S.D. LUKEFAHR

BREEDING AND IMPROVEMENT OF FARM ANIMALS

BREEDING AND IMPROVEMENT OF FARM ANIMALS

EIGHTH EDITION

J. E. Legates

William Neal Reynolds
Professor of Animal Science and Genetics
Dean, Emeritus, College of Agriculture and Life Sciences
North Carolina State University, Raleigh

Everett J. Warwick

Retired Staff Scientist
Science and Education Administration
United States Department of Agriculture

McGRAW-HILL PUBLISHING COMPANY

New York St. Louis San Francisco Auckland Bogotá Caracas
Hamburg Lisbon London Madrid Mexico Milan Montreal New Delhi
Oklahoma City Paris San Juan São Paulo Singapore Sydney Tokyo Toronto

This book was set in Times Roman by the College Composition Unit
in cooperation with General Graphic Services, Inc.
The editors were Denise T. Schanck and Scott Amerman;
the production supervisor was Denise L. Puryear.
The cover was designed by Tony Paccione.
New drawings were done by J & R Services, Inc.
R. R. Donnelley & Sons Company was printer and binder.

BREEDING AND IMPROVEMENT OF FARM ANIMALS

1 2 3 4 5 6 7 8 9 0   DOC DOC 8 9 4 3 2 1 0 9

ISBN 0-07-068376-X

Library of Congress Cataloging-in-Publication Data

Legates, James Edward.
    Breeding and improvement of farm animals/J.E. Legates, Everett
J. Warwick.—8th ed.
        p.   cm.—(McGraw-Hill publications in the agricultural
sciences)
    Warwick's name appears first on the earlier edition.
    Includes bibliographical references.
    ISBN 0-07-068376-X
    1. Livestock—Breeding.   2. Livestock—Genetics.   I. Warwick,
Everett James, (date).   II. Title.   III. Series.
SF105.L425   1990
636.08'2—dc20   89-13038

CONTENTS

PREFACE

Professor Rice began "Breeding and Improvement of Farm Animals" in 1928. Six decades later we are pleased to continue the eighth edition of his popular undergraduate classroom text. Both the knowledge and substance of animal breeding have changed markedly since the first edition. In the eighth edition the authors have focused on principles but hopefully have suggested sufficient examples to show where these principles can find application.

Students undertaking a study of animal breeding represent so many different backgrounds that it is not possible to fashion a text that will optimally match their prior preparation. It has been assumed that most students will have had previous instruction in genetics and reproductive physiology. Since this assumption will not be true for all students, Chapter 2 is intended to provide an update and review of genetic background and principles. Students without a previous course in genetics should, with careful study, gain sufficient undergirding for understanding the subsequent chapters. Chapter 3 has been revised and much reduced to focus primarily on the important function of reproduction in the generational transfer of inheritance.

Chapter 4 on qualitative genetics covers concerns with various polymorphisms and undesirable recessives which are frequently a matter of interest and question. While not considered a major thrust, they are deserving of consideration to provide a balanced view of animal breeding.

It is difficult to chart a course that is acceptable by all when the integration of genetics, mathematics, and statistics is sought. We have attempted to chart a middle road. Chapters 5 and 6 are intended to provide a beginning appreciation. Rigorous proofs are not provided; although, it is hoped that certain of the appendixes will provide stimulus to students with a mathematical bent. Those without such interest should be able to gain a working comprehension of the subject matter, even without an intricate understanding of these developments.

The next six chapters dealing with selection, mating systems, and performance and progeny evaluation are intended to give the student a feeling for what can be expected from the application of animal breeding in the changing of populations. Although these chapters are devoted to the genetic aspects of improvement, it is hoped that the student will sense the underlying concept that progress and profits come from a balanced consideration of genetic, environmental, and economic considerations.

In a book of this nature it is not possible to recognize all persons who have contributed in one way or another. However, we must mention the following reviewers: David S. Buchanan, Oklahoma State University; Richard R. Frahm, Virginia Polytechnic Institute and State University; Truman G. Martin, Purdue University; Robert R. Shrode, University of Tennessee; and Charles W. Young, University of Minnesota. We know that we could not have prepared this book without the assistance and encouragement of many colleagues and friends. Their advice and counsel have assuredly been responsible for many improvements in this edition, although the authors accept the responsibility for any shortcomings. Many individuals and organizations have graciously provided illustrative materials for which we are much indebted.

Our present effort could not have been accomplished without the encouragement and support of our wives. We extend to them a special measure of gratitude.

J. E. Legates

Everett J. Warwick

BREEDING AND IMPROVEMENT OF FARM ANIMALS

1

ANIMAL BREEDING: FOUNDATIONS AND CONTRIBUTIONS

The primary goals of humans since their emergence on earth has been to secure sufficient food to satisfy hunger and nutritional needs. During early times food was obtained by hunting and fishing and by gathering the fruit and seed of plants in their natural habitat. Sufficient area was roamed near critical sources of water to provide the needed food and the hide or fur for clothing. Truly, the human was "a hungry animal in search of food."

Much of human success in improving the quality of life can be attributed to the animals that have provided food, clothing, and power. Animal agriculture is a most advanced and specialized form of agriculture. Historically, the leading nations and societies have had well-developed livestock enterprise for their times. Three-fourths of the protein, one-third of the energy, most of the calcium and phosphorus, and a substantial share of essential vitamins and other minerals in the American diet come from animal products.

Animal agriculture has been supportive of the conservation and fullest utilization of our natural resources for the human good. It contributes to maintaining the delicate balance of nature. It provides an opportunity for the addition of increments of value in three ways. First, from the production of feed grains, by-products, and forages, a direct contribution from the land and related natural resources is realized. Second, by processing feed grains, by-products, and other feedstuffs through our animals, an additional increment of value is added in refining these raw materials. A third increment of value to society is added by the processing of our animals and animal products for distribution and utilization by the consumer.

DOMESTICATION OF ANIMALS

With the domestication of animals and plants the early nomadic tribal life gradually evolved to a more stationary culture as domestication and cultivation progressed. These developments have been of reasonably recent origin. Most records and archaeological evidence suggest that the domestication of animals, with the possible exception of the dog, began approximately 10,000 years ago. Prior to domestication, natural selection chose those individuals which were best adapted for survival and reproduction. Those types which survived most effectively reproduced more abundantly and dominated the local or environmental niches available to them.

The domestication of animals and plants provided the foundation on which civilization could be built. Without a productive agriculture it seems reasonable to conclude that social and scientific progress would have been extremely limited. Domestication very likely began at the end of the New Stone Age. Each domestic animal has had a rather restricted and humble origin, yet through constant selection for one or another character numerous forms have evolved—some useful, some ornamental, some having both qualities, and some neither.

The Horse

The name *horse* is equivalent to the Anglo-Saxon *hors,* which means "swiftness," and it is logical to suppose that this genus was able to survive the vicissitudes of time and enemy attack chiefly because of its speed.

The horse was probably the last animal to be domesticated, but its immediate ancestry as well as the date of its domestication is still a matter of dispute. It seems probable that at least two, or perhaps three, wild types have made their contribution to our domestic horse. One of these was probably the *steppe horse,* now known as the fossil representative of Przhevalski's horse. This was a small, sturdy, short-legged horse with a moderately long, heavy head. Another was the so-called *desert horse,* standing, as did the *steppe horse,* about 13 hands, or 130 centimeters (cm), high and corresponding closely to the now-extinct tarpan, or Mongolian horse. This strain was somewhat more slender than the steppe horse and had a shorter head. The third contribution was that of the *forest horse,* a type standing about 15 hands (150 cm) high with longer but stout limbs and having a long, narrow head and long body. It seems probable that all three types made a contribution to our modern breeds. The horse was apparently domesticated separately in Asia and Europe, probably earlier in Asia. A Turanian folk tamed the Przhevalski horse around 3000 B.C.

The earliest record of the horse dates back to Paleolithic times, about 25,000 years ago. Around an open camp at Soultré in France are found the remains of several thousand horses, indicating that horses may have served as a source of food. In later Paleolithic times, rock carvings of the horse were made, but they do not show it harnessed, ridden, or attached to any sort of vehicle; so we assume the horse was not yet fully domesticated.

The earliest trace of the horse hitched to a chariot goes back to about 2000 B.C. in Greece, whereas the first Egyptian records of the domestication of the horse date from about 1600 B.C. These were small horses, about 13 hands high, similar to Przhevalski's horse. The horse evidently grew in size and importance in Persia and Mesopotamia during the intervening years, and after about 750 B.C. it began to serve as a mount. Mounted riders were first given a place in the Olympian games in 648 B.C. The Arabs did not use horses until after the time of Christ.

The ancestry of the horse has been traced back about 55 million years by means of skeletons found in descending layers of the earth's crust. The forerunner of our present horse was an animal about 25- to 50-cm tall with four toes on the front feet and three on the back. Twenty million years later this horse stood about 60- to 65-cm tall and had three toes fore and back. It has since been reduced to but one toe (with two splits), but the overall size of the horse has increased, and its teeth have developed into more efficient tools for grinding feed.

Cattle

It seems probable that cattle were domesticated during the New Stone Age in both Europe and Asia. There are two types of domestic cattle: *Bos indicus,* the humped cattle of tropical countries, and *Bos taurus,* the cattle of the more temperate zones. Humped cattle were domesticated as early as 2100 B.C. Cattle played an important part in Greek mythology; they were sacred animals in many older civilizations, and their slaughter was therefore forbidden. The great ox, or aurochs, *Bos primigenius,* which Caesar mentioned in his writings, is generally considered to be one of the progenitors of our modern-day breeds. This was a very large animal, described by Caesar as ''approaching the elephant in size but presenting the figure of a bull.'' The wild park cattle of Britain are considered by some authorities to be the direct descendants of *B. primigenius.* Another progenitor of our modern breeds is *Bos longifrons,* a smaller type, with a somewhat dishlike face. This is the Celtic Shorthorn, which has been found only in a state of domestication. It was the only ox in the British Isles until 500 A.D. when the Anglo-Saxons came and brought the great ox, or aurochs, of Europe.

It is doubtful whether any of our present-day European or American breeds trace back solely to either one of these ancient types. It seems much more probable that our present breeds are the results of various degrees of crossing between them. The cattle of India and Africa, *B. indicus,* are characterized by a lump of fleshy tissue over the withers sometimes weighing as much as 15 to 25 kilograms (kg). They also have a very large dewlap, and the voice is more of a grunt than a low. They are thought to be descended from the wild Malayan banteng.

It seems probable that primitive humans first used members of the family Bovidae as a source of food. Domestication perhaps began when these animals were used as draft animals, probably in the first steps of the tillage of the soil. In their wild state there was little tendency to store excess fat on the body, as this would have been a hindrance rather than a help under the conditions then existing. Milking qualities also were most sufficient for the rearing of the young. As civilization developed, feed became more abundant, methods of caring for livestock improved, and the latent possibilities for rapid growth and milk production began to be realized under selection by humans.

That the ox played an important part in aesthetic development is attested by its use in architectural and interior mural decoration as well as by its frequent use as a subject of poetic fancy. The ox assumed great religious importance in many ancient civilizations; the best members of the breed were sacrificed to propitiate the gods. They were crowned with wreaths and honored in other ways during pageants and holidays. To some extent we perpetuate this custom in our fairs and expositions today. The Romans' term for money was *pecunia,* a word derived from *pecus* meaning "cattle," and in ancient times wealth was measured in terms of the cattle one possessed.

Swine

Swine are ungulates belonging to the suborder ariodactyls (even-toed). They belong to the family Suidae. The Dicotylidae, or peccaries, and the Hippopotamidae, or hippopotamuses, are closely related families, these three families comprising the Suina. These animals have tubercles on the molar teeth, and there is not a complete fusion of the third and fourth metapodials to form a cannon bone. The nose is elongated into a more or less mobile snout.

It appears that our modern breeds, *Sus domesticus,* have descended from at least two wild stocks: the Northern European breeds from the wild boar *Sus scrofa,* and those of Southern Europe, Asia, and Africa from one of the Malayan pigs, possibly the collared pig *Sus vittatus.* The former was a larger, coarser animal throughout than the latter, and it had a denser covering of hair.

Present-day breeds are no doubt the result of varying degrees of crossing between the parent stocks and their offspring. It seems probable that the pig was domesticated later than cattle and sheep and earlier than the horse. Selected for its ability to grow rapidly and economically, the pig is foremost in converting feed into flesh. Several of the breeds of hogs found in America are of our own breeders' making; for example, the Duroc, the Poland China, and the Chester White breeds are strictly American creations.

Sheep and Goats

These two genera of the family Bovidae are very closely related, so closely, in fact, that a naturalist never speaks lightly of "separating the sheep from the goats." The genus *Ovis* includes the sheep and its wild relatives, whereas goats and their kind make up the genus *Capra.* Sheep are distinguished from goats by glands in both forefeet and hind feet, by the absence of a true beard, and by the absence of the strong goaty odors in males. There are also marked differences in the skulls; and the horns generally spiral in opposite directions, the right horn of the sheep to the right like a corkscrew, and the goat's to the left. The sheep gets its Latin name *Ovis* from the Sanskrit *avi,* signifying "to keep" or "to guard."

Sheep probably originated in Europe and in the cooler regions of Asia in the Pleistocene or later Pliocene era. Remains of sheep or a goatlike animal have been found at the sites of lake dwellings from Neolithic times in what is now Switzerland. Sheep are thought to have been derived from the antelopelike animals allied to the gazelles because of certain similarities of the molar teeth. It seems certain that our modern breeds trace back to at least two remote ancestors, the mouflon of Europe (*Ovis musimon*) and the Asiatic urial (*Ovis vignei*).

The sheep was originally a hairy animal with an underfur of wool. No doubt people living in cold climates who used skins as clothing were the first to begin the selection of sheep for wool production. As in all our domesticated animals, there is wide variation among sheep. Some, like the African long-legged and Abyssinian maned sheep, bear hair instead of wool; some have spiral horns 50 cm or more in length, and others have no horns at all. The tail of the common domesticated sheep is long and slender; in some other strains it is a fat depot about 30 cm in width; whereas still others have merely a vestige of a tail. The last sort often carry huge patches of fat on their rear quarters, the stored fat in all cases serving to tide the animal over long periods of food shortage. The hunia, a tall, long-legged sheep, is used in India as a fighting animal.

Goats are also versatile in characteristics, yielding the underfur for Cashmere shawls and providing mohair, milk, meat, and draft power; they also provide one means of clearing up brush land because of their fondness for all sorts of tender shoots.

The Chicken

Chickens came from Southeast Asia and were kept in China as early as 1400 B.C. The most authentic information about the origin of the chicken, according to Jull[1] suggests that the existing breeds have descended from one or possibly four subspecies.

1 *Gallus gallus,* the Red Junglefowl
2 *Gallus lafayette,* the Ceylon Junglefowl
3 *Gallus sonnerati,* the Gray Junglefowl
4 *Gallus various,* the Java Junglefowl

Charles Darwin studied the origin of the chicken and in 1868 concluded that all the domestic stocks descended from the *Gallus gallus,* Red Junglefowl species. Darwin observed that in this species the voice was more similar to the domestic chicken. Further, when mated with domestic stock it produced offspring more freely, and these hybrids were more likely to be fertile than crosses with any of the other species.

[1] Jull, M. A. 1940. "Poultry Breeding," John Wiley & Sons, Inc., New York.

Several other researchers have investigated the origin of the domestic chicken. The predominant conclusion of their finding is that the domestic chicken owes most of its genetic origin to the Red Junglefowl, but some lesser contributions may have come from the other subspecies.

ANIMAL BREEDING FOUNDATIONS

As domestication progressed, animals were selected for special purposes in addition to reproductive fitness. The desire of humans for food, draft, or clothing began to influence the choice of animals to leave offspring. Thus selection for human requirements was gradually imposed upon natural selection for reproductive fitness. As these additional requirements were imposed, the heredity of the animals was modified. Environmental and managemental adjustments, such as aid in the procurement of fodder and protection from predators, were gradually implemented to permit individuals with the most desired expression of the productive function to survive and reproduce.

Practical Beginnings

Long before Mendel's principles were rediscovered in 1900, the mystery of the transmission of hereditary material from parent to offspring was recognized but not understood. Robert Bakewell (1725–1795) of Leicestershire in England has been identified as the first notable pioneer in animal breeding. He is credited with conducting the first systematic progeny tests of bulls and rams. Males of his choosing were leased out to other breeders, with Bakewell retaining the right to inspect all progeny. Males with the best progeny were returned for use by Bakewell. He is credited with laying the foundations for Leicester sheep, Longhorn cattle, and Shire horses.

The Colling Brothers, Charles and Robert, who are often referred to as the founders of the Shorthorn breed, studied the breeding procedures of Bakewell closely beginning in 1783. Favorite, one of the great sires of the day that had been bred by Charles Colling, was born in 1783. He and his son Comet were used extensively in the Collings brothers' herds. Charles purchased the cow Duchess in 1784; her influence continued with "Duchess" Shorthorns gaining popularity in the United States, reaching a peak in the 1870s.

Scientific Advances

Parallel with the developments among livestock breeders, scientists of this period sought to understand the origins of animals and plants. Many persons, beginning with early Hindu and Greek philosophers, have contributed to the accumulation of knowledge on the variation in and evolvement of existing species. Space will permit mention of only a few of these.

Buffon (1707–1788), a French naturalist, greatly enlarged the concept of mutability through direct environmental means, an idea which he developed in later life, having earlier subscribed to the idea of immutability. He was a precursor of Lamarck and perhaps also of Darwin in regard to pangenesis as well as the struggle for existence and the survival of the fittest. Coupled with his idea of change due to environmental influences was his further implied belief that these so-called *acquired* characters were heritable.

Erasmus Darwin (1731–1802), an English physician and poet and grandfather of Charles Darwin, is one of the most imposing figures in the field of human inquiry. He borrowed and enlarged on many old ideas concerning evolution and made distinctive new contributions. He was the first to stress the idea that evolution has been operating from the time of the first primordial life. To him new forms were but the flowering of potentialities originally deposited in the life stream by the Creator and called into being by the necessity for adaptation to the environment.

Lamarck (1744–1829), a French naturalist, can be rightfully called the father of the theory of evolution. He was the first to devise a classification scheme or phyletic tree to include all plants and animals, and he was the first to state that all animals formed a branching series of related forms, shading into one another by very gradual steps. His theory of evolution, propounded in 1809, had three main points of which only the first has survived in part.

1 The environment, directly in plants but indirectly in animals through the medium of the nervous system, causes changes in the organism.

2 The use or disuse of parts leads, respectively, to their further development or to their atrophy.

3 These so-called *acquired* characteristics are inherited.

Charles Darwin (1809–1882), the best-known proponent of the theory of evolution, was an English naturalist. Darwin was not the father of the idea of a gradual progressive change. He might perhaps more correctly be called the attending physician who brought the concept safely into the world, or, as Butler put it, "Darwin's chief glory is not that he discovered evolution but that he made men believe in it, and what glory," he added, "could be greater than this?"

Darwin sailed on the *HMS Beagle* as naturalist on a round-the-world trip lasting from 1831 to 1836. On this voyage he had the opportunity of studying the rich fauna and flora of South America and several island archipelagoes. He was struck by the manner in which animal types shaded into one another and by the distinctive forms found on separate islands. He had begun his journey with a belief in separate creation, but during the course of the voyage he became convinced of the mutability of species. On his return to England in 1837 he began to organize the known facts on plant and animal variation and to try to discover new ones. During the next year, while reading Malthus's "Essay on Population," the idea of the struggle for existence that is constantly going on in nature as an explanation of the great variety of plant and animal species came to him.

Darwin's theories involved four points:

1 Organisms vary, i.e., they are not exactly like their parents.

2 These variations are or may be hereditary, i.e., they may be passed along to descendants.

3 Due to the dynamics of reproduction in all species there is a continual struggle for existence in nature.

4 Those best adapted to survive in a given environment will be the most likely to survive and should therefore leave the most descendants.

Darwin's extensive and systematic observations provided a body of evidence that supported the significance of hereditary variation. His writings and the statistical developments of Galton shaped much of the thinking of animal breeders in the later portion of the nineteenth century.

Galton, without the benefit of Mendel's findings, added quantitative precision to the characterization of the variation which Darwin had drawn upon to formulate his theory of natural selection. Galton's law stated that given a correlation of .50 between parent and offspring in a population, with minimal inbreeding and for a highly hereditary trait, the correlation between an animal and a more remote ancestor is halved for each preceding generation. Lush pointed out that practical breeders in Britain had used this concept as early as 1815.

The rediscovery of Mendel's research in 1900 provided the beginning of the scientific foundation for breeding investigations and practice. Most of current animal breeding theory is based on the merging of genetic and statistical fundamentals first developed by Fisher and Wright and extended by Lush to animal breeding practice. The influence of the writings and thought of Lush and his students has shaped animal breeding investigations and practice for the past half century.

What Is Animal Breeding?

Animal breeding, which is the application of scientific knowledge to the genetic improvement of animals, has evolved from these beginnings. Genetics provides the foundation principles which should guide animal breeding practice. However, plans and programs to improve the genetic merit of livestock must draw heavily on contributions from statistics, biochemistry, physiology, economics, and other disciplines.

Genetic principles are utilized and synthesized into breeding programs. The task in animal breeding is twofold: (1) to select the most desirable animals based on the prediction of genetic merit and (2) to produce superior genotypes by the combination of genetics through breeding plans and systems of mating.

Livestock production is an economic enterprise, and animal breeding recommendations must be examined in the light of economic, as well as genetic considerations. Breeding programs involve systems of management as well as systems of breeding. The basic objective of animal breeding is to enhance the efficiency of production and the quality of the product for the ultimate consumer through planned genetic change.

Productive efficiency and product quality can be enhanced by both genetic and environmental improvements. Environmental influences, such as improved nutrition, disease prevention, or housing, usually have a more immediate and rapid influence on production. Genetic improvement is usually slower and often less dramatic, yet it is more permanent in that it alters the animal's heredity and remains to influence performance for a lifetime. A portion of this genetic change is also retained in succeeding generations as gametes are transmitted from parent to offspring. Genes do not express themselves in a vacuum, the final expression of a trait represents the joint action of the heredity of the animal and the environment provided for its expression. Understanding the relative importance of the direct and joint contributions of each of these factors is an important part of animal breeding research.

CONTRIBUTIONS OF ANIMAL BREEDING

Animal breeding has contributed much to the total improvement of livestock production. In certain instances, where performance records have been available to direct selection and to assess change, the results have been dramatic. Combined genetic, nutritional, and other advances have led to remarkable success in certain areas. A striking example is the increase in milk production over the past 60 years, as shown in Figure 1-1. Recognizing that many factors have contributed to this increase, performance records through Dairy Herd Improvement have provided unselected data to choose herd replacements and particularly to guide sire selection. Analyses of production records have provided genetic parameters for developing effective progeny testing and breeding plans. National programs of sire and cow evaluation provide computerized listings of sires for use in artificial insemination and cows to be mated as dams of young males in planned progeny-testing programs. Frozen semen has markedly extended the use of outstanding sires, some of which have sired over 200,000 offspring. Embryo transfer has been perfected to the point where outstanding females may now leave 100 or more progeny. For the past 20 years the rate of increase in milk yield has been about 1 percent of the mean yield per year. Analyses indicate that approximately half of this is due to genetic improvements. The highest production by an individual cow is now 25,248 kg of milk in 365 days, milked twice daily.

Chicken used to be a delicacy that was relished primarily for Sunday dinner, but changes in broiler production since the early 1930s have made poultry one of our most available meats. In fact, its consumption per capita has been increasing more rapidly than any other meat in recent years. Dramatic genetic changes in the broilers and in the systems of rearing and marketing have contributed to this remarkable change. In the 1940s it took about 12 weeks to de-

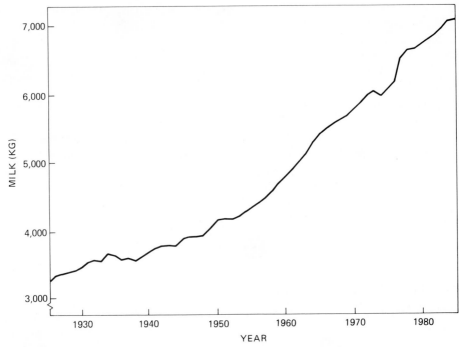

FIGURE 1-1
Increase in average milk yield per cow from 1925 to 1985 for all cows on Dairy Herd Improvement test in the United States. (*From DHI Summary, Animal Improvement Program Laboratory, USDA.*)

velop a broiler weighing 1.6 kilos; whereas, today a 1.75-kilo broiler can be developed in only 6 weeks. Many nutritional and housing innovations have contributed to this remarkable change. Feed requirements per kilo of gain have been reduced from 3.5 to 1.9 kilos, largely by improved genetic efficiency and ration formulation. Breeding improvements from the early use of the available heavy egg-laying strains to the specialized highly selected cross-combinations of Cornish males and White Rock females have been striking. Larger-breasted, faster-growing, feed-efficient birds have made chicken one of our most readily available sources of animal protein.

Early animal husbandry texts include many photographs of what was then known as the "lard" type of hog. Animals were carried to slaughter weights of 200 to 250 kilos, as lard for cooking was one of the primary products. The change to sedentary occupations, the availability of cheaper sources of fats, and a weight-conscious public have reduced the dietary demand for lard. A dramatic change in the conformation and leanness of our swine has taken place as selection for the "leaner-meat" type of swine has progressed.

The development of the *probe* technique for determining back-fat thickness in the live animal permitted measuring a trait closely related to carcass values

FIGURE 1-2
Holstein cow. Beecher Arlinda Ellen, "Excellent" at 91 points. She is the World Record
Leader for milk yield with a 365-day, twice daily milking, lactation of 25,248 kg. In 1 year
Ellen produced enough milk to provide a liter a day for one person for 69 years and 2
months. (*Courtesy Select Sires, Inc., Plain City, Ohio.*)

FIGURE 1-3
Contrast of the growth attained by
broilers for market, 1.4 kg in 1940 and
2.0 kg in 1987, with the same approximate quantity of feed. Intensive
selection, choice of specialized strains
in crossbreeding, plus ration and
nutritional advances have contributed to
this marked improvement in efficiency.
(*Courtesy of Department of Poultry
Science, N.C. State University.*)

and still allowed the use of the superior animals for breeding. Previously the
fatness of the carcass could be determined only after the animal had been
slaughtered. Then sibs or littermates had to be used for breeding. The probe
technique has been widely employed in swine improvement programs. It is relatively simple, involving cutting a small slit in the skin in the back and thrust-
ing a thin metal rule down through the soft fat to the firmer lean to read the fat

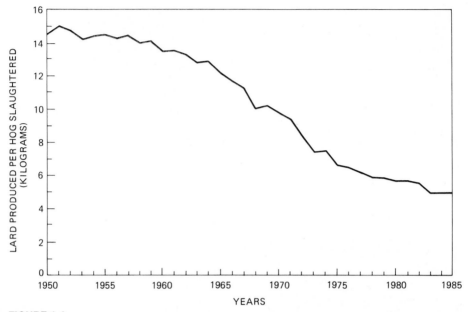

FIGURE 1-4
Trend in amount of lard produced per hog slaughtered in the United States, 1945–1985.

thickness directly. Other, more intricate, devices have been since developed, but the probe measurement has provided an accurate live-animal assessment of back-fat thickness which is correlated with the carcass quality.

CHALLENGES

The above are examples of the results of animal breeding in practice. Genetic engineering, meshing with the long-established and developing breeding principles, should offer numerous fruitful future opportunities. Genes for rat growth hormone have been successfully transferred to fertilized eggs of mice. The mice that developed from the eggs varied from normal to twice-normal size. More recently, similar transfer of genes into swine has been successful. These findings demonstrate the potential for genetic engineering, but much research must still be done to make possible direct genetic improvement using such new techniques.

Apart from the interest of fanciers, animal breeding is an economic undertaking. As such, the breeds, breeding goals, and programs need to be determined with proper consideration of the economic concerns dictated by the system of management and market for the end products. In dairy cattle, in situations where cheese rather than fluid milk is of primary economic concern, strains or breeds with higher protein and fat production should be considered. Measures of net economic efficiency have been developed for dairy cattle, beef cattle, and swine.

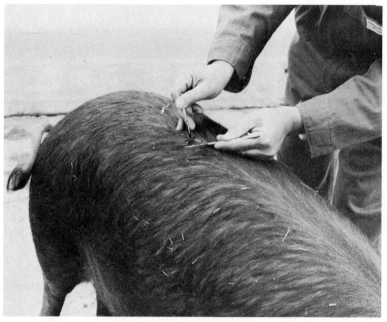

FIGURE 1-5
Measuring back-fat thickness using the probe technique. (*Courtesy of Dr. H. O. Hetzer formerly of USDA.*)

FIGURE 1-6
Duroc gilts at about 80-kg live weight from the 13th generation of a selection experiment for high fat (left) and low fat (right). Note differences in conformation. (*Courtesy of U.S. Department of Agriculture.*)

Basically, the task of the animal producer is to convert vegetable products (some edible by humans, and some not) and inedible animal by-products into palatable, nutritious human food. Other animal products important in our everyday life include fiber for clothing, leather, pharmaceuticals, and a wide array of other materials. In a world with constantly increasing human population and a consequent narrowing of the margin of safety between world food needs

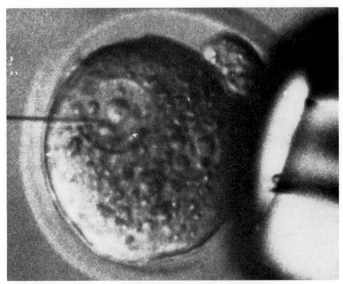

FIGURE 1-7
Injecting genes into pronucleus of mouse egg with a micropipette to produce transgenis mice. (*Courtesy Robert M. Petters, North Carolina State University.*)

FIGURE 1-8
Transfer of growth hormone genes to this piglet has been accomplished, a technique that offers much promise for use with farm animals. Direct injection of growth hormone in swine produces much leaner growth. (*Courtesy of Vernon G. Pursel, USDA.*)

and potential world production, it is essential that the conversion of vegetable to animal products be achieved efficiently.

This book addresses the genetic factors affecting the efficiency of animal production. It is, however, essential that the student fully realize that this is only one aspect of overall production efficiency. Many animals are inefficient and unprofitable producers because they are poorly fed and poorly managed. An optimum environment also involves the control or elimination of the various livestock pests and diseases which take an annual toll of millions of dollars. The problem facing animal producers consists in feeding and managing more efficiently what we have, as well as raising the level of potential performance through selective breeding.

SUMMARY

Humans are in a constant struggle to produce sufficient food for their present needs and for those of an ever-expanding population. Animals were an important part of the human food chain, even before their domestication. In their natural habitat, animals were selected on the basis of their ability to survive and reproduce. Since that time they have been chosen for special productive purposes in addition to reproduction. The rediscovery of Mendel's research in 1900 provided the beginning of a sound foundation for planned genetic change. Animal breeding is the application of scientific knowledge to the genetic improvement of animals. Genetic principles are used to predict the genetic merit of available breeding animals, and the potential for productive efficiency is improved by selection and systems of mating. Since livestock production is an economic enterprise, animal breeding practices and recommendations must be sound economically as well as genetically. The ultimate goal is to increase productive efficiency in order to raise the net return to the producer and to provide an abundant, economical supply of animal products for the consumer.

SUGGESTIONS FOR FURTHER READING

Books

Barker, J. S. F., K. Hammond, and A. E. McClintock (eds.). 1982. "Future Developments in the Genetic Improvement of Animals," Academic Press, New York.

Brackett, B. G., G. E. Seidel, Jr., and S. M. Seidel (eds.). 1981. "New Technologies in Animal Breeding," Academic Press, New York.

Darwin, C. 1859. "Origin of Species," A. L. Burt Company, New York.

Darwin, C. 1868. "The Variation of Animals and Plants under Domestication," 2 vols., John Murray (Publishers) Ltd., London.

Davenport, E. 1910. "Domesticated Animals and Plants," Ginn and Company, Boston.

Mason, I. L. (ed). 1984. "Evolution of Domesticated Animals," Longmans, Green & Co., Ltd., London.

Smith, C., J., W. B. King, and J. C. McKay (eds.). 1986. "Exploiting New Technologies in Animal Breeding: Genetic Development," Oxford University Press, London.

Articles

Craft, W. A. 1958. Fifty Years of Progress in Swine Breeding. *J. Anim. Sci.* 17:960–980.

Warwick, E. J. 1958. Fifty Years of Progress in Breeding Beef Cattle. *J. Anim. Sci.* 17:922–943.

White, J. M., W. E. Vinson, and R. E. Pearson. 1981. Dairy Cattle Breeding and Genetics. *J. Dairy Sci.* 64:1305–1317.

Terrill, C. E. 1958. Fifty Years of Progress in Sheep Breeding. *J. Anim. Sci.* 17:944–959.

CHAPTER 2

GENETIC BASIS FOR ANIMAL BREEDING

The science of genetics is concerned with the inheritance or the transmission of biological material (genes) from parent to offspring and hence from generation to generation. A comparatively young science, its foundation was laid by Gregor Mendel, who did experimental crossing studies with garden peas at the monastery in Brunn, Austria, and published the results in 1865. The importance of his research was not recognized until the principles he developed were independently rediscovered by three scientists in 1900. Developments since then have been rapid, and genetics has evolved to include a number of special fields involving almost all scientific disciplines.

MECHANISMS OF INHERITANCE

Bodies of higher organisms are composed of many billions of microscopic cells. From a genetic standpoint at least two types are distinguishable: *somatic*, or body, cells and *gametes*, or germ cells.

Each cell nucleus contains a number of bodies known as *chromosomes*. These stain darkly with basic dyes in microscopic preparations; hence the reference to *chromo*, or "colored" and *soma*, or "body." They are composed principally of nucleoproteins which are complexes of deoxyribonucleic (DNA) acid and histones. Chromosomes carry hereditary material and provide for its transmission from parent to offspring. The mitochondria found in the body of the cell also possess DNA (mitochondrial DNA).

Chromosomes in Reproduction and Inheritance

Chromosomes are usually present in pairs in the cells of higher animals and plants (eucaryotes). The number of pairs characteristic for a given species is

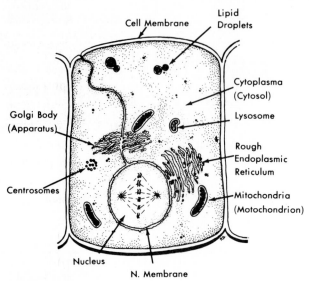

FIGURE 2-1
Generalized drawing of a cell showing present-day concepts.
(*From Campbell and Lasley. Original drawing by Dr. Robert Godke, Louisiana State University.*)

given in Table 2-1. Individual pairs differ in size and other morphological features that permit their identification in appropriately prepared and stained preparations. Cells with two members of each chromosome pair are said to be *diploid,* or to have the $2n$ chromosome number, where n equals the number of chromosome pairs in a species. Figure 2-2 shows the chromosome complement of domestic swine (*Sus scrofa*), which normally has 19 pairs of chromosomes.

During development some cells become specialized for reproductive purposes and serve as progenitors for the gametes, or germ cells. In animals, the progenitors of the gametes are known as *spermatogonia* in males and *oogonia* in females. In plants the corresponding cells are pollen, or *microspore,* cells and embryo-sac, or *megaspore,* mother cells. These cells have the ability to divide by *mitosis* to provide additional cells of the same types. In mitosis, the diploid chromosome number is maintained. From a genetic viewpoint the key step in this process is the reduction, or *meiotic* cell division, in which only one member of each chromosome pair goes to each new cell. In animals, the cells resulting from meiotic division subsequently undergo one mitotic division during the process of differentiation into gametes. They contain only one member of each chromosome pair and are *haploid,* or have only the n number of chromosomes. When the male and female gametes unite at fertilization, one member of each chromosome pair is transmitted to the new individual by the male and one by the female. Thus, the diploid, or $2n$ chromosome, number is restored.

TABLE 2-1
DIPLOID CHROMOSOME NUMBERS OF SOME MAMMALS*

Common and scientific names	Chromosome number
Humans (*Homo sapiens*)	46
Horse (*Equus caballus*)	64
Ass (*Equus asinus*)	62
European cattle (*Bos taurus*)	60
Zebu cattle (*Bos indicus*)	60
American Bison (*Bison bison*)	60
Domestic buffalo (*Bubalus bubalus*)	48
Musk Ox (*Ovibos moschatus*)	48
Reindeer (*Rangifer tarandus*)	70
Sheep (*Ovis aries*)	54
Goat (*Capra hircus*)	60
Swine (*Sus scrofa*)	38
Dog (*Canis familiaris*)	78
Cat (*Felis catus*)	38
Domestic rabbit (*Oryctolagus cuniculus*)	44
Mouse (*Mus musculus*)	40
Rat (*Rattus norvegicus*)	42
Chicken (*Gallus gallus*)	36

*From Hsu and Benirschke (1967–75).

The process of gamete formation is illustrated diagrammatically in Figure 2-3 for a hypothetical animal species with three chromosome pairs. In the figure, chromosomes of paternal origin have been depicted in black and those of maternal origin in white. Any combination of chromosomes of maternal or paternal origin could go to any secondary spermatocyte or oocyte. The combinations in Figure 2-3 are for illustration only and do not include all possibilities.

In the male, each primary spermatocyte has the potential for forming four spermatozoa. In the female both the meiotic and the subsequent mitotic divisions are unequal as regards the *cytoplasm,* or nonchromosomal material. The result is a concentration of food material for the embryo in the *ovum,* or egg, and the formation of three nonfunctional polar bodies. It is a matter of chance whether a cell with a particular chromosome complement develops into an ovum or a polar body. Thus, an ovum could have any one of the combinations of chromosomes possible in the primary oocytes. The number of possible combinations for any species is equal to $2n$.

Biochemical Basis of Heredity

The discovery of nucleic acids was reported by Miescher in 1869, only 4 years after Mendel's experimental results were published. Miescher's findings, like the laws of inheritance, remained unused in genetic developments for decades. Tremendous progress has been made since the 1950s in gaining an understand-

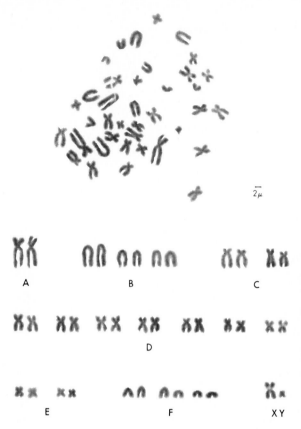

FIGURE 2-2
The 38 chromosomes (19 pairs), stained by conventional methods, of a male Yorkshire pig (*Sus scrofa*) at metaphase, at which stage each chromosome appears to be doubled. *Above,* a photomicrograph of the chromosomes of a single cell. *Below,* chromosomes of a similar photo cut out and rearranged by pairs in a *Karyotype.* It will be noted that all pairs are very similar in size and form except for the X and Y—the sex chromosome pair in which differences are readily apparent. *(Courtesy of Dr. Richard A. McFeely, University of Pennsylvania.)*

ing of the biochemical aspects of heredity. As has already been mentioned, the basic hereditary material in higher animals and plants is deoxyribonucleic acid, or DNA. DNA is a chemically complex substance with large molecules capable of virtually an infinite number of structural variations. It provides the fundamental information at the molecular level for growth and development.

Watson and Crick in 1953 developed a model for the structure of the DNA molecule for which they later received a Nobel Prize. Long spiraling double-stranded threadlike molecules of DNA are located in the nucleus of cells of higher animals and plants. DNA is composed of a linear sequence of basic units of *nucleotides* with each of these containing an organic base, a pentose sugar, and a phosphate. Four different nitrogenous bases, adenine, thymine, cytosine, and guanine, result in four kinds of nucleotides. The molecule is composed of two *polynucleotide* chains, or strands, arranged in a coiled spiral, or helix, manner. The name *double helix* is often applied to the structure. The strands are held together by hydrogen bonds between pairs of bases with the phosphate and sugar forming the outside of the helix. The two strands are

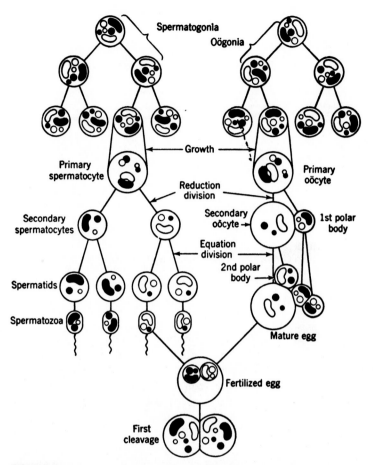

FIGURE 2-3
Diagrammatic representation of cell divisions during gamete formation
in animals. Bodies within the cells represent chromosomes, with those
of the same sizes and shapes being of the same pair (homologous).
Chromosomes are shown in black and white to represent paternal and
maternal origin, respectively. Only a portion of the chromosome
combinations possible in the gametes are shown. (*After Shull, 1938.
Heredity. 3d ed.*)

complementary to each other, with adenine always paired with thymine and
guanine with cytosine.

In replication, the two strands of the molecule come apart and each serves
as the template for the synthesis of a new complementary strand, thus maintaining genetic continuity. Replication must precede cell division in order for
the full complement of genetic material to be present in each new cell.

DNA has a four-letter alphabet, A, T, C, and G, comprised of linked mol-

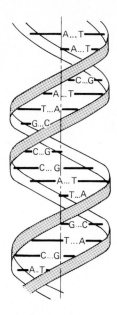

FIGURE 2-4
Diagram of a portion of the DNA molecule showing the coiled, double helix arrangement with bases as crosslinks. The letters represent: A, adenine; T, thymine; G. guanine; C, cytosine. (*From Office of Technology Assessment.*)

ecules of adenine, thymine, cytosine, and guanine, respectively. Triplets of these sets of four letters form the "words" of a message to direct the formation of specific amino acids. However, protein molecules are not made directly from DNA. Rather, the message from the sequence of linked A, T, C, and G molecules that comprises a gene is first copied by a process called *transcription* into single-stranded threadlike ribonucleic acid molecules called messenger RNA (mRNA). RNA is closely related to DNA, with three letters (A, C, and G) of its four-letter alphabet being the same as those in DNA, but with U (uracil) replacing T (thymine). Messages from words formed by the DNA alphabet are transcribed as RNA messages in the A, U, C, G alphabet. Messenger RNA molecules move through the nuclear membrane into the cytoplasm of the cell, where by *translation* their individual messages lead to the formation of specific protein molecules. This process utilizes the numerous biomanufacturing elements called ribosomes in the cytoplasm of the cell.

Translation is a much more complex process than transcription. The letters in the messenger RNA must dictate the correct sequence of the 20 different amino acids in the developing protein chain. This is accomplished by the genetic code, which consists of the 64 possible three-letter combinations of the four RNA letters. Each three-letter combination, or word, is termed a *codon*. Each codon directs the insertion of one of the 20 amino acids into a growing protein chain or directs the termination of the chain. Only codon AUG can initiate a protein chain; however, codons UAA, UAG, or UGA can terminate a protein chain. Some amino acids are inserted by the direction of more than one codon. For example, six different codons are capable of inserting the amino acid arginine into protein sequences.

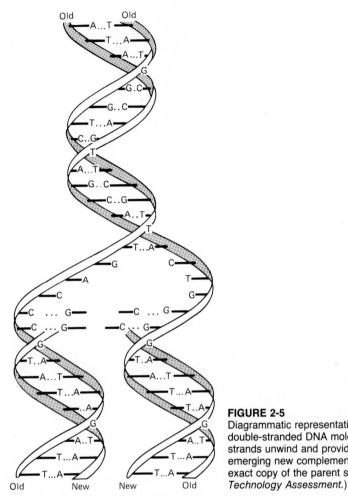

FIGURE 2-5
Diagrammatic representation of duplication in the double-stranded DNA molecule. The original strands unwind and provide templates for the emerging new complementary strands, each an exact copy of the parent strand. (*From Office of Technology Assessment.*)

The code for amino acid insertion is essentially the same for all life forms. This important fundamental principle underlies the success achieved in incorporating genes from one organism into the strands of DNA in another. The incorporation has been accomplished by using a combination of restriction enzymes or endonucleases which cut DNA strands and other enzymes (DNA ligases) that catalyze the joining of the cut ends. A restriction enzyme cuts a DNA strand on when it locates a specific sequence of nucleotides. For example, one restriction enzyme might cut between adenine and thymine in the sequence GAATTC, while a different restriction enzyme might cut between guanine and cytosine in the sequence GGCC. These sequences are called *restriction sites*. If a restriction site is repeated along the strand of DNA, the DNA would be cut wherever the particular site recurs. This is the basis for recombinant DNA (rDNA) techniques.

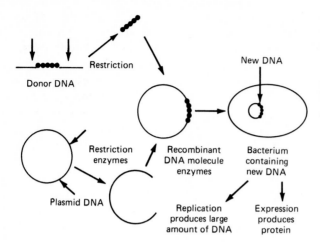

FIGURE 2-6
Recombinant DNA in bacteria. Restriction enzymes recognize specific sites on the DNA strand and cut the DNA at that point. Selected gene(s) from donor DNA molecules are removed and inserted into plasmid DNA molecules to form recombinant DNA. This rDNA can then be cloned to make many copies in its bacterial host which can produce large quantities of the desired protein. (*From Office of Technology Assessment.*)

Genes and Mutation

The *gene* is the unit of inheritance. This term was coined by Johannsen in the very early days of genetics to describe the basic, particulate unit of inheritance. It was assumed (1) to represent a unit of chromosome which was not physically subdivisible, (2) to be a unit of physiological function or expression, and (3) to be the smallest unit which could undergo genetic change, or mutation. Genes were often thought of as being on the chromosome in much the same manner as beads on a string.

With development of knowledge of the biochemistry of the hereditary material, together with some experimental observations which were not completely compatible with classical concepts, additions and redefinitions were in order. We previously related that a *codon* was a combination of three nucleotides which specifies the addition of a particular amino acid in the molecule. The term *muton* has been suggested for the smallest genetic unit capable of change, or mutation, and the term *recon* for the smallest indivisible unit of DNA capable of recombination. Both the muton and the recon represent single bases in a nucleotide. *Cistron* is the term used to represent the unit of DNA carrying the information necessary for the formation of one polypeptide chain in protein formation. It is equivalent functionally to the classical gene. The term gene continues in use but is now generally thought of as a functional rather than a structural entity.

Mutation is a change in a functional hereditary unit or gene. The usual definition of mutation is that it is a sudden, heritable change in the genetic material. Normal development of any form of life is dependent upon the functioning of a large number of genes. Each species throughout evolutionary history has developed a characteristic gene complement. For genes with visible effects or with major effects upon survival and on adaptation of the species to its environment, most individuals are eventually pure for the favored genes, carrying the same gene in each member of a chromosome pair. These genes are often referred to as "wild type." However, variant forms of these genes, having no apparent detrimental effects on the organism when present in only one member of a chromosome pair of an individual, are often present at low frequencies in wild populations. Genes of this type are termed *recessive. Dominant* genes are those with an observable effect when present in only one member of a chromosome pair. Degree of dominance can vary widely. In some cases it is complete, and the outward effect is the same as if the dominant gene were present in both members of the chromosome pair. In other cases dominance is incomplete with some level of intermediate expression.

Variations brought about by gene mutations are the materials with which the geneticist works and upon which the evolutionary process depends. Mutations may be recessive or completely dominant, or they may exhibit some intermediate degree of dominance. Recessive mutations are by far the most frequently observed type, partially because they can be carried for many generations in a hidden form. Often they are brought to light in a species only under laboratory or domesticated situations in which there is some degree of inbreeding.

Dominant mutations are observed much less frequently. Those with favorable effects are presumably rather quickly incorporated in a species and thus become the wild type. Those with unfavorable, but nonlethal effects, are rather quickly lost from populations due to natural selection. Those with lethal effects do not appear as a phenotypic type and are thus difficult to study.

Chemically, mutation is a change in DNA at a particular point in a chromosome. Formerly, efforts were made to distinguish between mutations of a single gene (so-called point mutations) and chromosomal changes. However, with increasing knowledge of the chemical nature of the hereditary material, it has become clear that distinctions between single-gene changes and chromosomal changes may be only those of degree. Thus, the term mutation has come to be used to include both.

It is thought that mutations can develop from mistakes in base pairing and to deletions or insertions of hereditary material. When it is realized that each gene may consist of up to several hundred base pairs, the possibilities for distinctly different changes are seen to be enormous. Thus, the concept of numerous chemically different variants of the same gene is realistic.

Mutation is always a relatively rare event but spontaneous rates differ widely between genes. In *Drosophila*, some loci may mutate as frequently as once in 20,000 individuals or as infrequently as once in 200,000. Some genes in humans may mutate as often as once in 10,000 to 100,000 persons.

Mutation rates can be increased by ionizing radiation, by ultraviolet radiation, and by a wide variety of chemicals. Radiation-induced mutations and those induced by many chemicals appear to be of a random nature. In other cases, however, chemical mutagens are specific for certain sites. It has always been a dream of geneticists to produce "directed" favorable mutations. The techniques of recombinant DNA may eventually bring this dream to reality.

Extrachromosomal Inheritance

Thus far, we have discussed inheritance based on materials carried in the chromosomes. Chromosomal inheritance is of major importance, but evidence also exists for extrachromosomal types of inheritance.

Many characters have initially been suspected of being nonchromosomal in inheritance. However, upon further examination many of these cases have been found to have a chromosomal basis. One of these was sex linkage which will be discussed in a subsequent section. In some cases, such as in the direction of coiling in snails, the apparently aberrant hereditary behavior has been found to be due to the fact that an individual's appearance is based on genes carried by the mother rather than on the genes of the individual. These cases are thus truly chromosomal.

There are, however, cases of extrachromosomal inheritance. The best known is that of *plastids* in plants. These are bodies found in the cells and which are involved in photosynthesis. Many characteristics of plastids are controlled by chromosomal hereditary material; but in some cases, such as in variegated plants having leaves varying greatly in color, inheritance seems to be due to factors transmitted in the maternal cytoplasm. *Mitochondria* are another type of small cytoplasmic bodies found in both plants and animals. They perform metabolic functions essential for survival. Both plastids and mitochondria are self-replicating particles and both have small amounts of DNA. It has been hypothesized that both were at one time free-living organisms which far back in evolutionary history became obligate endosymbionts of larger cells. Recently, evidence has been reported of the maternal inheritance of mitochondrial DNA (mtDNA) in humans. Statistical analyses of animal production data have suggested that cytoplasmic maternal genetic inheritance influences growth, milk yield, and composition.

CLASSICAL GENETICS

Neither Mendel nor his successors for many years had an understanding of the chemical nature of hereditary material. Their studies were, of necessity, limited to determining the behavior or mode of transmissions of particulate units of heredity or genes. The term *classical genetics* is often used to describe studies of this type. A knowledge of the mode of transmission of genes is basic to all genetic specialties. The methodology of classical genetics is to make matings among individuals of a species having apparent differences in such ways that the mode of transmission can be determined. Basic laws were determined by observing and counting individuals with sharply differentiated characters. Mendel's success was due to his choice of material and his mathematical treatment of the results.

Two major principles or laws were postulated by Mendel from his data. The first involved *segregation* and *recombination*. Individual genes were viewed as discrete units which maintained their identity without blending with other genes in the zygote. These genes, which are present in duplicate, separate (segregate) in the formation of gametes and recombine as discrete units at fertilization. The second is the principle of *independent assortment*. In regard to this, Mendel stated: "The relation of each pair of different characters in hybrid union is independent." The genes influencing the traits in his experiments sorted out independently. Later we will see that this second postulate was modified by linkage. Mendel was apparently not aware of this, since knowledge that most genes are linked together on the chromosomes was not published until 1903.

One-Factor Inheritance

One trait Mendel studied was plant height in peas. He had two varieties, one growing to a height of 180 to 210 cm (tall) and the other to only 22 to 45 cm (dwarf). When crossed, all the offspring were tall. When the offspring of the crosses were allowed to self-pollinate, they produced offspring in the approximate ratio of 3 tall to 1 dwarf. The results are what would be expected if (1) each individual carried two hereditary factors affecting plant height, (2) the two original varieties were pure for tallness or dwarfness, respectively, (3) the hereditary factors did not blend but retained their identities even though tallness obscured or dominated the factor for dwarfness when both were in the same individual, (4) the offspring produced germ cells of two kinds in equal numbers with half carrying the factor for tallness and half the factor for dwarfness, and (5) the two kinds of germ cells combined at random, i.e., with those carrying the factor for tallness having an equal probability of combining with one carrying the factor for tallness or with one carrying the factor for dwarfness, and vice versa.

If we let T represent the factor for tallness and t the factor for dwarfness, the cross with the resulting 3:1 ratio can be represented schematically as follows:

Parents:

Phenotype	tall	×	dwarf
Genotype	TT		tt
Gametes	all T		all t

Offspring, or F_1 generation:

Phenotype	tall	×	tall
Genotype	T		T
Gametes	½ T ½ t		½ T ½ t

Offspring of F_1 mated to F_1, or the F_2 generation:

	tall	tall	tall	dwarf
Phenotype	tall	tall	tall	dwarf
Genotype	TT	Tt	Tt	tt
Phenotypic ratio		3	:	1

Several new terms used above plus a few others illustrated by the example are defined as follows:

Hybrid[1] The offspring of parents which are genetically pure for one or more pairs of different hereditary factors.

F_1[1] The hybrid, or first filial generation from a given mating. Offspring from inter-mating of the F_1 generation are the F_2, etc.

Phenotype The external appearance or some other observable or measurable characteristic of an individual. In the above example, tall and dwarf are the phenotypes.

Genotype The genetic constitution of an individual. In the case of height of pea plants the genotypes are *TT*, *Tt*, and *tt*.

Dominant One member of a pair of hereditary factors or genes whose effect is manifested in the phenotype wholly or partially, regardless of which other member of the pair or series is present. In the pea plant size example, *T*, the factor for tallness, is dominant to *t*, the factor for dwarfness.

Recessive An hereditary factor whose effect is not observable when present with the dominant member of the pair or series. In the genotype *Tt*, the factor *t* is not phenotypically expressed, hence, it is recessive.

Homozygote (adj. homozygous) Individuals which are genetically pure for a given pair or series of hereditary factors. The genotypes *TT* and *tt* are homozygous. A homozygous individual will produce only one kind of gamete as regards this particular factor pair or series.

Heterozygote (adj. heterozygous) Individuals which carry unlike members of a given pair of hereditary factors, i.e., *Tt*. Heterozygous individuals will produce two kinds of gametes in equal proportions for the factor pair or series involved.

Segregation Separation of members of a pair of factors at gamete formation. The genes remain as constant entities through the generations; they separate at gamete formation rather than combine or blend. This is the first principle of Mendelian heredity.

Alleles Members of a pair (or series) of hereditary factors at a locus or location on a chromosomal pair which segregate in the formation of gametes.

The simple cross of tall and dwarf pea varieties illustrates behavior in characters controlled by only a single allelic pair of genes or hereditary factors. All the F_1's are genetically *Tt* and phenotypically tall. Each parent was genetically pure or homozygous and produced only one type of gamete for this particular character.

When the F_1's are mated with each other to produce the F_2 generation, the situation is more complicated. Since the F_1's are genetically impure, or *heterozygous*, they each produce two kinds of gametes. These are produced in equal numbers, and fertilization occurs at random with gametes of the other parent. That is, a *T* gamete of one parent has an equal probability of uniting with a *T* or a *t* gamete of the other parent. The four possible F_2 genotypic combinations as shown in the diagram occur with equal frequency. Thus the F_2 generation is composed of ¼ dwarf and ¾ tall individuals. Of the talls, ⅓ (¼ of all offspring) will be homozygous. They are thus genetically like the original tall parent. Two-thirds will be heterozygous like the F_1's. If these

---

[1] In common usage, these terms have assumed somewhat broader meanings than given here and are applied to offspring of species crosses, to progeny of crosses of inbred lines within a species, and even in some cases to offspring of breed crosses.

heterozygotes are used for breeding, their genetic performance will be exactly like that of the F_1 generation.

It should be emphasized that the F_2 ratios are what would be expected on the average if large numbers of offspring were produced. With only a few offspring there could be major deviations from the expected ratio. Male gametes are produced in large numbers in most species, and it is a matter of chance which ones eventually fertilize female gametes. This chance at fertilization is governed by the laws of probability, which are discussed in Chapter 5.

Literally thousands of characters have been identified in plants, insects, and higher animals which behave in inheritance as one-factor pairs with dominance. The first report of such a trait in cattle, the polled and horned condition, was in 1902. In other monofactorial cases the genes exhibit only partial dominance. This is illustrated diagrammatically in the following case of a cross of pure red and pure white Shorthorn cattle.

Parents:						
Phenotype	red		×		white	
Genotype	RR				rr	
Gametes	all R				all r	
Offspring, or F_1 generation:						
Phenotype	roan		×		roan	
Genotype	Rr				Rr	
Gametes	½ R ½ r				½ R ½ r	
Offspring of F_1 mated to F_1, or the F_2 generation:						
Phenotype	red		roan		white	
Genotype	RR	Rr		Rr	rr	
Phenotypic ratio	1	:	2	:	1	

It can readily be seen that the genes behave exactly the same in this cross as in the cross of tall × dwarf peas. However, the heterozygote is intermediate in color, roan.

The degree of dominance can vary greatly. In some cases it is complete or almost complete so that the heterozygote gives no phenotypic evidence of its genetic constitution. In other cases the heterozygote is a phenotypic intermediate.

Two-Factor Inheritance

A more complex type of inheritance than monofactorial inheritance occurs when two pairs of factors are considered concurrently. They may affect different phenotypic characters or some single character.

This can be illustrated by two factors segregating in Holstein cattle. Most members of the breed are horned and black and white, but some individuals are naturally polled and some are red and white in color. When a pure polled (*PP*) individual is mated to a horned individual (*pp*), all the offspring (F_1) are polled (*Pp*), since the gene for polledness is dominant to the gene for horns. When the F_1's are intermated, a ratio of 3 polled to 1 horned offspring is expected. When a pure black-and-white (*BB*) individual is mated to a red-and-white (*bb*) individual, all the offspring are black (*Bb*), since the gene for black pigment is dominant to the gene for red. Intermating the F_1 would give an expected ratio of 3 black to 1 red offspring.

Now if pure black-and-white and horned (*BBpp*) Holsteins are crossed with red-and-white and pure polled (*bbPP*) Holsteins, all the offspring are expected to be black-and-white and polled (*BbPb*). When these F_1 individuals are intermated the following results are expected:

$9/16$ polled and black-and-white offspring
$3/16$ polled and red-and-white offspring
$3/16$ horned and black-and-white offspring
$1/16$ horned and red-and-white offspring

This can be represented diagrammatically:

Parents:
Phenotype	horned, black	×	polled, red
Genotype	*pp BB*		*PP bb*
Gametes	*pB*		*Pb*

F_1 Generation
Phenotype	all polled, black
Genotype	all *Pp Bb*

F_1 Gametes ♂

	(PB)	(Pb)	(pB)	(pb)
♀ (PB)	Polled, black *PPBB*	Polled, black *PPBb*	Polled, black *PpBB*	Polled, black *PpBb*
(Pb)	Polled, black *PPBb*	Polled, red *PPbb*	Polled, black *PpBb*	Polled, red *Ppbb*
(pB)	Polled, black *PpBB*	Polled, black *PpBb*	Horned, black *ppBB*	Horned, black *ppBb*
(pb)	Polled, black *PpBb*	Polled, red *Ppbb*	Horned, black *ppBb*	Horned, red *ppbb*

Since the F_2 phenotypic ratio in a monohybrid showing dominance is 3:1, in a dihybrid (if the respective sets of genes are in different chromosomes) we would expect the ratio in the F_2 to be $(3:1)^2$, or 9:3:3:1.

These results demonstrate the second law of inheritance, namely that of *independent assortment*. Independent assortment simply means that one member of a pair of genes going to one gamete has no influence on which member of any other pair that goes to that cell.

With independent assortment, the four possible types of gametes are formed in equal proportions. As you can see this has been assumed in the above 4 × 4 "checkerboard," or table. By inserting the fraction ¼ for each gamete and multiplying the ¼ for any gamete by the ¼ for any other, you can see that $1/16$ of the total offspring will represent each genetic combination.

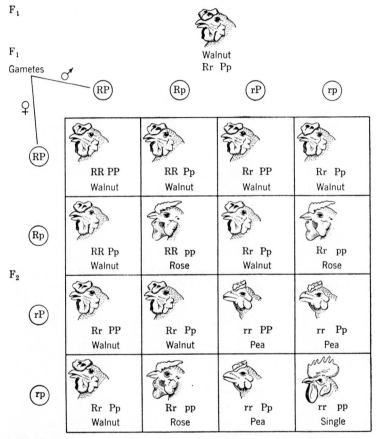

FIGURE 2-7

Diagram showing interaction of factors for comb form in fowls. The cross of a pure rose-comb bird with a pure pea-comb one gives all walnut-combed offspring. The 16 possible combinations of the F_1 gametes, with their genotypes and the phenotypes resulting from gene interaction, are shown in the F_2 checkerboard. (*From Sinnott and Dunn.*)

Depending upon dominance or lack of dominance in the gene pairs, the F_2 of a two-factor cross can deviate from the 9:3:3:1 ratio found when both pairs exhibit dominance. If we cross a polled, red Shorthorn with a horned, white one, we would have:

$$PPRR \times pprr$$

giving *PpRr*, a polled roan offspring. If we make the F_2 from this, we expect the following ratio since color exhibited only partial dominance.

3 polled, red:1 horned, red
6 polled, roan:2 horned, roan
3 polled, white:1 horned, white

If we had a case in which both characters lacked dominance, our phenotypic ratio would become 1:2:2:4:1:2:1:2:1.

The above dihybrids involved two pairs of genes determining different qualities. Cases are also known in which two pairs of genes act upon the same character. Comb type in poultry is an example. If a bird has the double recessive *rrpp*, it is single-combed; if it is *RRpp* or *Rrpp*, it is rose-combed; if *rrPP* or *rrPp*, it is pea-combed; and if it has at least one *R* and one *P*, it will be walnut-combed.

Linkage

In 1906 it was found that not all pairs of hereditary factors assorted independently. Bateson and Punnett in England were breeding sweet peas. Crosses of purple (dominant) and red had given a 3 purple:1 red ratio, and long pollen grains (dominant) and round ones also had given a 3 long:1 round in the F_2. It

FIGURE 2-8
Summary of phenotypic classes in two-gene F_2, resulting from dominance and different types of interallelic interactions or epistasis.

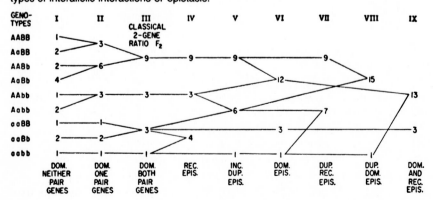

was expected that a cross of purple-long with red-round would give a 9:3:3:1 ratio in F_2. Instead of this, Bateson and Punnett obtained the following results:

Characteristics	Actually obtained			Expected	
	Number	**Proportion**	**Ratio**	**Number**	**Ratio**
Purple, long	4,831	.694	11.1	3,910.5	9
Purple, round	390	.056	.9	1,303.5	3
Red, long	393	.057	.9	1,303.5	3
Red, round	1,338	.192	3.1	434.5	1
Totals	6,952	.999	16.0	6,952	16

It is evident that these two characters of sweet peas did not follow the second Mendelian law, since they did not assort into all the possible combinations in an independent fashion. The two genes that were derived from the grandparents, purple and long from one grandparent and red and round from the other, tended to be held together. Hence, there were more of these combinations than expected in the F_2 and fewer than expected of the new combinations (purple-round and red-long). Since the characters tended to stay together, this feature of heredity was called *linkage*.

Linkage was not total since some new combinations were formed. Instead of the expected total of 2,607 new combinations, only 783 were found. The formation of new combinations in situations such as this is known as *recombination*.

Genetic evidence of the types reviewed thus far, reinforced by much detailed evidence on the behavior of chromosomes, provided early indications that hereditary factors were located in the chromosomes. In one-factor inheritance, the hereditary factors segregated independently. Microscopic studies of chromosomes indicated that they behaved similarly. In many cases of two-factor inheritance, the factors assorted independently. Here, too, behavior of hereditary factors parallels that of chromosomes since the members of different chromosome pairs assort independently. Obviously, each chromosome pair must contain more than one pair of hereditary factors since there are thousands of the latter and only a relatively small number of pairs of chromosomes. Table 2-1 gave the chromosome numbers of several animal species.

In the case of linked hereditary factors, their behavior in transmission can be explained on the basis of their being located in the same pair of chromosomes, thereby preventing independent assortment. However, since the linkage is not complete, there must be exchange of hereditary material between the two members of the same chromosome pair. This exchange of material, or *crossing over*, does occur at meiosis, and its frequency has been proved to parallel the recombination frequency of hereditary factors.

Studies of the genetic transmission of many allelic pairs in many combinations with one another has permitted the mapping of chromosomes in some intensively studied species. The most completely mapped species are the fruit fly (*Drosophila melanogaster*) and corn or maize (*Zea mays*). Among mammals, the mouse (*Mus musculus*) has been most intensively studied and reasonably extensive chromosome maps developed.[2] In farm animals less is known about linkage groups and the specific location of individual genes.

Linkage is a conservative force in heredity. It does not prevent the formation of new genetic combinations but reduces their frequency.

Epistasis

Interactions between nonallelic genes are termed *epistasis*. Several types of epistasis are known for specific cases of two-factor inheritance. It is probably a widespread phenomenon. Indeed, in a broad sense, the expression of any gene is dependent upon interactions and interrelationships with others.

If one gene of a pair masks the presence and prevents the manifestation of its allele, we say it is *dominant*. Likewise, a gene or genes of one allelic pair may mask the presence and manifestations of those of another pair. Several kinds of epistatic gene action are known, and the epistatic genes themselves may be either dominant or recessive.

If we mate a black mouse *AABB* (gene *A* for color, gene *a* for diluted color, gene *B* for expression of any color, and genes *bb* masking all color, that is, epistatic to *A*) to an albino *aabb*, all the F_1 are black, *AaBb;* but the F_2 will appear as 9 blacks, 3 creams, and 4 albinos. This is due to the fact that the presence of at least one *A* and one *B* produces black, two *a*'s and at least one *B* produce cream, but either *AA, Aa,* or *aa,* together with *bb,* result in albino. This is because the *b*'s mask the expression of *A* or *a;* that is, *bb* is epistatic to *A* or *a* so that the last two portions of the usual 9:3:3:1 ratio are thrown together phenotypically.

The foregoing illustrates the nature of epistasis or factor interaction. Accompanying Figure 2-8 gives in diagrammatic form various modifications of the classical 9:3:3:1 F_2 ratio which can occur. Names have been given to the different kinds of two-factor epistases. These are rather infrequently used now, but are included in the figure for their historical interest.

All modifications of the 9:3:3:1 ratios found in F_2's point to interactions in development between independently inherited genes. Many epistatic interactions have been studied biochemically, particularly several which control flower color in plants, coat color in mammals, and eye color in insects. It has been found in all cases studied that genes are involved in the specific chemical reactions necessary for development of color pigments.

At present we have very few clear-cut examples of epistasis for production traits in farm mammals, but it is not difficult to imagine that there may be almost limitless numbers of possible epistatic reactions actually present. Dominance can be a hindrance in breeding because it makes it impossible to sepa-

---

[2] See pp. 206–207 in King (1975). "Handbook of Genetics," vol. 4, Plenum Press, New York.

rate the homozygotes and heterozygotes by inspection. Likewise, epistasis can be a hindrance in breeding since, if any desirable qualities of an animal are due to epistatic combinations, they may not be passed on intact to offspring because of the halving nature of heredity.

Inheritance Involving Three or More Gene Pairs

Phenotypic and genotypic ratios have been determined for cases of inheritance involving three or more pairs of genes influencing different qualitative characters. Such studies can become tedious because the number of possible gametes, genotypes, and phenotypes increases rapidly with increases in number of gene pairs. The number of individuals required in controlled matings for firm establishment of ratios becomes astronomical if many gene pairs are involved. (These facts are illustrated in Table 5-1 in Chapter 5.)

Most traits of a quantitative nature are under the influence of several pairs of hereditary factors. Individually their effects are small, and most such traits are also influenced by environmental factors. These facts, together with the large number of gene pairs, make it impossible in most cases to identify individual genes and determine specific genetic ratios. However, the principles of segregation and recombination apply in such cases just as in the genes for the qualitative traits we have been using as illustrations thus far.

Multiple-Gene Heredity

Early in this century it was often believed that Mendelian-type inheritance was limited to qualitative traits such as color or horns. Hereditary differences in quantitative traits such as size or milk yield were believed to depend upon some other mechanism. It has been found, however, that characters such as those are dependent upon the action of many pairs of genes, each of which behaves in a Mendelian fashion. Usually we are not able to identify individual genes, but relatively simple cases have been established. These establish the basic principle that the mechanism of inheritance is the same whether few or multiple genes are involved.

The multiple-gene hypothesis grew out of the work of Nilsson-Ehle in 1908 with wheat color and East's 1910 work with ear length of corn. The former crossed several strains of red and white wheats. In general the red color was only partially dominant over white, for the F_1 was not as dark-red as the red parent. In the F_2 of one cross there were 3 reds (1 dark red, 2 lighter red) to 1 white. This indicated a one-gene pair situation. Another cross of red and white wheats gave a similar F_1 but an F_2 of 15 reds (of varying shades) to 1 white, thus indicating a two-gene pair situation. Still a third cross gave the usual F_1 but an F_2 of 63 reds (again of varying shades) to 1 white. This indicated a three-gene pair situation. Breeding tests revealed that there were one, two, and three pairs of genes segregating in these cases. In the two-gene-pair cross it was shown that a wheat with four genes for red was redder than one with three; this latter one was redder than one with two; and this, in turn, was redder than a wheat with only one gene for red. In these cases there is not complete dominance, but the genes act in a cumulative or additive fashion.

The mode of inheritance involving several gene pairs with cumulative but individually indistinguishable effects is now thought to be operative (along with dominance, overdominance, epistasis, and other types) in the inheritance of most of the commercially important animal characters. In other words, instead of color of wheat we might substitute size, weight, yield, or many other quantitative traits. This explains the difficulties encountered in attempting to unravel the exact mode of inheritance of many characters in farm animals.

We can seldom identify single-gene effects for quantitative traits in farm animals. However, statistical methods are available for estimating average effects of genes. These methods are covered in Chapter 6. Essentially, they estimate the average effect of substituting a gene say A for a in a genotype.

Multiple Alleles

In the foregoing examples the different alleles were considered as pairs, or alternate forms of a gene which could occupy a certain location in a given pair of chromosomes. Alleles also occur in series of three or more genes which can occupy a given chromosome locus. We call these *multiple alleles*. Normally individuals can carry no more than two members of a multiple allelic series, but in some cases large numbers of alleles are present in different individuals in a population. As discussed earlier, genes are composed of deoxyribonucleic acid (DNA), which can occur in an almost infinite variety of chemical combinations. Alleles and multiple alleles are presumably the result of rather small chemical differences.

One of the best-known series of multiple alleles occurs in humans as regards blood type. We do not inherit blood from our parents (it is made by the embryo, just as are bone and muscle), but we do inherit genes which control the type of our blood. There are four blood types in the human which are derived from the interaction of three allelic genes. The first gene is A, the second A^b, and the third a. Genes A and A^b are both dominant over gene a, but there is a lack of dominance between genes A and A^b, thus giving a new type, just as genes R (red) and r (white) Shorthorn cattle give a new type, Rr (Roan).

Thus, we can write the genetic makeup of humans for this particular series of blood types as follows:

Type A	AA or Aa
Type B	A^bA^b or A^ba
Type AB	AA^b
Type O	aa

The knowledge of the inheritance of these blood types (together with that of other types determined by genic series of other loci) is used in the courts to establish possible paternity in disputed cases.

The large number of loci in each chromosome, coupled with the fact that most species have several pairs of chromosomes, provides for the existence of very large numbers of genetically different types in any species even if only two genes could occupy any one locus. This range of genetic variability is further extended by the existence of multiple alleles. Several very extensive series of multiple alleles affecting qualitative characters are known in both plants and animals. Several of these which control blood antigens and other biochemical polymorphisms of farm animals are discussed in Chapter 4. Multiple alleles may be important for quantitative characters, but identification of individual series would be very difficult.

Sex Determination

Looking at the entire universe of animals and plants, a wide variety of genetic mechanisms determine sex. In addition, environmental factors often influence sexuality. In some cases environmental effects determine sex, in some cases they can reverse sex, and in many others they can influence the normality of sexual development. It is beyond the scope of this book to cover the subject in detail, but for an excellent review the inquiring student is referred to Srb et al. (1965).

In mammals, birds, many insects, and in many other forms of animal and plant life, chromosomes are of two types. Most pairs are alike in males and females. These are termed *autosomes*. There is an additional unlike pair known as *sex chromosomes*. In mammals, normal females have two so-called X chromosomes while males have one X paired with an unlike mate called the Y chromosome. The female is termed the *homogametic*, the male the *heterogametic* sex. Many insects, including *Drosophilia melanogaster,* the fruit fly which has contributed so extensively to genetic knowledge, have the same XY arrangement as mammals. In birds, reptiles, some insects, and some fish and amphibia, the situation is reversed. Males have two X chromosomes while females have either one X and one Y or an X with no pair mate. The latter case is called the XO condition. In these species the male is the *homogametic* and the female the *heterogametic* sex.

Sex chromosomes segregate at meiosis. The mechanism serves to maintain nearly equal numbers of the two sexes as diagrammed in Figure 2-9.

It is more accurate to say that the XX and XY conditions are associated with femaleness and maleness, respectively, than to say that sex is determined by these chromosomes. The same is true for species in which males are XX and females XY or XO.

The reason for caution in this regard is that the situation is considerably more complicated in at least some species. In *Drosophila* the Y chromosome seems to be neutral as regards sex determination, with the X chromosome exerting an influence toward femaleness and the autosomes an influence toward maleness. By experimentally producing individuals with variable numbers both of X and Y chromosomes and of sets of autosomes, it has been shown

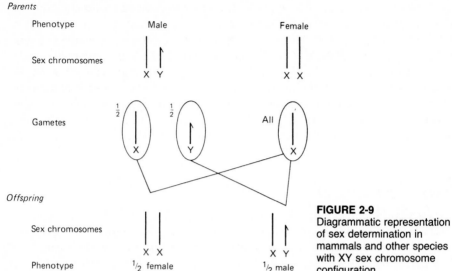

FIGURE 2-9
Diagrammatic representation of sex determination in mammals and other species with XY sex chromosome configuration.

that individuals with one X and the normal two sets of autosomes are males regardless of whether a Y is also present. Similarly, individuals with two X chromosomes and two sets of autosomes are females regardless of whether one or more Y's are also present. Other combinations have confirmed this *balance theory* of sex determination.

In mammals, at least in mice and in humans, the Y chromosome appears to be decisive with the presence of a Y resulting in a male and its absence a female.

Sex Linkage

In some of the examples used thus far the genes involved were in the autosomes and behaved the same in both sexes; in other examples we were dealing with monoecious plants in which both sexes are represented in the same individual and no sex chromosomes exist.

Chromosomes or portions of chromosomes are termed *homologous* if the allelic genes are in the same loci on each chromosome pair and if they pair at meiosis. They are *nonhomologous* if these criteria are not met. In species with the XY sex chromosome type, major portions of the X are nonhomologous with any portion of the Y, whereas certain portions are homologous. The relative extent of homologous and nonhomologous portions differ widely between species.

Genes located in the nonhomologous portion of the X chromosome behave in a characteristic fashion in inheritance and are termed *sex-linked*. Sex-linked characters had been observed in humans from ancient times, but the mode of inheritance had not been understood until the white-eyed character of *Drosophila* was studied by Morgan around 1910. *Drosophila* normally have red eyes. A white-eyed type appeared in a culture. When white-eyed males were mated to red-eyed females the two eye colors behaved in crosses as shown in Figure 2-10. In the diagram the + symbol is used to represent the wild type or normal allele of the gene for eye color. As can be seen, all the F_1's are red-eyed. In the F_2 all females are red-eyed whereas half the males are red-eyed and half the mutant white type. The student can readily diagram the cross of a white-eyed female and a red-eyed male and show that all daughters will have red eyes and all sons white eyes.

FIGURE 2-10
Diagrammatic representation of sex-linked inheritance using the white-eye gene of *Drosophila melanogaster* as an example. The + symbol represents the gene for wild type (or red-eyed) and *w* the recessive mutant for white eye.

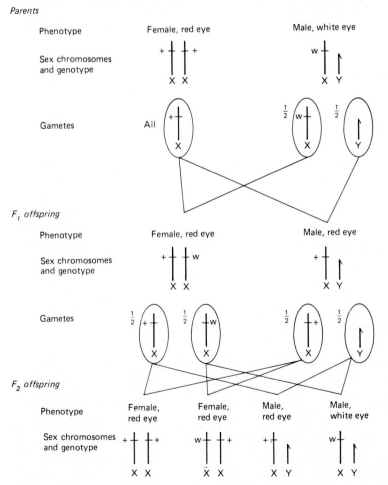

In sex-linked inheritance a trait depending upon a single recessive gene will always be more frequent in the heterogametic sex since the effect of the recessive gene has no possibility of being obscured by its dominant allele.

In humans, red-green color blindness is a sex-linked recessive and behaves in the same manner in inheritance as white eye in *Drosophila*. About 8 percent of the men in the United States but less than 1 percent of women are color-blind. The disease hemophilia in which the blood does not clot normally and affected persons bleed seriously from slight injuries is also a sex-linked recessive. Several other sex-linked recessives are known in humans.

Curiously, relatively few sex-linked genes have been found in mammals other than humans. Several have been identified in mice and a few in the dog. Few are known in cattle, sheep, swine, or horses.

A relatively larger number of sex-linked genes have been identified in poultry. The gene for barring of feathers is a dominant gene located on the sex chromosome. Since in birds the male is the homogametic sex, mating a barred Plymouth Rock male to a nonbarred or solid-colored female will produce only barred males and nonbarred females. This was a technique for genetic autosexing used early in the development of the broiler industry. Sex-linked genes also have been observed to influence body size, the rate of sexual maturity, and broodiness.

Sex-Influenced Heredity

In addition to the direct impact of sex-linked genes, several other influences related to genes in the autosomes are expressed differently in the two sexes. A classic example of this situation is a type of pattern baldness in humans. The gene is dominant in men but recessive in women. In crosses between a horned breed of sheep, the Dorset, and a hornless breed, the Suffolk, horned males and hornless females are produced indicating a difference in dominance between the sexes.

CHROMOSOMAL ABERRATIONS

Deviations from the chromosome number and morphology (karyotype) characteristic of a species have long been known. Indeed, in the early development of genetics some of the crucial evidence that genes were located in the chromosomes came from relating aberrant chromosome segregation or morphology to behavior of hereditary traits. Chromosomal changes or aberrations are often brought to light by cytological studies after disturbances in normally expected Mendelian ratios or after changed linkage relationships have been noted. Different types of chromosome aberrations are listed and briefly discussed in the material which follows.

Inversions

If a portion of a chromosome has become reversed and the normal order of a portion of the genes is changed, the change is termed an *inversion*.

Duplication and Deficiency

A break may occur in a chromosome with the part breaking off being lost and creating a deficiency or becoming attached to the other member of the homologous pair to form a duplication in the chromosome. The latter could occur as a result of unequal crossing over between homologous chromosomes.

Translocation

It sometimes happens that a chromosome becomes broken and one of the pieces attaches to another chromosome. If the piece becomes attached to a nonhomologous chromosome, a new linkage relation is set up and a translocation is produced. If, for instance, a small piece of the X chromosome in *Drosophila* were to become attached to one of the two large pairs or to the small IV chromosome, then the genes which were contained in the broken piece of X chromosome and which had formerly shown sex-linked inheritance would no longer do so. Likewise, if a piece of one of the autosomes were to break off and become attached to an X chromosome, those autosomal genes that had formerly shown ordinary Mendelian inheritance would now show sex-linked inheritance.

Reciprocal Translocation

In ordinary translocation, there is no trading between the chromosomes or chromatids; a section of one chromosome simply goes over to another chromosome without any reciprocity on the part of the latter. In *reciprocal translocation,* there is reciprocity with exchanges of parts between nonhomologous chromosomes.

Changes in Chromosome Number

Normally, a new organism arises from the union of an egg and a sperm, each of which has the reduced or haploid number of chromosomes. Their union re-

FIGURE 2-11
Diagrammatic representation of the process of inversion in which the normal order of some of the genes in a chromosome is changed.

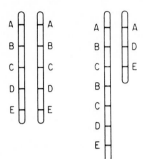

FIGURE 2-12
Diagrammatic representation of duplication and deficiency in a chromosome pair.

sults in restoration of the *2n* or diploid chromosome number. Sometimes due to error in the meiotic process, the new individual receives one (or occasionally more than one) too many or too few chromosomes, thus being *2n* + 1 or *2n* − 1 in number. Those with one extra chromosome are termed *trisomics* and those with a chromosome missing are called *monosomics*. *Aneuploidy* is a general term applied to individuals with an abnormal number of chromosomes. Some organisms with abnormal chromosome numbers are viable and can be studied cytologically. Down's syndrome in human beings is due to a *2n* + 1 chromosome number. The human conditions of Turner's syndrome in which the Y chromosome is missing and Klinefelter's syndrome in which there is an extra X chromosome are referred to in Chapter 3 in connection with sex ratios. There are counterparts of Klinefelter's syndrome in several farm animals.

Polyploidy is the term applied to conditions in which one or more entire extra complements of chromosomes are found in an individual. Polyploids can be formed in a number of ways including nonexpulsion of the first or second polar body in germ cell formation, fertilization of an ovum by two sperm cells (dispermy), or by a doubling of the chromosome number without cell division at an early embryonic stage. Individuals with three sets of chromosomes (*3n*) are known as *triploids* and *4n* individuals as *tetraploids*. Such polyploid individuals are rare in higher animals and so far as is known are not viable. In lower forms such as *Drosophila* they can be viable.

Chromosome Abberations in Farm Animals

Most chromosome aberrations tend to be present in populations at only low frequencies, since affected individuals are often sterile or have low reproductive rates. As a result the aberrations usually disappear from a population rather rapidly. Many chromosomal aberrations result in meiotic irregularities as a result of genes pairing or attempting to pair with alleles even though their normal positions in chromosomes have been disturbed.

Over the years much had been learned about the chromosomes of farm animals, but only by means of the painstaking methodology that was available until a series of new techniques developed in the 1950s and 1960s which revolutionized the field of animal cytology. These new methods facilitated chromosome studies and have resulted in a virtual explosion of knowledge since that time (see review by Fechheimer, 1974).

A number of chromosomal aberrations have been studied in farm animals. These include several trisomics associated with phenotypic abnormalities. A number of abnormalities, including polyploids, have been found in pig embryos at the blastocyst stage but not at later stages, suggesting that early embryonic mortality could be due, in at least some degree, to chromosomal aberrations (see review by Bruere, 1974).

Robertsonian Translocation

Of unique interest is the finding of many Robertsonian translocations in domestic animals (see reviews by Gustavsson, 1974, and Bruere, 1974). Translocations of this type were first described by Robertson in insects in 1916. They involve chromosomes which undergo centric fusion. Basically, this means that two chromosomes which have their centromeres (area for spindle fiber attachment) near the ends (acrocentric) fuse end to end. They then behave as a single chromosome with a central or metacentric centromere.

Animals heterozygous for a Robertsonian translocation have one less than the normal chromosome number and, if homozygous, two less. In both cases the total amount of DNA and the complement of genes is the same or nearly the same as in normal $2n$ individuals. The first report of a Robertsonian translocation in a farm animal was in 1964 in the Swedish Red and White breed of cattle. Since that time they have been identified in a number of breeds of cattle, and in sheep, swine, and goats. In cattle, the most frequent of these has become known as the $1/29$ translocation, with the numbers indicating the numbers of the pairs of chromosomes which fuse. The abnormality is present in unexpectedly high frequencies in some cattle breeds. More than 14 percent were found to be carriers of the translocation in a sample of over 2,000 animals of the Swedish Red and White breed. Translocations involving other chromosome pairs have also been described in cattle. At least three Robertsonian translocations have been found in sheep. In swine, chromosome polymorphism was first noted in European wild pigs in East Tennessee. In this population, a translocation of the Robertsonian type was present in such frequency that 31.5 percent of the individuals studied were homozygous for it (having 36 chromosomes), 53.7 percent were heterozygous (having 37 chromosomes), while only 14 percent had the normal 38 chromosomes.

As stated earlier, most chromosome aberrations do not persist in populations due to lower-than-normal viability or reduced reproductive rates of affected individuals. Translocations of the Robertsonian type may be an exception to this generalization. Reports differ in the apparent effects on reproductive rates, productivity, and viability. In some cases it is probable that reproductive rates are somewhat reduced, but in others there has been no apparent effect. Thus, as of this writing, the reasons for the occurrence, the

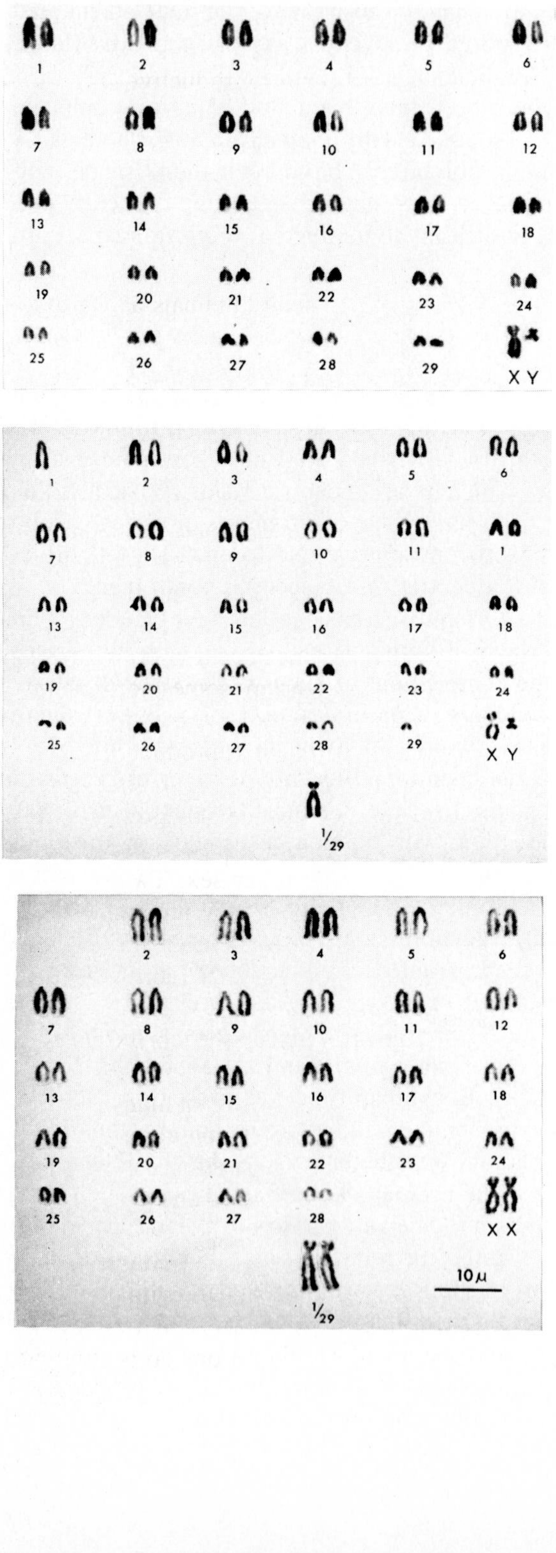

FIGURE 2-13
Metaphase chromosomes of cattle (*Bos taurus*) arranged in karyotypes. *Above,* normal male; *center,* a male heterozygous for the 1/29 Robertsonian translocation; and *below,* a female homozygous for the 1/29 Robertsonian translocation. (*Courtesy of Dr. Ingemar Gustavsson, Royal Veterinary College, Uppsala, Sweden.*)

relatively high frequency of occurrence in some populations, and the persistence of these translocations are a mystery. It does appear that their occurrence and normality or near-normality in viability and reproductive rate provide a mechanism by which changes in karyotypes can be introduced and retained in populations as part of the evolutionary process.

SUMMARY

Hereditary factors (genes) are contained in the chromosomes in the nuclei of microscopic cells which make up the bodies of all higher animals and plants. Hereditary processes are dependent upon deoxyribonucleic acid (DNA) which is a major component of chromosomes. DNA molecules are very large, of a double-helix form, and are capable of virtually an infinite number of variations. Each is made up of a series of nucleotides, each of which is composed of a base, a sugar, and a phosphate. The bases are paired in differing sequences which provide the genetic code. The functional units of heredity are specific segments of DNA called genes (or cistrons). DNA has the ability to replicate and serves as the template for formation of messenger ribonucleic acid (mRNA) which travels to the ribosomes of the cell carrying the genetic code for protein formation. Proteins in turn function as enzymes in controlling cell development and function. Genes and chromosomes occur in body cells in pairs, with each genetic pair occupying a specific chromosome locus. There may be multiple allelic series of genes which can pair, but only two normally occur in each individual. In the formation of gametes (reproductive cells), chromosome and gene numbers are reduced with only one member of each pair or allelic series going to each gamete. Random chance at segregation determines which member of a pair goes to each gamete. Chance also determines which gamete fertilizes a particular gamete from the other sex. Two or more pairs of genes (or allelic series) assort independently if they are in different pairs of chromosomes. They are said to be linked if located in the same chromosome pair. Linked genes remain together at gamete formation more frequently than would be the case if they assorted independently. Frequency of recombination is dependent upon the distance apart in the chromosomes. Qualitative characters often are under the control of a single pair of genes, whereas quantitative characters are usually under the control of many pairs of genes which may act additively or multiplicatively, may exhibit dominance, or may interact epistatically with each other. Most chromosome pairs are alike in the two sexes and are known as autosomes. In higher animals one pair, known as the sex chromosomes, is composed of unlike members. One sex (females in mammals, males in birds) has two like members, whereas the other sex has one of each kind and is known as the heterogametic sex. Hereditary factors located in the sex chromosomes are labeled sex-linked. Although mammals normally have two of each autosome, chromosomal aberrations do occur, and some individuals may lack a chromosome or may have one or more extra chromosomes.

SUGGESTIONS FOR FURTHER READING

Books

Ayala, F. J., and J. A. Kigh, Jr. 1984. "Modern Genetics," 2d ed., Benjamin/Cummings Publishing Co., Menlo Park, California.

Burns, G. W. 1980. "The Science of Genetics," 4th ed., The Macmillan Co., New York.

Eldridge, F. E. 1985. "Cytogenetics of Livestock," AVI Publishing Co., Inc., Westport, Connecticut.

Friefelder, D. 1983. "Molecular Biology," Jones and Bartlett Publishers, Portola Valley, California.

Goodenough, U. 1978. "Genetics," 2d ed., Holt, Rinehart and Winston, Inc., New York.

Strickberger, M. W. 1985. "Genetics," 3d ed., The Macmillan Co., New York.

Srb, Adrian M., R. D. Owen, and R. S. Edgar. 1965. "General Genetics," 2d ed., W. H. Freeman and Co., San Francisco.

Watson, J. D. 1976. "Molecular Biology of the Gene," 3d ed., W. A. Benjamin Co., Menlo Park, California.

Proceedings, "First World Congress on Genetics Applied to Livestock Production," Madrid. October 7–11, 1974. Graficas orbe, S. L., Padilla 82, Madrid (3 vols). Volume 1 contains articles in English relative to subjects in this chapter:

Basrur, P. K. Innovations in Cytogenetics and Their Applications to Domestic Animals. Pp. 215–227.

Bomsel-Helmreich, O. Experimental Induction of Chromosome Abnormalities. Pp. 201–213.

Bruere, A. N. The Discovery and Biological Consequences of Some Important Chromosome Anomalies in Populations of Domestic Animals. Pp. 151–175.

Fechheimer, N. S. Cytogenetics (Introduction). Pp. 109–125.

Gustavsson, I. Chromosomal Polymorphism. Pp. 191–199.

Makino, D. General Aspects of Chromosomes in Domestic Animals. Pp. 127–133.

CHAPTER 3

BRIDGING THE GENERATIONS

The reproductive process provides the critical mechanism for transmitting genes from one generation to the next and retaining their availability for further selection. Animal genetics and reproductive physiology are uniquely interwoven in attempts to improve animal populations. Animals with superior genotypes for production traits cannot contribute to improvement unless their reproductive capacity is maintained at satisfactory levels. High reproductive performance is also essential for an economically efficient commercial animal enterprise. Developments such as artificial insemination, embryo transfer, and cloning can markedly increase the number of offspring per parent; hence the influence of genetically superior individuals.

REPRODUCTIVE RATES

Maintaining a high reproductive rate is important economically; it is also crucial in providing maximum selection pressure for desired traits. It determines the percentage of animals which must be retained merely to replace those lost from the herd due to death, sterility, or old age. As a consequence, the intensity of selection for desired traits is directly dependent on the reproductive rate.

The length of the reproductive cycle and its components differ for various classes of farm animals. This establishes a biological time constraint to the length of the generation interval for a given species and influences the genetic change that can be expected for a specific time interval. The duration of components of the reproductive cycle in our major farm mammals is provided in Table 3-1.

The number of young born at each gestation also varies among species and classes of animals, because some have single offspring and others have litters of varying number. Table 3-2 gives what can be considered practical rates of reproduction. These optimal rates are rarely achieved in practice. In the United States it is estimated that the number of young born per year for each breeding female averages about 0.85 in beef and dairy cattle, 0.7 in horses, 1.05 in sheep, and 16 to 18 in swine. Many males are used in small herds or flocks, and they do not have opportunities to produce maximum number of offspring. The impact of the reproductive rate on selection intensity is given further consideration in Chapter 7.

Lowered reproductive rates can be due to many causes. Some are well understood and some are still subjects for research. Some animals are completely sterile and produce no offspring. Suboptimal numbers of offspring per year may result from failure to conceive at a given service thus requiring repeated matings, or from failure to maintain a pregnancy, low litter size, or low rates of twinning.

Under natural breeding systems, with typical reproductive rates and normal attrition, approximate percentages of animals reared in a given generation which must be retained for breeding to maintain constant herd or flock size are given in Table 3-3.

AUGMENTING REPRODUCTIVE RATES

Reproductive rates higher than those suggested in Table 3-2 have much potential for increasing rates of improvement in seed-stock populations such as a large herd or pure breed. The potentials for effecting rapid positive change in commercial populations is also of critical economic importance.

TABLE 3-1
DURATION OF COMPONENTS OF THE REPRODUCTIVE CYCLE IN FARM ANIMALS

Class of livestock	Age at puberty, months		Age at first breeding, months	Estrus cycle, days		Average gestation length days
	Avg.	Range		Avg.	Range	
Cattle	11	8–14	13–16	21	16–24	283
Horse	14	10–24	24–36	21	10–37	336
Sheep	7	6–9	16–20	16	14–20	150
Swine	7	5–8	7–9	21	18–24	113

TABLE 3-2
PRACTICAL MAXIMUM REPRODUCTIVE RATES UNDER NATURAL MATING AND COMMON
SEASONAL BREEDING PATTERNS

Class of livestock	Ovulation rate	Young per gestation	Gestations per year	Young per year
Cattle	1	1	1	1
Horse	1	1	1	1
Sheep	1–2*	1–2*	1–2†	1–2
Swine	12–30	6–15	2	12–30

*Extremely variable among breeds, with some releasing three or more ova and averaging two or more lambs.
†Depending upon whether or not accelerated lambing is used.

FIGURE 3-1
A 35-month-old Finnsheep ewe with her litter of six lambs. This breed is noted for its
prolificacy. A litter of six is exceptional, but this ewe produced triplets at 12 and 24 months
of age. (*Courtesy of Finnsheep Breeders Association.*)

Artificial Insemination

Males produce large numbers of spermatozoa—more than are needed for fertilization at a single service. Artificial insemination takes advantage of this
fact. Semen is collected artificially, subdivided into amounts required for inducing pregnancy, and mechanically introduced into the reproductive tracts of
a number of females. Species vary as to the number of females which can be
successfully inseminated with the spermatozoa in a single ejaculate. Averages
are about 600 in cattle, 15 in sheep and goats, 30 in swine, 110 in horses, and

TABLE 3-3
APPROXIMATE PERCENTAGES OF ANIMALS REQUIRED FOR BREEDING TO MAINTAIN
CONSTANT NUMBERS WITH NATURAL BREEDING

	Percentages needed	
Class of livestock	Males*	Females
Beef cattle	4–5	40–50
Dairy cattle	4–5	50–70
Horses	3–5	30–50
Sheep and goats	2–4	40–50
Swine	1–2	10–15

*Could be reduced to 0.5 percent or less with artificial insemination.

50 in turkeys. In practice, reproductive rates of this magnitude could seldom if ever be attained, but they do illustrate possibilities. Official data are not available regarding maximum numbers of progeny which have been produced by individual sires through use of artificial insemination. A few bulls are thought to have sired 200,000 or more offspring. Man (page 298 in Cole and Cupps, 2d ed., 1969) quotes data on a 5-year-old Caucasian ram which yielded enough semen in a single breeding season to inseminate 17,681 ewes and sire 18,414 lambs. Table 3-4 gives estimates of the number of progeny which could potentially be obtained per year per breeding male in farm animals through full use of all semen that could be produced.

Obviously, it would be very difficult to have as many females as could be inseminated from a single ejaculate in estrus at a given time and in the immediate vicinity of the male at the time of semen collection. Therefore, for artificial insemination to be practical in any except very special circumstances, it

TABLE 3-4
POTENTIAL NUMBER OF OFFSPRING PER MALE PER YEAR IN ARTIFICIAL
INSEMINATION

Class of livestock	Sperm per ejaculate, in billions	Motile sperm per yr,* in billions	Motile sperm required† per insemination, in millions	Potential pregnancies per year‡	Potential progeny raised per year§
Bull	9	1217	10	73,020	65,718
Ram	2.5	1950	120	9,750	13,160
Boar	50	5850	1200	2,925	21,940
Stallion	18	1825	100	10,950	9,855

*Based on regular schedules of semen collection at maximum frequency believed compatible with long-time maintenance of semen volume and concentration at normal levels.
†As estimated by Gomes in Cole and Cupps (1977) to represent minimum doses for normal fertility.
‡Assuming a 60 percent conception rate for females inseminated.
§Assuming a 10 percent death loss of embryos and offspring born in case of bull, ram, and stallion and 25 percent in case of boar.

was necessary to develop procedures for preserving and storing semen during transport and until females were in estrus. Practical and usable procedures for accomplishing this have been developed for cattle. Techniques are less satisfactory for other species, but progress is being made and artificial insemination programs are possible with all classes of farm animals. The effectiveness of semen preservation methods and the number of females which can be inseminated per semen collection have been major factors in the differential use of artificial insemination with different species.

In cattle, successful methodology was developed in the late 1930s for diluting and preserving semen at temperatures slightly above freezing for periods of up to 3 or 4 days. These techniques, with some subsequent improvements, provided a basis for initial large-scale artificial insemination programs with dairy cattle. In 1952 Polge found, rather accidentally, that with the inclusion of small amounts of glycerol in dilutors, cattle semen could be frozen and stored for long periods at temperatures of $-79°C$ (Celsius) (dry ice) to $-196°C$ (liquid nitrogen) with retention of viability adequate for successful use in artificial insemination when thawed. Frozen semen procedures are now very widely used. Liquid nitrogen is the preferred refrigerant. The ultimate time limit for successful preservation of semen, or indeed if there is a limit, is not known. At this writing, conception is known to have been obtained with semen stored over 30 years. Frozen semen has introduced heretofore undreamed-of flexibil-

FIGURE 3-2
Mother and young bull calf born 30 years after his sire's death. Semen from the calf's sire, Cottonade Emmet, was collected November 19, 1953, just prior to his death, and frozen in liquid nitrogen. (*Courtesy of National Association of Animal Breeders.*)

ity into breeding programs. Semen from a given male can be used concurrently nationwide and worldwide, and even long after a bull is dead. It has made it possible to collect from bulls on a regular year-round basis and make maximum use of superior individuals.

First records of the use of artificial insemination of dairy cattle in the United States show that 7,358 cows were inseminated in 1939. The number inseminated increased rapidly during the 1950s, with an average of 6,580,236 dairy cows inseminated each year during the period 1956–1960 and 7,220,614 during the period 1966–1970. Acceptance of artificial insemination came later in beef cattle, and an average of 850,057 were inseminated each year during the period 1966–1970. Within-herd insemination has replaced specialized technicians, and records of the number of animals inseminated have not been available since 1970.

Table 3-5 gives information on use of artificial insemination with cattle in the United States. A slight increase for dairy and a slight decline for beef usage is evident. Exports of dairy semen have increased over fourfold since 1970. Data are not available on the extent of use with other species, but the figures are comparatively low. In some other countries artificial insemination is used more widely than in the United States. In Denmark, for example, nearly 100 percent of all cattle are artificially inseminated. In the Soviet Union, artificial insemination is used extensively with sheep.

In sheep and goats, progress in developing methods for long-time semen preservation has lagged. However, in recent years some success has been attained in long-time storage by deep freezing. Artificial insemination can be quite successful using semen immediately after collection or during the first few hours. Considerable labor is required to detect ewes in estrus. This, coupled with low values per animal, has limited use of artificial insemination with sheep in the United States. In this species, as with beef cattle, development of an effective, low-cost procedure for estrus synchronization would undoubtedly stimulate the use of artificial insemination. However, universal use undoubtedly depends upon improving methods for long-term semen storage.

In swine, relatively large volumes of semen and large numbers of spermatozoa are required for conception with artificial insemination. Even so,

TABLE 3-5
ARTIFICIAL INSEMINATION IN CATTLE IN THE UNITED STATES

Year	Units of semen sold or custom-frozen*	
	From dairy bulls	From beef bulls
1970–74	11,707,398	3,754,213
1975–78	11,650,634	2,486,005
1979–82	13,891,580	2,443,122
1983–86	13,352,162	2,288,760

*Data, courtesy of National Association of Animal Breeders, and represent averages on a yearly basis for the years indicated.

FIGURE 3-3
A Nichols Large White turkey male and a Broad-Breasted Bronze turkey hen. Males reach weights of 30 to 35 kg at breeding age and are unable to mate naturally. At maturity, males of some strains attain a weight of up to 45 kg. (*Courtesy of Department of Poultry Science, North Carolina State University.*)

there is much potential for extending the use of superior sires. Development of methods for long-time storage of semen has been slow. However, in recent years swine semen has been successfully stored in the frozen state. Commercial use of artificial insemination is growing but the extent of its potential acceptance is still problematical. With present artificial insemination techniques, there is evidence of slightly lower average litter size than following natural service.

Legend tells us that artificial insemination was first used successfully with horses. An Arab chieftain is said to have stolen semen, kept it warm in a bag under his arm, and inseminated his prize mare! Regardless of the authenticity of this tale, artificial insemination has long been practiced to some extent with horses. Pioneer artificial insemination research was done with horses in Europe in the late 1800s and early 1900s before its use had been seriously considered for other species of farm animals. Oftentimes, artificial insemination with horses has consisted merely of collecting the semen which escapes during normal mating and introducing it into the reproductive tract of the mare after service to "reinforce" natural mating. It has also been used with horses as a disease preventative—usually collecting semen by artificial vagina and immediately introducing it into an estrual mare.

Use of artificial insemination to extend the use of outstanding stallions has to date been very limited. In part this is due to extremely restrictive rules on artificial insemination by many breed associations. Long-time semen storage is still being perfected as numbers of sperm required for fertilization are larger than for cattle. In spite of these difficulties, horse artificial insemination as a commercial enterprise has been developing.

It is of interest that turkeys are now produced almost exclusively by artificial insemination. Selection of heavy broad-breasted types has made natural mating almost impossible. This is an example of how selection for productive traits has impaired mating.

EMBRYO TRANSPLANTATION AND CLONING

As early as 1929 subcutaneous implantation of pituitary glands was shown to increase the number of ova ovulated in laboratory animals. By the late 1930s partially purified pituitary gonadotropic hormones had been developed. Both these and the gonadotropic hormone found in the blood serum of pregnant mares were shown to increase ovulation rates in farm animals. The ova were capable of fertilization and normal early embryonic development, but when there had been several ovulations in cattle or sheep, nearly all embryos died during the first 1 to 2 months of gestation. The result was no increase, or in many cases a decrease, in net reproductive rate. Embryonic and fetal death was apparently due to inability of the uterus to sustain the additional embryos. Transplantation to recipient females when in the two-cell to blastocyst stages resulted in normal development, thus proving that the embryos themselves were normal.

This work stimulated interest in the idea of increasing progeny numbers from females of outstanding merit by (1) using gonadotropic hormones to stimulate ovulation of large numbers of ova and then (2) transplanting fertilized ova or early embryos to recipient females for pregnancy and parturition. The resulting offspring would genetically be progeny of the outstanding female, the recipient having served merely as an incubator.

FIGURE 3-4
A set of triplet calves produced and raised by their mother following superovulation as a result of treatment with a gonadotropic hormone. The Angus-Holstein crossbred mother raised all three calves to 205 days at which time their total weight was 491 kg and the weight of the dam 479 kg. One calf was a male, the other two freemartin females. (*Courtesy of Dr. E. J. Turman, Oklahoma Agricultural Experiment Station.*)

The first live calf from embryo transplantation was born in 1953. By the late 1960s, research—much of it in England by Rowson, Polge, and collaborators—had reached a stage suggesting the practicality for commercial application with cattle. At about the same time, a great deal of interest developed in Canada and the United States in the use of several continental European cattle breeds for beef production. Importation of females of these breeds to North America was very costly, and only limited numbers could be imported through existing quarantine facilities. Embryo transplantation provided a method for increasing numbers of these breeds in North America.

The procedure for cattle involves treatment of the donor female to induce superovulation and breeding or artificially inseminating her in order for fertilization to occur. This is comparable to the treatment used to induce multiple births without embryo transfer. The embryos are removed about 4 days after estrus and transplanted to recipient females in the same stage of the estrual cycle as the donor. In some cases the recipients are synchronized with hormone treatments, but some firms maintain large-enough herds of potential re-

FIGURE 3-5
Five ⅞ Maine-Anjou heifers resulting from transplantation of fertilized ova from a single donor cow to five recipients in an Ohio commercial beef herd. (*Courtesy of Dr. Donald R. Redman, Ohio Agricultural Research and Development Center, Wooster.*)

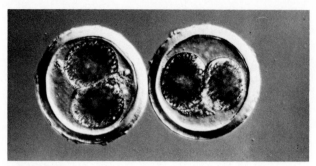

FIGURE 3-6
Very early swine embryos at the two-cell stage (X approximately
190). (*Courtesy of Dr. Vernon G. Pursel, U.S. Department of
Agriculture.*)

cipients to have the requisite numbers of females in estrus on the same day as
the donor.

Numbers of offspring per cow secured by transplants have been variable.
Two to eight or, in exceptional cases, more normal embryos may be secured
from a cow at one time and on average about 50 percent of them result in successful pregnancies after transplantation. One case of 52 pregnancies in recipient cows in a 9-month period using embryos from a single donor cow has been
reported.[1]

As used in most cases to date, the procedures are expensive and are economically feasible as a means of increasing numbers of rare exotic breeds and
possibly for increasing numbers of offspring from outstanding females.
Nonsurgical methods for embryo recovery and transplant are being used commercially. Research continues in long-time storage of embryos, although it
also has been perfected for routine use.

The high cost of embryo transfer will likely continue to limit its usefulness
in the production of herd replacements. It will continue to be used to produce
young bulls from elite dams for progeny testing. Theoretical considerations indicate that the rate of genetic improvement could be increased by 10 to 15 percent with the use of multiple ovulation and embryo transfer.

Splitting, or cloning, embryos into two or more segments, with each developing into an offspring, is now being used commercially for special circumstances. Currently such a division can be carried out only a limited number of
times, but techniques are developing to obtain more clones from a single embryo. Such individuals are monozygotic since they are derived from a single
fertilized egg, and they are the same genetically as the infrequent "identical,"
or monozygotic, twins that occur naturally.

---

[1] *Livestock Breeder Journal,* June 1977, p. 27.

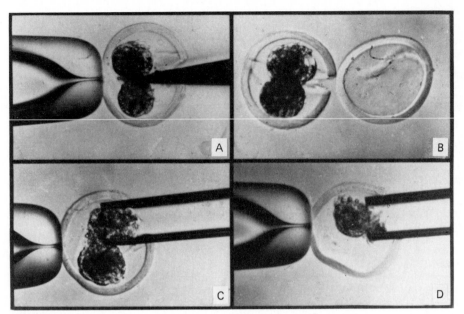

FIGURE 3-7
Bisection of a bovine embryo. A. With a microsurgical blade, an embryo is divided into two groups of cells. The embryo is held stationary on a holding pipette by suction. B. The bisected embryo alongside a second capsule that has been prepared to receive one of the embryo sections by cutting open an unfertilized egg and removing the original contents with a suction pipette. C. One of the embryo sections is removed from the original capsule. D. The embryo section removed from the original capsule is inserted in the second capsule. The result is two embryos ready for transfer to properly synchronized host females to produce genetically identical offspring. (*Photographs courtesy of George E. Seidel, Jr., Colorado State University.*)

SEX RATIOS AND SEX CONTROL

In farm animals, the sex ratio at birth is rather close to 1:1 with some species having a slight excess of males and others females. Further, in spite of apparent deviations from a 1:1 ratio in the progeny of some sires and dams, it appears that most if not all deviations can be accounted for by the laws of probability as discussed in Chapter 5. In practice, it sometimes seems that there are more deviations than could be accounted for by chance, but it should be remembered that we usually talk only about the exceptional cases. Thus, the cases we hear about represent very incomplete samples of the total population.

There could be many advantages in both improvement of seed-stock herds and in commercial animal production if means for controlling sex were available. In dairy cattle, "planned matings" of the best progeny-tested bulls are made to cows with high levels of production and other desired characteristics for the purpose of producing bull calves of superior potential genetic merit. The resulting bull calves are prime candidates for progeny testing at young ages. However, the cows in such matings cooperate by producing bull calves

FIGURE 3-8
Genetically identical twin calves resulting from bisection of a bovine embryo and placement of the embryos in host or surrogate mothers, who carry the calves to term. (*Photograph by Robert Harvey. Courtesy of George E. Seidel, Jr., Colorado State University.*)

only about half the time on average. Superovulation, sexing, and embryo transfer are about to overcome this limitation.

There are often advantages in commercial production of having females of a specific breed or cross which can be mated to males of a "terminal sire" breed or cross for production of market animals. Presently, ease of producing these special females is reduced by the fact that an equal number of males are born. In beef and sheep, males usually are more efficient meat producers than females. Therefore, there would be advantages in producing only males in herds devoted purely to the production of slaughter animals.

As presented in Chapter 2, sexuality in farm mammals is associated with chromosome differences. There is one pair of chromosomes called the *sex chromosomes*. For this pair, the female has two like chromosomes called X chromosomes. The male has one X chromosome and one different in size and shape called the Y chromosome. The female produces only one type of gamete, or germ cell, and it contains one X chromosome plus one member of each pair of autosomes. The male produces two types of germ cells, half containing one X and the other half one Y chromosome, in addition to one member of each pair of autosomes in each case. These unite randomly with female germ cells at fertilization to produce approximately half females (XX) and half males (XY) in the next generation. In poultry, sex is also related to chromosome complement, but females have the unlike and males the like pair of sex chromosomes.

Recognizing that sex chromosome differences are probably the primary factor in sex determination in farm animals, we must also recognize that the development of functionally normal males and females is a complicated process involving the coordinated interaction of genetic, hormonal, and environmental factors. Many of these are incompletely understood. Further, there are deviations from a 1:1 sex ratio at different life stages which pose interesting and as-yet unanswered questions.

In bovine twins of unlike sex, it has been observed from antiquity that there is sterility in the female (associated with major abnormalities in the development of the reproductive tract) in about 90 percent of the cases. The same condition also occurs in females born with males in triplet or higher-order birth sets. The sterile females are termed *freemartins*. Lillie hypothesized many years ago that the freemartin condition was due to vascular anastomosis of placentas with interchange of blood between twins and that hormones of the male exerted a deleterious influence on the early development of the female reproductive system. Later, Owen's work with blood antigens proved that vascular interchanges occur. Subsequent studies have shown that the freemartin condition occurs only when blood admixture has also occurred. Both XX and XY chromosome complements have been found in leukocytes and in a number of other tissues in both the freemartin and the apparently normal male co-twin. This condition is termed *sex chromosome chimerism*. The word *chimera* refers to "an individual in which two or more genomes coexist."

The possibility of controlling sex has intrigued humanity from antiquity, and literally hundreds of methods have been proposed to accomplish it. Since the two types of sperm produced by mammals contain different members of the sex chromosome pair, there is a possibility of discovering a physical or chemical difference between them which would permit their separation or the inactivation of one type. If this could be accomplished, only X- or only Y-bearing sperm cells with X or Y chromosomes could be used for insemination and only the desired sex produced.

Many proposals for sex control are based on folklore or superstition. Studies aimed at separating sperm types by centrifugation, electrolysis, and antigenic approaches have been unsuccessful or at least have not proved suitable for commercial use. Assuredly, efforts to separate X- and Y-bearing sperm will continue, and at this time some laboratories have advertised that they can already make a practical and useful separation.

While success in sexing sperm has been elusive, early sexing of embryos before transfer has been possible at a high level of accuracy. Bovine and ovine embryos have been accurately sexed at the 12- to 15-day stage by chromosome analysis. However, the procedure is time-consuming and requires highly trained personnel, making it unlikely for commercial embryo sexing. Antigenic methods using the H-Y antigen offer a method that can become successful as a method for sexing embryos prior to transfer. The H-Y antigen is in the histocompatibility group which is present in the mammalian male. Experimental data using monoclonal antibody techniques have approached an accuracy of 90 percent.

Sex ratios continue to be of interest to both basic scientists and livestock breeders. At this time it is worthy of emphasis that most deviations from a 1:1 ratio at birth are relatively small. It is also worthy of repeating that, in spite of the fact that breeders often feel that some sires tend to beget an undue proportion of one sex or the other, there is little evidence to substantiate this. Thus, breeders are to be discouraged from eliminating otherwise highly desirable sires because they have in the past produced a preponderance of one sex, and they should be advised not to make other changes in breeding programs based on past sex ratios.

SUMMARY

Successful reproduction is essential to the transmission of the genes from one generation to the next. Maximum reproductive rates are essential to maximize progress in genetic improvement and to maintain profitable commercial production. Female reproductive processes are cyclic in nature with periods of estrus or heat accompanied by ovulation occurring at regular intervals. Lengths of both estrus and estrual cycles are characteristic of each species. Augmentation of reproductive rates of the male through artificial insemination has permitted rapid genetic advances in dairy cattle and to a lesser extent in beef cattle. It has potential for improvement in other classes of animals, but with some limitations in swine and horses due to large semen volumes required for fertilization. Embryo transplantation has a potential for enhancing the rates of genetic improvement through improving reproductive rates of superior females, but currently it is a comparatively expensive process. Cloning of embryos offers a further opportunity to extend the impact of superior animals. Sex ratios are near 1:1 in farm animals. Successful methods of sex control could have important implications for both genetic improvement and reduced production costs. However, the search continues for a practical and effective method.

SUGGESTIONS FOR FURTHER READING

Books

Cole, H. H., and P. T. Cupps (eds.). 1977. "Reproduction in Domestic Animals," 3d ed., Academic Press, New York. (A reference is made in text to 2d edition of this book, published in 1969.)

Hafez, E. S. E. (ed.). 1984. "Reproduction in Farm Animals," 5th ed., Lea & Febiger, Philadelphia.

Kiddy, C. A., and H. D. Hafs (eds.). 1971. "Sex Ratio at Birth—Prospects for Control," American Society of Animal Production, 113 North Neil Street, Champaign, Illinois.

Maule, J. P. (ed.). 1962. "The Semen of Animals and Artificial Insemination," Technical Communication 15, Commonwealth Agricultural Bureaus, Farnham Royal, Bucks., England.

Nalbandov, A. V. 1976. "Reproductive Physiology of Mammals and Birds," 3d ed., W. H. Freeman and Co., San Francisco.

Perry, E. J. (ed.). 1968. "The Artificial Insemination of Farm Animals," 4th ed., Rutgers University Press, New Brunswick, New Jersey.

Proceedings, "First World Congress on Genetics Applied to Livestock Production," Madrid, October 7–11, 1974. Graficas Orbe, S. L., Padilla 82, Madrid (3 vols.). Volume 1 contains many review articles. Two of interest relative to this chapter are:

Bruere, A. N. The Discovery and Biological Consequences of Some Important Chromosome Anomalies in Populations of Domestic Animals. Pp. 151–175.

Young, G. B. Mendelian Factors and Reproductive Criteria. Pp. 57–63

Salisbury, G. W., N. L. VanDemark, and J. R. Lodge. 1978. "Physiology of Reproduction and Artificial Insemination of Cattle," W. H. Freeman and Co., San Francisco.

Sorensen, A. M., Jr. 1979. "Animal Reproduction," McGraw-Hill Book Co., New York.

Articles

Britt, J. H., N. M. Cox, and J. S. Stevenson. 1981. Advances in Reproduction in Dairy Cattle. *J. Dairy Sci.* 64:1378–1402.

Gordon, I. 1976. Controlled Breeding in Cattle. Part 1. Hormones in the Regulation of Reproduction, Oestrus Control, and Set-Time Artificial Insemination. Part 2. Pregnancy Testing, Control of Calving, Reduction of the Calving Interval, Induction of Twinning, Breeding at Younger Ages, and Future Developments. *Anim. Breed. Abs.* 44:265–275 and 451–460.

Maijala, K. 1987. Genetic Control of Reproduction and Lactation in Ruminants. *J. Anim. Breed. Genet.* 104:53–63.

Marcum, J. B. 1974. The Freemartin Syndrome. *Anim. Breed. Abs.* 42:227–242.

Powell, R. L., H. D. Norman, and F. N. Dickinson. 1975. Sire Differences in Sex Ratio of Progeny. *J. Dairy Sci.* 58:1723–1726.

CHAPTER 4

QUALITATIVE GENETICS IN ANIMAL BREEDING

Most traits of economic importance in farm animals are quantitative in nature. Most are genetically influenced by many pairs of genes. However, many simply inherited traits of a qualitative nature are important in animal breeding. They may be of direct economic importance, or they may exert harmful effects which completely overshadows or prevents the expression of genes controlling desired quantitative traits.

A qualitative trait is defined as one which can be characterized categorically for such things as a particular color or for the presence or absence of horns, a certain defect, a given blood antigen, or a specific type of protein in blood serum or other blood constituents. This contrasts with quantitative characters in which a gradation of measurable variation exists.

The expression of many, but not all, qualitative traits can be related to one or to a few pairs of genes. Numerous qualitative traits have been identified as having a genetic basis in farm animals. For many, the mode of inheritance is well established. In other cases a hereditary basis is known to exist, but the exact mode of inheritance is not known with certainty.

We often speak of a gene or of genes for a trait such as horns, color, or a physical abnormality. Strictly speaking, this is not correct, since the gene we are talking of is only one of many necessary for development of the trait. It is, however, one of the gene pair, or an allelic series in which variation has occurred and enabled us to identify an effect. By analogy, it may be correct for us to say that a chip is responsible for a computer's failure to operate, but we can hardly say that the replacement chip is responsible for the successful operation of the computer. Rather, it operates as the result of the successful, coordinated function of this chip and many other components.

In this chapter it is not our purpose to attempt a cataloging of genetically controlled qualitative traits.[1] We hope, rather, through photographs to illustrate some of the many forms which can occur and to discuss some of the problems a breeder may face in connection with them. Some useful aspects of genetic variants will also be considered.

LETHALS AND GENETIC ABNORMALITIES

Although wide anatomical and physiological differences occur within species, there is an eventual limit of deviation from the norm beyond which the organism cannot survive. Death of the organism may occur at any stage of development—immediately following fertilization, during embryonic differentiation, at parturition, or postnatally. Death may be due to a variety of causes, such as injury, disease, malnutrition, and harmful irradiations such as x-rays and gamma rays. Any cause of death is termed a *lethal effect*. Among the many causes of death are gene changes which are incompatible with development or survival. These genes are known as *lethal* genes. Some genes are deleterious to the organism but not lethal, provided environmental factors are especially favorable; these genes are called *semilethals*. There are undoubtedly many undetected lethal genes, but there are likewise many unexplained deaths due to environmental factors.

The development of a new individual from a tiny fertilized egg is one of the most interesting and complicated processes in all of nature. What any individual will eventually become is obviously dependent on both heredity and environment. Heredity provides the basic specifications; the environment, both internal and external, provides the wherewithal for fulfilling the specifications. When an organism succumbs before the usual time for its species, it is often exceedingly difficult to determine whether the cause of its early demise is hereditary or environmental, or perhaps some combination of both.

Is an Abnormality Heritable?

Typically, a lethal or other abnormality will first come to the attention of a breeder when one or more defective individuals appear in the herd or flock. There are no absolute rules for determining whether the abnormality is hereditary or environmental in origin, whether it is due to some combination of hereditary and environmental influences, or whether it is merely an "accident of development."

The following would indicate a hereditary basis:

1 If previous studies on a scale large enough to be conclusive had shown a hereditary basis for a phenotypically similar condition in the same species or

[1] See Bogart, R. 1959. "Improvement of Livestock," pp. 71–111, The Macmillan Co., New York, for a tabulation of numerous genetic defects.

breed. This would not necessarily be a completely reliable guide because some characters, almost or completely phenotypically indistinguishable, may be caused by either heredity or environment. Hydrocephalus in cattle is a probable example.

2 If the condition appeared only in some breeding groups or families. An example would be a condition appearing in some sire progenies but not in others.

3 If it occurred in herds where there had been inbreeding. Inbreeding does not create genetic abnormalities, but since most such abnormalities are recessive, it tends to bring them to light as a result of increased homozygosity resulting from inbreeding.

4 If it occurred in more than one season when rations and environment differed.

The following would indicate an environmental basis:

1 If it had previously been reliably reported as due to ration or environment.

2 If it occurred when the ration of the dam was known to have been deficient or when she had been under other stress.

3 If it did not recur after rations or environments were changed or improved.

If a condition met none of the above criteria and occurred only once or very rarely, it should be looked upon as an accident of development. No action to prevent recurrence should be undertaken until facts are available to provide a sound basis for action.

The foregoing represents an oversimplification in that we have arbitrarily classified abnormalities into those due to (1) genetic causes, (2) environmental

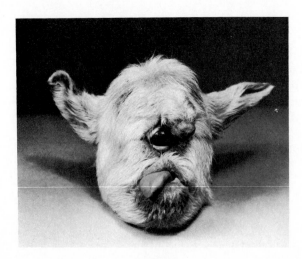

FIGURE 4-1
Head of a cyclopian type of malformed lamb with a centrally located single eye, a proboscislike protuberance originating just above the eye and absence of the premaxilla bone. This condition and variations of it have been produced experimentally by feeding the weed *Veratrum californicum* early in gestation. The sensitive period to the embryo is the 14th day of gestation. (*Courtesy of the late Dr. Wayne Binns, Poisonous Plant Research Laboratory, U.S. Department of Agriculture, Logan, Utah.*)

effects, and (3) accidents of development, i.e., those for which we have no explanation. In a strict sense every abnormality is the product of heredity *and* environment. If the gene or genes conditioning an abnormality are uniformly expressed over a range of environments which includes the ''normal'' range, we see only the genetic effects and think of it as genetic.

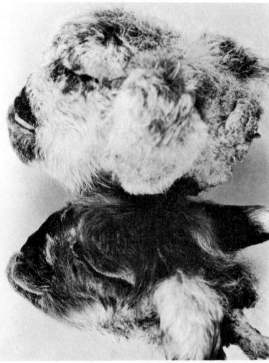
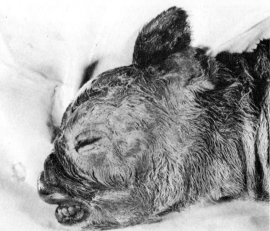

FIGURE 4-2
Above, two ''monkey-faced'' lambs thought to be less severe expressions of the same environmentally induced condition shown in Figure 4-1. *Below,* a calf homozygous for the gene which conditions the trait adenohypophyseal aplasia. It was born dead after a gestation period 100 days longer than normal. Note the striking resemblance in profile of the lambs and the calf.
(Courtesy of Dr. Clyde Stormont, University of California.)

Genetic Abnormalities Affected by Environment

The expression of several abnormalities in laboratory animals and plants varies within the range of normal environments. The bar-eye condition in the fruit fly *Drosophila melanogaster* is one of the best known of these. The normal compound eye of this insect has many subunits or facets—usually 800 or more. In bar-eye individuals the number is much reduced, but the reduction is much larger at high rearing temperatures than at low. Expression of abnormalities of this type are said to exhibit "genotype-environmental interactions." In swine, scrotal hernia has an heritable basis, but its incidence is also influenced by a maternal effect. The nature of the effect is not known, but it appears to have differential effects on different genotypes.

Several defects in farm animals are conditioned partially by hereditary variations of a quantitative nature and partially by environmental factors. The best known of these is cancer eye of cattle. It occurs more frequently in Herefords than in other breeds. Although it occurs in most geographic locations, within this breed it is more frequent in locations with high-average annual hours of sunshine. Latitude and altitude are also related to incidence, probably as a result of differences in the ultraviolet component of sunlight. It is usually an affliction of older cattle. Its incidence increases, and it occurs at younger ages in cattle maintained on high levels of nutrition. Hereditary variation affects the age of cancer development as well as occurrence or nonoccurrence. Hereford cattle with pigmented eyelids and corneoscleral areas are less susceptible than white-eyed types. Preference in the United States often has tended to be for white-eyed types. This is probably a case of selection for a "fancy point" having harmful effects.

Selection against defects of these types is difficult because they are expressed more frequently at advanced ages. A sire may have left many offspring before it is discovered that his daughters have an unusually high susceptibility to a condition such as cancer eye. Selection for pigmented eyes should be an effective indirect method of selecting against cancer-eye susceptibility, but unfortunately, pigment in the corneoscleral areas does not develop fully until the age of 5 years or more. Thus, selection for it cannot be fully effective at young ages.

Losses from Genetic Abnormalities

No accurate data are available on the total losses from hereditary abnormalities. Usually, any one abnormality, although it may be serious in a given herd, will not be a serious industrywide problem. Total losses from all such conditions may be more serious than generally realized. In most farm animals, prenatal death losses range from 20 to possibly over 50 percent of all fertilized eggs. Causes are not well understood, but increases with inbreeding suggest that some of the loss may have a genetic basis.

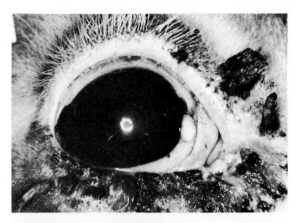

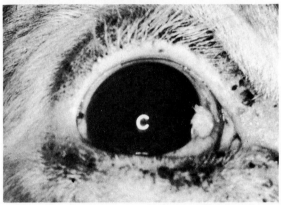

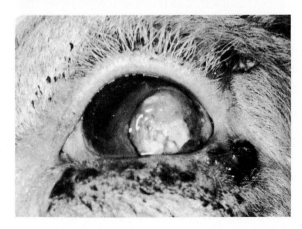

FIGURE 4-3
Three stages in the development of "cancer eye" in cattle.
Above, a plaque.
Middle, a slightly more advanced but still benign stage of papilloma.
Below, a carcinoma. (*Courtesy of Dr. David E. Anderson, University of Texas Medical School, Houston.*)

FIGURE 4-4
A "crooked calf." Conditions similar to this have been observed in many Western states. Deformities of the front legs are most consistent, but the neck, back, and hind legs can be affected. The condition has been produced experimentally by feeding lupine (*L. sericeus*) during the 40th to 70th day of gestation. It is believed that other nutritional imbalances or toxic materials may also cause it. (*Courtesy of the late Dr. Wayne Binns, Poisonous Plant Research Laboratory, U.S. Department of Agriculture, Logan, Utah.*)

Recessive Genetic Defects with Some Expression of Heterozygote

Most lethals and abnormalities are recessive in inheritance, whether due to one or several genes. The death or culling of affected individuals usually keeps the frequency of the gene or genes at low levels and in equilibrium with mutation rate as discussed in Chapter 6 on population genetics. A few cases are known in which recessiveness is not complete, and the heterozygotes, or "carriers," have characteristics which make them more favored or desired by breeders than the homozygous normals. The classic example of this is the Dexter breed of cattle. Cattle of this type are always heterozygous for a semidominant gene which when homozygous produces a lethal achondroplasia (bulldog calves). Dexters themselves show the effects of this gene by shortness of leg. When intermated, Dexters produce about ¼ long-legged individuals known as Kerrys, ½ short-legged Dexters, and ¼ bulldog calves.

Apparently the hereditary situation is as follows:

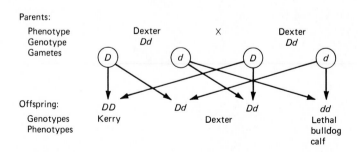

Preference of British breeders for the short-legged Dexters has resulted in the development of a breed carrying a lethal gene at a frequency of 0.5. Since the heterozygous individuals are easily identified, it would be an easy matter for breeders to cull them if they so desired. The Dexter type can be propagated without the production of lethals by avoiding the intermating of Dexters. Kerry × Dexter matings give 50 percent each of Kerry and Dexter types.

A more insidious situation occurred in a Swedish dairy breed. A type of infertility characterized by gonadal hypoplasia was found to be highly hereditary but not a clear-cut case of a single pair of genes. Although it was not demonstrated beyond all reasonable doubt, evidence indicated that unilaterally affected cows, on average, produced milk with a higher-than-average fat percentage and were favored in selection. The gene or genes responsible for the condition attained such frequency in the breed that bilateral gonadal hypoplasia (with resultant sterility) reached a level constituting a serious prob-

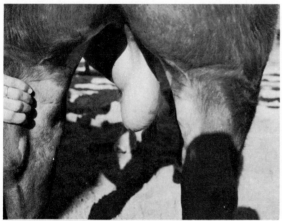

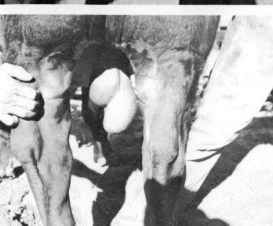

FIGURE 4-5
Two yearling Hereford bulls showing incomplete descent and imperfect development of left testicles. This defect appears to have a heritable basis since it has occurred in one inbred line and not in other Herefords maintained in the same environment. It is not, however, simply inherited, and it also varies in expression from cases like those shown to cryptorchidism. In the latter condition, the testicle is retained in the body cavity. (*Courtesy of Dr. R. S. Temple, formerly of the U.S. Department of Agriculture.*)

lem to the breed. Intentional selection against the condition subsequently reduced the frequency markedly.

In the midyears of the twentieth century, a type of dwarfism characterized by small size, high mortality, bulging foreheads, undershot jaws, difficult breathing, a tendency to bloat, and poor coordination (Figure 4-7) reached a frequency in at least two breeds of beef cattle in the United States high enough to constitute an economic problem. The common name of "snorter" dwarf was applied to these animals. Initially the condition appeared to be inherited as a simple recessive. Although definite proof is lacking, the apparent increase in frequency strongly suggested that the gene was not completely recessive, but had some effect in the heterozygous condition making animals shorter-bodied

FIGURE 4-6
Top left, a tailless Angus Bull. *Top right,* a Holstein calf with no anal opening (atresia ani). *Below left,* a Hereford calf born tailless, eyeless (microophthalmia), and with multiple genital abnormalities. *Below right,* a calf with a short lower jaw (brachygnathia inferior). The latter condition has a hereditary basis in at least some breeds. The others are of unknown etiology. These and other photos of abnormalities are printed to illustrate the wide variety of things which can go wrong in developmental processes. (*Courtesy of Drs. H. S. Leipold and Keith Huston, Kansas State University.*)

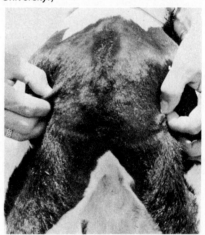
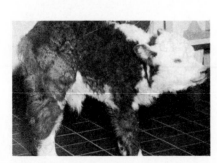

FIGURE 4-7
Above, a "snorter" or brachycephalic dwarf Hereford calf, 9 months of age, with a normal Jersey heifer of the same age. *Below,* a crossbred Hereford X Angus dwarf of the same type with her Angus mother. Since this type of dwarfism is inherited as a one-gene recessive, the occurrence of a crossbred dwarf indicates that both the parental breeds include animals carrying the same recessive gene. (*Courtesy of Drs. H. W. Leipold and Keith Huston, Kansas State University, and Dr. K. P. Bovard, Virginia Polytechnic Institute and State University.*)

and lower-set. This led breeders to select heterozygous, or "carrier," animals for breeding purposes in more than a random proportion of cases.

Situations such as the foregoing are difficult to cope with. The heterozygous animals apparently have an *average* superiority, but they overlap so much with homozygous normals that they cannot be accurately identified phenotypically. In some cases research has developed methods for identifying them or aiding in their identification. An example is the procedure of evaluating x-rays of lumbar vertebrae which was of some usefulness in identifying carriers of the gene for snorter dwarfism.

A partial deficiency of uridine monophosphate (UMP) synthase has been discovered in Holstein cattle. This enzyme is responsible for the conversion of orotic acid to UMP, the precursor of all other pyrimidine nucleotides. Affected animals have half the normal activity for this enzyme when heterozygous for the condition. Homozygous recessive genotypes apparently are lethal in utero. Enzyme assays provide an accurate identification of carriers, all of which thus far have been traced to one male.[2] Advances in molecular genetics may permit identification of carriers through DNA probes. More often, however, breeders will have to use predigree information and progeny test methods for reducing the frequency of such conditions.

Semilethal Recessives Related to Economically Desirable Traits

Fortunately, few characters of this kind are known. However, one is of sufficient interest and importance to be worthy of some discussion.

In several breeds of swine in both the United States and Europe, a pork quality problem became apparent in the 1950s and 1960s in types selected for muscularity and thin back fat. It is known as a pale, soft, exudative pork (abbreviated PSE). In carcasses of affected individuals the lean tissue is light in color and lacks firmness; fluids may seep from cut surfaces.

A second condition, highly associated with PSE, but not completely linked, is the so-called pork stress syndrome (PSS) in which affected pigs are unable to withstand even short periods of strenuous physical exercise and may die quickly when subjected to such stress. In some cases susceptible animals may live to market weights or even to reproductive ages if not subjected to undue stress. However, death rates may be high, particularly during marketing, and for this reason PSS can be a highly important economic problem.[3]

A test involving exposure of pigs to a standard level of halothane anesthetic for a prescribed period was devised to identify stress-susceptible animals. In this test, stress-susceptible pigs develop muscular rigidity, while normal animals are unaffected. PSS has been found to be inherited as a simple recessive. Interestingly, it is related to certain blood groups.

Perhaps the most dramatic example of the relationship of PSS to economic traits is a report from Switzerland.[4] Two lines of pigs were selected from the same

[2] Shanks, R. D., et al. 1987. J. *Dairy Sci.* 70:1893–1897.
[3] See review of Campion, D. R., and D. G. Topel. 1975. J. *Anim. Sci.* 41:779–786.
[4] Vogeli, P., et al., 1984. J. *Anim. Sci.* 59:1440–1450.

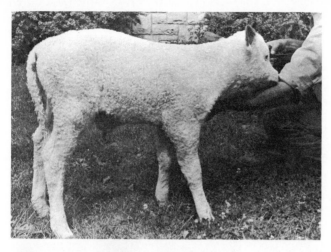

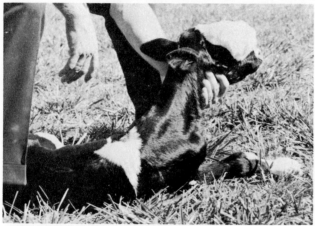

FIGURE 4-8
Above, an Ayrshire calf homozygous for a recessive gene, resulting in an extremely curly hair coat known as "Karakul curl." *Below,* a Holstein calf with brain hernia—thought to be caused by a recessive lethal gene. (*Courtesy of Drs. H. W. Leipold and Keith Huston, Kansas State University.*)

base population, one for superiority and one for inferiority of an index based on daily gain and back-fat thickness. After six generations, 42 percent of the superior line were halothane-susceptible whereas none of the inferior line were affected.

In the United States the condition has apparently been greatly reduced in frequency in recent years as a result of breeders selecting against it. Probably, the industry is tolerating more fatness in market animals as a result of selection against this problem.

Strategy for Genetic Defects

The action to be taken if a lethal or genetic abnormality is discovered in a herd depends upon the type of herd and the seriousness of the abnormality. In a commercial herd, the only action usually necessary is to cull the sire or sires

FIGURE 4-9
Upper left, a yearling Angus bull affected with a muscular hypertrophy condition often referred to as "double-muscled" or "dopperlender." *Upper right,* the carcass of an affected newborn calf showing extreme muscle development in the rear quarters. *Lower left,* an affected and a normal calf, each 6 weeks of age. *Lower right,* the carcasses of

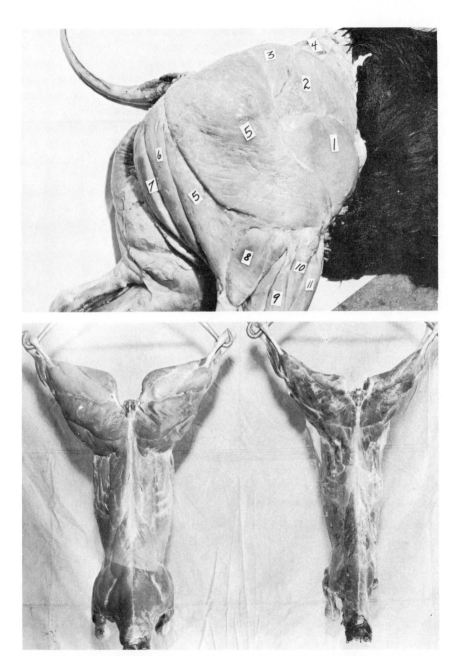

the calves shown at left. This condition is found in several cattle breeds. In most cases it behaves as a recessive with some expression often apparent in the heterozygote. Other cases, possibly involving modifying genes, do not satisfactorily fit a single-gene hypothesis. (*Courtesy of Prof. Walter H. Smith, Kansas State University.*)

which produced the defective offspring and replace them with unrelated males. For most traits, the frequency is so low that the probability of obtaining new sires which also carry the undesired gene or genes is very low. Only normal offspring will likely be produced from new sires even though some females in the herd still carry the defective gene. The probability of acquiring replacement sires which do not carry the deleterious gene or genes can be increased by a knowledge of the pedigree lines to avoid. Some breed associations are now making such information available on a restricted basis for certain defects.

Corrective measures may need to be more drastic in seed-stock herds since the owner has an obligation to provide stocks which will perform well for future customers. Further, not all buyers make rational choices in deciding which herds to patronize, and seed-stock herds sometimes must not only provide productive stocks but must be above suspicion as regards genetic defects. For seed-stock herds the following should be considered as possible measures for elimination of the defect or for reducing its frequency:

1 Cull all sires which have produced defective offspring.

2 Replace the herd sires culled with animals whose pedigrees indicate there should be only minimal probabilities of the new sires being heterozygous for the defect.

3 Remove all females which have produced defective offspring from the seed-stock herd itself. They may be placed in an auxiliary herd and used to progeny-test future herd sires to determine whether they are heterozygous for the gene responsible for the defect.

4 Cull other close relatives of affected individuals including normal offspring of sires and dams which have produced defective individuals.

5 If the affected individuals are viable and fertile, retain them for progeny-testing prospective breeding animals.

6 Progeny-test prospective herd sires before using them extensively in the herd (see Chapter 6 for more details).

How far to pursue the foregoing program will depend upon a number of factors including, but not necessarily limited to (1) economic importance of the defect (i.e., does it affect productivity or is the defect merely esthetic?), (2) frequency and distribution in the breed, and (3) importance attached to presence or absence of the defect by other seed-stock breeders or commercial producers to whom breeding stock might be sold. Thus, each case should be studied carefully and procedures adapted to the specific circumstances.

COLOR, HORNS, AND FLEECE

Most livestock breeds have standard colors or color patterns which serve as breed trademarks. Most of these are unrelated to productivity, but a few are of economic worth. As an example, white udders in beef cows may lead to "snow burn" if cows calve in the spring before snow is gone. As discussed

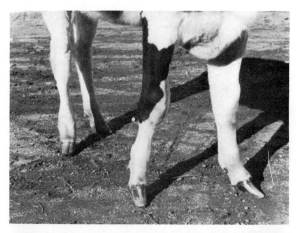

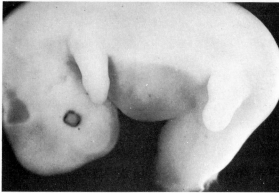

FIGURE 4-10
Above, a Holstein with the single-toe condition, syndactylism, often popularly called "mule foot." It is inherited as a simple recessive. *Below,* a 37-day embryo removed by Caesarean section which already shows the syndactylous condition. Evaluating embryos at this stage or later permits a rapid progeny test. (*Courtesy of of Drs. H. W. Leipold and Keith Huston, Kansas State University.*)

earlier, pigment in and around the eyes of white-faced cattle reduces the incidence of eye and eyelid cancers. In hot areas with intense sunlight, light coat colors in cattle reflect more heat and thus are an aid in maintaining normal body temperatures. Black wool will not take dyes of other colors and is of very low value.

Most basic color variations are inherited in a fairly simple fashion, and in many cases maintenance of the breed trademark constitutes no special problem. However, modifications of basic patterns sometimes depend upon additive-type gene action with considerable nonhereditary variation also occurring. Examples of this are the color patterns in breeds such as Hereford cattle and Hampshire and Poland China swine.

Shade of color (i.e., light or dark) has been a concern in several breeds of both cattle and swine—especially red breeds. In spite of the fact that there is no known relationship between shade of color and productivity, fads for certain shades have sometimes developed. Selection for unfixable color patterns, specific shades of color, or other details of appearance reduces the intensity of selection for important traits and is to be avoided if at all possible.

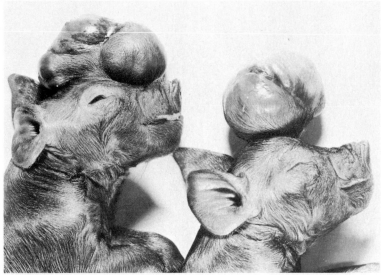

FIGURE 4-11
Pigs homozygous for a recessive gene causing "brain hernia." This gene is usually lethal when homozygous, but by special care a few affected animals have been successfully reared and used in breeding tests. (*Courtesy of Dr. Clyde Stormont, University of California.*)

Some standard colors in breeds are dominant. The black color of the Angus is an example already mentioned. Complete elimination of the recessive gene for red in the breed is very difficult (actually almost impossible) without a general system of a progeny-testing. This would probably not be economically feasible. In spite of many years of eliminating reds as they occur, and probably some culling of bulls and cows producing reds, various estimates are that 1 in 200 to 1 in 400 births are of red calves. Red is also present as a recessive in the Holstein breed. Currently both the Angus and Holstein breeds have registries for animals with red.

In most breeds of sheep and cattle, the presence or absence of horns depends upon fairly simple genetic patterns. The size and shape of horns are apparently modified by many pairs of genes, each with minor effects.

Breed standards often specify horn size and shape as well as presence or absence. Horns are of economic importance in cattle, being a detriment under farm conditions because of the trouble of dehorning. They sometimes have advantages under range conditions.

In European breeds of cattle, the presence or absence of horns usually (there are some unexplained exceptions) behaves as if under the control of a single pair of allelic factors with the dominant allele resulting in the absence of horns, or *polledness*. Some of the traditionally polled breeds such as the Angus are apparently homozygous, or nearly so, for the gene for polled and

FIGURE 4-12
An Ayrshire calf homozygous for a recessive gene resulting in
congenital dropsy. Most cases are lethal but a few survive.
(*Courtesy of Drs. H. W. Leipold and Keith Huston, Kansas State
University.*)

FIGURE 4-13
A bovine monster. Abnormalities of this type
occur rather rarely and are not known to have
hereditary bases.

produce essentially 100 percent polled calves in crosses with other European
breeds. Polled strains of other breeds developed in the United States during
this century from polled mutants have been forced to draw heavily on horned
stocks for foundation material. They are still far from pure. Making them
homozygous for polled will be a difficult task—having the same problems as
eliminating genes for red from a black breed.

In most fine-wool sheep, the presence or absence of horns depends upon a
single pair of alleles with heterozygotes being horned in males. Females of all

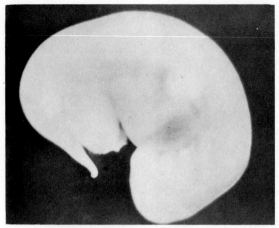

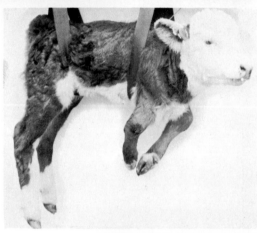

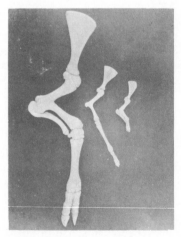

FIGURE 4-14
Under experimental conditions irradiation at early embryonic stages can result in anatomical abnormalities. *Above,* a normal 32-day calf embryo—note the developing front limb bud; *center,* a calf with malformed shortened forelegs born to a cow given a 400-r air-dose of gamma radiation from Co^{60} on the 32nd day of gestation; *below,* front leg bone (left) of the calf shown in center photo and leg bones of a lamb and a pig whose dams were irradiated on the 23rd and 21st days of gestation, respectively. Note ankylosis of radiohumeral joint in all three sets of leg bones. (*Courtesy of Dr. R. L. Murphree, University of Tennessee.*)

genotypes are polled or have a slight amount of horny growth known as knobs. The polled gene is related in some way in breeds of this type to cryptorchidism, a defect in which testicles are retained in the abdominal cavity rather than descended into the scrotum. Whether the cryptorchidism is due to a pleiotropic effect of the same gene or to a closely linked gene is not known with certainty. A few normal rams have been progeny-tested at high levels of probability and are apparently homozygous for polled. This would indicate either (1) close linkage which has been broken or (2) presence of modifying genes which prevented the expression of cryptorchidism even though this is a normal pleiotropic effect of the gene.

An interesting case of a new industry being built upon a single mutant gene is the *N* gene discovered in New Zealand Romneys. This gene is nearly, but not completely, dominant with newborn lambs which carry the gene and which have a high level of long, coarse, medullated halo-hairs in their fleeces. In mature sheep the *N* gene results in fleeces with coarse primary fibers. The wool is useful for carpet manufacture. It has advantages over many carpet wools from other breeds in that it is entirely white and therefore can take dyes of any color, and it usually has greater tensile strength and a lower incidence of kemp than other breeds. A new breed, the Drysdale, has been developed from Romneys carrying the *N* gene. A carpet-wool industry based upon it has developed in New Zealand. In size and fertility the Drysdale is essentially the same as the Romney, but it produces 0.5 to 1.0 kg more fleece.

GENETIC DIFFERENCES IN DISEASE AND PARASITE RESISTANCE

The success of plant breeders in developing crops which are highly resistant to specific diseases has long been an important factor in maintaining yields at high levels. Genetic differences in resistance to specific diseases in laboratory animals and in poultry have been repeatedly demonstrated. Further, some of the physiological characters apparently associated at least to some degree with disease resistance have been identified. Examples are the relation of number of leukocytes to the resistance of mice to *Salmonella typhimurium* and the genetic control of the thermoregulatory mechanism to resistance of chickens to *Salmonella pullorum*.

In chickens, genetic differences in resistance to lymphomatosis have been exploited in commercial stocks. However, as yet this approach has not provided a complete answer to the problem.

In larger farm animals genetic differences in resistance to specific diseases undoubtedly exist. They have been demonstrated conclusively for mastitis and much suggestive evidence is available on other diseases. Genetic differences have been demonstrated in the resistance of sheep to the common stomach worm *Haemonchus contortus*. Tick resistance or tolerance has been noted in *Bos indicus* cattle, and milk fever incidence is generally higher in Jerseys than in Holsteins.

In plants, resistance to disease has often been found to depend upon a single gene. Introduction of these single desired genes into commercial varieties can be accomplished rather rapidly when needed. To date, genetic differences found in resistance to disease or parasites in animals have not been of this type. Apparently many gene pairs are involved. In other words, effective use of genetically controlled disease resistance appears to present many of the same problems as do breeding for improvement of quantitative traits, such as milk production or growth rates. This situation, together with the fact that low reproductive rates (as compared to plants) would make it difficult to incorporate genes for disease resistance into entire populations quickly, has not encouraged research on the problem.

Philosophy to date has generally been that diseases in large farm animals can be controlled more effectively and economically by vaccines, sanitary procedures, test and slaughter, etc., than by genetic approaches. Thus, relatively little study has been given to the potential effectiveness of genetic procedures.

Breeding for disease resistance would likely be very expensive. Unless physiological indicators of resistance could be used in selection, a system of challenging animals of every generation through use of a live, virulent organism would be necessary. High death or debility rates would be almost certain until resistant animals were located. An indirect expense would be the fact that selection for disease resistance would reduce or eliminate possibilities of selecting for other desirable traits.

In recent years much research has been done on the major histocompatibility complex (MHC) of mammals with most emphasis on humans and mice. This is a complex of genes which influences immune responses and thus resistance to certain diseases. All farm animals have an MHC, but these are given different names in different species. In cattle it is BoLA, in the pig SLA, in the sheep OLA, in the horse ELA, and in the chicken B. Enough research has been done to prove that variations in the MHC are related to disease resistance. As yet the only practical use of this information has been in the chicken. With more research information, it is possible that useful tests can be developed that will indicate resistance to specific diseases in other species. Such tests, if developed, could eliminate the need for challenges with live organisms each generation, but would have the same disadvantages as other means of selecting for disease resistance, reducing intensity of selection for other economically important traits.

In spite of the apparent difficulties, the field of genetic resistance to disease and parasites is one which would probably justify more time for research with farm animals than has been set aside for it to date. It might be especially fruitful for disease and parasites for which other control methods are unsatisfactory. Chronic, nonlethal conditions such as mastitis in dairy cattle are sometimes thought to be a potentially fertile area for genetic selection. In most cases, however, until proven methods of selecting for disease resistance have been developed, breeders should depend upon other methods of control and concentrate on selection for traits on which progress is more probable.

BLOOD GROUPS AND BIOCHEMICAL POLYMORPHISMS

Literally hundreds or perhaps thousands of genetically controlled, chemical variants have been discovered in cells and body fluids of farm animals. Some of these are antigens or antigenic factors carried on the red blood cells. They are usually referred to as *blood groups*. They are identified by immunological procedures involving antibodies. In some cases naturally occurring antibodies are found in animals not having a particular antigen. These can be used as a test substance for identifying individuals carrying the antigen. This parallels the situation in humans as regards the ABO blood group system. In other cases, including the vast majority of blood antigens in farm animals, antibodies are not known to occur naturally. Antibodies for use in test reagents are produced by transfusing the blood of individuals with a given antigen into animals not carrying it. Antibodies are produced and obtained from the blood serum of transfused animals for testing. It would be unusual if the blood of the donor and recipient animals differed in only one antigen. Thus, in most cases a given antisera will have antibodies for many antigens. Therefore, the antisera must be fractionated by a series of absorptions to produce a reagent specific for only one or a desired combination of antigens.

Determination of whether the blood of an individual being tested carries a particular antigen is based on either the agglutination (clumping) of red blood cells or lysing (cells ruptured with release of hemoglobin) when cell suspensions are mixed with an antisera containing the specific antibody and an appropriate complement such as rabbit serum to enhance the reaction. Most farm animal blood group identifications are based on lysis rather than on agglutination.

Development and maintenance of a complete range of antisera for a species requires painstaking effort and is a costly undertaking. As a result relatively few laboratories in the world have the capability for making blood group determinations in farm animals. For example, to our knowledge there are currently only four laboratories in the United States with capability for making cattle blood group determinations.

Biochemical polymorphisms is the term often applied to molecular variants of protein, although the term is broad enough to cover other types of variants as well. Protein variants or polymorphisms are detected by biochemical procedures, principally electrophoresis. Starch gel electrophoresis is used in most cases. Chromatography can also be used in some instances and as a supplement to electrophoresis. Electrophoretic separation is based upon the fact that different proteins have different electrical charges. When proteins are placed in an appropriate buffer solution and subjected to an electrical potential, they migrate toward the anode or the cathode depending upon their charge and at rates influenced by the intensity of their electrical charge and by molecular size and shape. As a result they are separated during migration. After appropriate staining they can be identified as bands in the starch gel or other medium in which the buffer solution is held. Identification of protein polymorphisms does not require the specialized reagents needed for blood typing. Thus, larger numbers of laboratories can make determinations.

Presently, in spite of a few demonstrations of apparent relationships of blood groups and biochemical polymorphisms to fertility, survival, or productivity, the existence of these polymorphisms remains one of the engimas of biology. It is difficult to conceive of their being maintained in populations if they have no biological significance or function. Determining their functions, or indeed determining whether they have any function, has in most cases proved elusive. A few have deleterious effects when present in specific combinations in families. The best-known of these negative effects in farm animals occurs in horses. Some mares not having certain blood group genes, but when pregnant with a foal which carries one or more of them from its sire, may develop antibodies to the blood of the foal and excrete them in the first milk or colostrum. If the foal nurses its mother immediately after birth, an acute, often fatal, hemolytic disease (neonatal isoerythrolysis) develops. By 48 to 60 hours after birth, the stomach of the foal loses its ability to absorb antibodies and transfer them to the bloodstream. Thus, if the breeder waits until the foal has attained this age, affected foals can safely nurse their mothers. This condition in horses is somewhat analogous to the Rh factor in humans. In a portion of the cases in which an Rh− mother is pregnant with an Rh+ baby, some admixture of fetal and maternal blood occurs and the mother develops antibodies to the baby. These antibodies adversely affect the baby, and it may be born with a severe hemolytic disease (erythroblastosis fetalis). If the condition is not handled cautiously, often with an exchange of fetal blood, it can be fatal.

Some cases have been reported in farm animals in which specific blood groups or protein polymorphisms have apparently been related to lowered fertility or reduced survival in at least some populations or breeds. These reports should be confirmed in other populations before conclusions can be drawn regarding the general applicability of the findings. In chickens, differences in one blood group locus are associated with differences in rejection of skin homografts and in another with susceptibility to infection by an avian leukosarcoma virus.

Many studies have been conducted to correlate production characters of farm animals with specific blood groups and biochemical polymorphisms. Some results have been suggestive of real relationships in at least some breeds. However, in most cases the percentage of phenotypic variance accounted for has been too low for the blood groups or polymorphisms to be useful as selection aids. This is a rapidly developing field, and it is quite possible that meaningful and useful relationships will be found. It has been reported that blood antigens in swine are related to the so-called pork stress syndrome (PSS).[5]

[5] Vogeli, P., et al. 1984. *J. Anim. Sci.* 59:1440–1450.

Newer genetic engineering techniques have given rise to much speculation regarding the use of DNA fragments where there is a high density of genetic variability identified by detectable polymorphisms. These are commonly termed restriction fragment length polymorphisms (RFLPs). Fragments, or segments, of the chromosome are identified by marker loci. These loci are more precisely identified by use of a copy of the DNA sequence which has been cloned. Such clones serve as probes to identify the presence of a specific RFLP without the need for a breeding test. When a large number of RFLPs are identified, most of the active DNA can be identified. Genes closely linked in the RFLP would be expected to be transmitted with a high level of probability, assisting, it is hoped, in the accumulation of a large proportion of desired genes for a quantitative trait. Much work is yet to be done to demonstrate the practical usefulness of this approach to improving quantitative traits where blood groups and protein polymorphisms have not been fruitful in the past.

To date, blood groups and protein polymorphisms have had their major usefulness as aids in determining accuracy of parentage records. To be most useful for parentage determinations, it is necessary that a character be expressed or fully developed at birth or at a young age and that it not change throughout life. Its mode of inheritance must be simple and well understood. Ideally, the character should be easily determined. Many blood groups and biochemical polymorphisms meet all the qualifications except the last. Unfortunately, most determinations require at least reasonably sophisticated laboratory procedures.

A second use made of blood groups and polymorphisms in farm animals is in the study of phylogenetic relationships among breeds and types. Genes for color, horns, or other visible characters and genes related to size or productive potential must be assumed to have had their frequencies altered by human selection. In contrast, while genes controlling blood groups and biochemical polymorphisms have been subject to whatever natural selection pressures which may exist, they are not known to have been subject to direct selection by humans. They thus provide a largely unbiased measure of genetic similarities of populations. Determinations of frequencies of many such genes in each of several breeds can be used to estimate the genetic similarity or dissimilarity of breeds, i.e., the "genetic distance" between them. One of the hazards facing animal production, looking a century or more into the future, may well be the loss of genetic variation due to extinction of less popular breeds and of less productive strains of breeds which survive. The resultant genetic uniformity could seriously hamper the ability of humans to adapt livestock to changed conditions at some future period. Information on genetic distances may be a valuable tool in selecting breeds to be preserved for the purpose of maintaining genetic variations in populations even though their current economic usefulness is limited.

Blood Groups

The first intensive, in-depth studies of blood groups in a farm animal were with cattle. Because of the large number of antigens discovered and their (to that time) unique modes of genetic determination, the results have been significant in extending the basic understanding of molecular genetics. The results have also had practical usefulness in serving as aids in determining the accuracy of parentage records.

Table 4-1 summarizes information on blood factors and the phenotypic systems of blood groups identified in several mammalian species. Twelve blood group systems have been identified in cattle. About 80 different serologically specific antigenic substances have been found. Similar, but less extensive, genetic systems controlling blood antigens have been found in other animal spe-

TABLE 4-1

APPROXIMATE NUMBER OF BLOOD FACTORS, RECOMMENDED TYPING PROCEDURES, AND THE PHENOTYPIC SYSTEMS OF BLOOD GROUPS IDENTIFIED IN CATTLE, DOGS, HORSES, PIGS AND SHEEP*

Species	Blood factors, approximate number	Typing procedures, in order of preference	Recognized phenotypic systems of blood groups, minimum number of alleles in parentheses
Cattle	80	Hemolysis only	Twelve systems: A (10), ‡B (500+), C (70 +), F-V (5), J or J-Oc (4 +), L (2), M (3), N (2), S or S-U (8), Z (3), R' − S' (3), + and T' (2)
Dogs	9	Agglutination Hemolysis (A₁ only)1	The nine canine blood factors—A₁, A₂, B, C, D, E, F, C, and Tr—are yet to be classified with respect to phenotypic systems.
Horses	20	Hemolysis Agglutination	Eight systems: A (5), C (2), D (6), K (2), P (3), Q (6),‡ T (2), and U (2)
Pigs	60	Agglutination† Hemolysis	Fifteen systems: A-O (A locus 2; S locus 2), B (2), ‡C (2), D (2), ‡E (13), ‡F (3), ‡G (2), ‡H (6), I (2), ‡J (3), K (5), L (6), ‡M (9), N (3), ‡ and 0 (2),‡
Sheep	50	Hemolysis Agglutination (D only)	Seven systems: A (2), B (60+), C (3), D (2), M (4),‡ R-P (R locus 2; I locus 2) and X-Z (2).‡

*Adapted from Stormont, C. J., pp. 505–513 in *Research Animals in Medicine,* USDHEW (NIH) 72-333, 1973.

†Some of the agglutinins act as so-called incomplete agglutinins and require the use of antiglobulin to bring about agglutination. Some of the others produce agglutination when the red cells are suspended in dextran solutions. The majority, however, act as saline agglutinins.

‡Closed systems—that is, there is no phenotype which is nonreactive with all of the reagents used in typing the particular system. All others are open systems. That is, there is one phenotype in the particular system which is characterized by the absence of reactions with all reagents used in typing that system.

cies. However, at the time of the initial discoveries, cattle were unique in that (1) large numbers of antigenic substances were found to be parts of single genetic systems and (2) inheritance appeared to be under the control of several extensive series of multiple allelic genes.

The B system is of particular interest in that many of the antigenic substances are almost always found only in combination with one or more other antigens of the group. Very rarely a change in an antigenic combination occurs. The B group system has over 500 genetic units which, except for the rare changes in combinations, behave in inheritance as multiple alleles. However, it is now generally believed that the B locus, as well as other blood group loci, each of which controls several antigens, is genetically complex. A number of theories have been advanced to explain the structure of these loci. One possibility is that each locus is made up of a series of subloci, each of which controls an antigen. For the most part they remain as units, but rare crossovers or recombinations do occur. The changed combinations, whether due to interallelic crossover or to some other type of change, do not interfere with use of the systems for parentage determinations.

In inheritance each combination of antigens is transmitted as if controlled by a single gene. There is no dominance. Thus, if an individual carries a given gene, its red blood cells will contain the antigen or antigenic combination determined by it. The identification system involves the use of letters to represent genetic systems and antigens within the systems. Letters representing the systems are given first, followed by those for the antigens as superscripts. For example, an animal could be found to have B system antigens B, O_2, A', E'_3, O_3, I' and K'. His genotype would be represented as $b^{BO_2A'E'_3}b^{O_3I'K'}$ since these combinations are known to be inherited as groups. In the B system it is sometimes possible to determine genotypes from phenotypes as in this example. In other cases this is not possible due to the fact that the same antigens could be present in more than one gene combination.

Blood group gene frequencies vary widely in different breeds or types of cattle and other farm animals. Many studies have been made of gene frequencies in different breeds and types of the various species in many parts of the world and more are being reported each year.

With the numbers of loci and the large numbers of alleles within many loci, the numbers of possible blood types in any species are so large as to defy cataloging. A number of years ago someone estimated that there could be more than 156 billion blood types in cattle.

Each farm animal species has been found to have several blood group systems, although none are as extensive as in cattle. Several of the systems of other species are apparently analogous to some of the cattle systems. Some of them include antigen complexes controlled by single genes as in cattle. Information on the systems in different species and references to earlier reviews on individual species can be found in general reviews by Stormont (1973) and Rasmusen (1975).

Blood Proteins

Large numbers of chemically different protein variants have been discovered in farm animals in the globulins (transferrins), albumins, and enzymes of blood and in hemoglobin. Fourteen of these have been described in the horse and are listed in Table 4-2 for illustrative purposes. Similar or more extensive tabulations could be developed for other species. The field is active, and new variants are discovered frequently. Most if not all the blood protein polymorphisms are genetically controlled by allelic pairs or allelic series with no dominance or, as is sometimes stated, with codominance. Thus, if an animal has a gene for a specific substance, the substance can be detected in the blood by appropriate procedures, and the presence or absence of specific substances is directly related to the genotype.

Milk Protein

About 80 percent of the protein in milk is in the form of casein. In cattle there are three major types of casein: α_{S1}, β, and K, plus several minor types which we will not consider here. The remaining protein is in the form of the so-called whey proteins. There are two types of these, α-lactoglobulin (the immune globulins) and β-lactalbumin (serum albumin). For detailed reviews on polymorphisms of milk proteins in cattle and their nomenclature see Thompson and Farrell (1974) and Whitney et al. (1976).

Each of the five major types of protein has two or more genetic variants. Each variant is related to a single autosomal gene. Starch gel electrophoresis is used to detect the variants of both casein and whey protein simultaneously.

TABLE 4-2
BLOOD PROTEIN SYSTEMS OF THE HORSE*

Protein/enzyme	Locus	Source	Number of alleles
Transferrin	Tf	Serum	10
Albumin	Al	Serum	3
Esterase	Es	Serum	7
Prealbumin	pR	Serum	8
Prealbumin	Xk	Serum	3
Postalbumin	Pa	Serum	2
Hemoglobin	Hb	Red cells	2
Carbonic anhydrase	CA	Red cells	5
6-phosphogluconate dehydrogenase	PGD	Red cells	3
Phosphoglucomutase	PMG	Red cells	3
Phosphohexose isomerase	PHI	Red cells	3
Acid phosphatase	AP	Red cells	2
Catalase	Cat	Red cells	2
NADH diaphorase	Dia	Red cells	2

*Adapted from Sandberg (1974) "Proceedings, First World Congress on Genetics Applied to Livestock Production," vol. 1, pp. 253–265. (Used by permission.)

The amino acid sequences of several of the genetic variants are now known. They are being studied intensively in efforts to relate their sequences to those of specific RNA molecules.

The major genetic variants in milk of cattle are:

α_{S1}-casein—Four variants, A, B, C, and D. Controlled by four allelic autosomal genes with no dominance. Phenotypes correspond to genotypes, i.e., A (A/A), AB (A/B), B (B/B), AC (A/C), etc.

β-casein—Four variants also with four allelic autosomal genes with no dominance.

K-casein—Two variants controlled by an allelic pair with no dominance.

β-lactoglobulin—Three variants controlled by an allelic series with no dominance.

α-lactalbumin—Two variants controlled by an allelic pair with no dominance.

The genes controlling the genetic variants of α_{S1}- and β-caseins are closely linked, and not all the possible combinations occur. The genes controlling variants in K-casein are linked to the α_{S1} and alleles, but not so closely. It has been estimated that they are less than 2.8 crossover units apart.

Functions and significance of the genetically controlled milk protein variants are not well understood. Thompson and Farrell (1974) give some interesting speculations. Others have stated that the variants are "just there" with no apparent function. α_{S1}-A-casein is considered to be inferior for cheese making. Apparently the other genetic variants have no known relationship to physical or nutritional qualities of milk.

Two studies (Ng-Kwai-Hang et al., 1986, and Lin et al., 1986) with Holsteins and Ayrshires and their crosses found relationships between milk protein types and yields of milk, fat, and protein. Four loci accounted for 8.9 percent of the variance in milk yield. Two studies with Guernseys and Holsteins (Haenlein et al., 1988, and Gonyon et al., 1988) reported little association of 14 polymorphic loci with milk yield. Significant associations of protein (Guernseys) and fat (Holstein) percentages were observed for specific loci.

The protein variants of milk constitute valuable genetic markers for basic studies and for estimating phylogenetic relationships. As aids to selection and parentage determination, they suffer from the disadvantage that determinations of type of milk protein can be made only after the first lactation begins—close to the time that full- or part-lactation records will be available.

Other Biochemical Polymorphisms

In addition to the red cell blood antigens and the blood and milk protein polymorphisms discussed, there are a large number of genetically controlled variants of other types, including hemoglobins, enzymes, and leukocyte antigens. Generally, these variants have been studied less extensively in farm animals than in laboratory animals, but much information on genetic variation is avail-

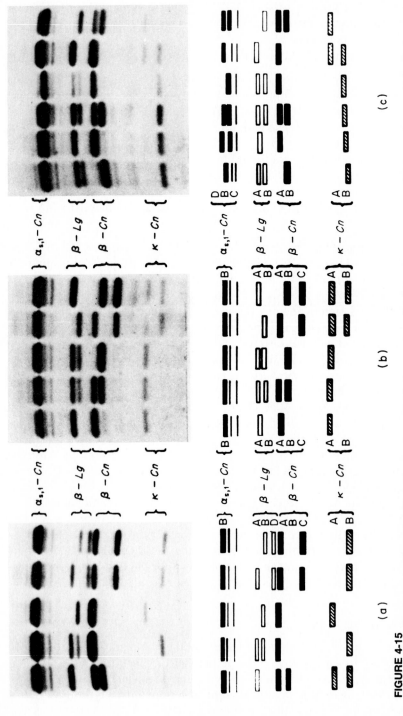

FIGURE 4-15

Simultaneous thin starch gel electrophoretic phenotyping of caseins and β-lactoglobulins. Top row of photographs shows patterns for each of 15 milk samples, while lower row of photographs gives diagrammatic representations of the same samples. Letters on lower row of photographs refer to genetic types of the proteins. (*Courtesy of Dr. Marvin P. Thompson, Eastern Regional Research Center, U.S. Department of Agriculture, Philadelphia.*)

able in specific cases. Several references listed at the end of the chapter will provide entrée to the scientific literature for several of these areas.

Parentage Tests

In using the term *parentage tests,* bear in mind that no blood group or biochemical polymorphism can *identify an individual as a parent.* They can only *exclude* individuals as parents. Put more simply, these tests indicate whether a given individual *could* be a parent, not whether the individual *is* the parent. Cases of uncertain parentage arise frequently in seed-stock herds and flocks for a variety of reasons. A female may have been accidentally mated to two males during an estrus period, or identification of vials of semen used for artificial insemination may be in doubt. Parturition may have occurred in two females in the same pasture or stall very close to the same time with no attendant present, and one female claims both offspring or an exchange of offspring is suspected. These are the more common types of problems but by no means exhaust the list of possibilities. Often, one parent is known, and the problem is one of determining whether one or more of the possibilities as to the identity of the other parent can be excluded. In addition to the accidental or unintentional occurrences leading to uncertainty as to parentage, there is always the possibility of fraud in registration of purebred animals. Breed registry societies often use these tests to resolve situations where fraud is suspected.

All parentage tests depend upon the fact that an individual cannot have a specific blood antigen or blood or milk protein variant unless one or both parents had it. Probabilities of solving cases of uncertain or disputed parentage upon the basis of any one blood group of biochemical polymorphism depend upon the number of alleles in the system, their frequency in the population involved, and whether genotypes are identifiable directly from phenotypes. The probability of excluding nonparents as possible parents increases as the number of genetic systems which can be studied increases. With the wide array of blood groups and biochemical polymorphisms now known in most species, the probabilities are relatively high. In two breeds of Swedish horses, Sandberg (1974) estimated that tests of six blood groups and nine blood proteins would resolve more than 90 percent of questionable paternity cases involving two stallions.

For most determinations it is desirable to have tests of the offspring, of the known parent, and of the possible or suspected other parent(s). However, it is not always necessary to have tests on the known parent. For example, if a male is homozygous for a specific transferrin or other variant and the offspring does not have it, that male is excluded as a possible parent. Currently, superovulated dairy cows may be inseminated with semen from more than one sire. The blood types of the bulls used in combination with the blood type of the cow must permit exclusive identification of the sire of this calf. Prior approval by the registry association is required for registration.

OUTDATED IDEAS IN ANIMAL BREEDING

The terms and expressions which follow are outmoded or not believed to be based on fact. However, many will be found in certain popular writing and older literature. We believe that a student should be aware of them.

Acquired Characters

This term refers to the possibility of a character induced in the body (soma) of an individual by environmental forces becoming part of its germ plasm and being transmitted to future generations. This topic was one of the historic battlegrounds of biology. Speaking generally, things which happen to the soma, either favorable or unfavorable, do not directly affect the genes, in which reside the potentialities for future generations. Environment does permit (or prohibit) the development of existing potentialities but in general does not alter the genes.

There is an abundance of examples showing that acquired environmental modifications have no influence on the hereditary make-up of the animal. When the tail of a sheep is docked, the sheep has acquired a character which is not passed on to its descendants. In succeeding generations it will be necessary to dock the lambs, notwithstanding the fact that docking has been practiced for hundreds of generations. In this category come also the dehorning of cattle, clipping of dogs' ears and tails, and so on. In none of these cases is the acquired character inherited by the offspring. In the human family, circumcision, which has been practiced for thousands of years, and the binding of feet to prevent full development, as practiced by Chinese women, might be mentioned. Here, as in the cases cited of farm animals, there is no evidence that these environmentally acquired characters eventually become inherited.

Weismann, in order to test this hypothesis, cut off the tails of mice for 19 generations in succession and secured no shortening of the tails or absence of tails in any of the descendants. As Walter remarks, "It is a good thing that children of warriors do not inherit their parents' honorable scars of battle else we would have long since been a race of cripples." The same author also remarks in this regard that "evidently wooden legs are not inherited, but wooden heads are."

Atavism, or Reversion

The terms *atavism* and *reversion,* meaning "the reappearance of some ancestral trait or character after a skip of one or several generations," are often encountered in the older literature on animal breeding. Such reappearances were mysterious before the physical basis of heredity was understood. The birth of a red Angus when the past few generations have been black is now understood by each parent supplying the gene for red. Recessives may be carried along, hidden by dominants, for any number of generations. Whenever two recessives come together, or, in other words, whenever the dominant gene is lacking, the "atavistic" character will be evident. There is nothing mysterious about it, for it is one of the normal manifestations of the hereditary mechanism.

Nick, or Nicking

The terms *nick* or *nicking* are sometimes used to describe an individual mating or the mating of members of a family with members of another family, in which the progeny are either better or worse than would have been expected from the merit of the parents. Usually the terms are used to describe better-than-expected results.

Genetically, a favorable nick *could* occur in an individual mating purely as a result of chance. The offspring could get a larger-than-average number of favorable genes from each parent as the result of chance at segregation and fertilization. If more productive offspring than expected were produced from matings between different families, chance could still be a factor if only small numbers were raised. However, if larger numbers were produced, the probability of this being chance would be reduced.

Likely explanations with large numbers are (1) the two families carried different allelic pairs of genes which complemented each other and produced a favorable epistatic combination in most offspring, or (2) with overdominance, favorable results could occur if the families were genetically so different that the offspring were above average in heterozygosity.

Nicking has not been demonstrated to be very important in several studies on outbred animals within pure breeds. It is almost certain that many breeders put more emphasis than can be justified in looking for favorable nicks. Families in such populations are usually only as much related to each other as half sibs, or perhaps a little more. Therefore, the proportion of their genes which are similar because of relationship could not be large enough to establish a very predictable breeding behavior when mated to animals of another such family. Sometimes ideas gain general acceptance among breeders that such and such a family "always" or "never" crosses well with another. Usually, there is little or no basis for such ideas.

With highly inbred lines, the possibilities of nicking being important are greatly increased, but it is then usually referred to as *specific combining ability*.

Prepotency

The term *prepotency* means the ability of an animal, either male or female, to stamp a given set of characteristics on its offspring to the exclusion of effects of genes from the other parent. The term is usually applied to males and is usually thought of in relation to a sire stamping his own characteristics on the offspring. It is also applied to cases in which the expected resemblance or uniformity among the offspring is more than expected even if the resemblance is not to the sire himself.

Genetically, prepotency depends primarily upon an animal being homozygous for dominant genes. The term is often misused, and exaggerated ideas are often held about the prepotency of particular sires. Sometimes a rugged, masculine appearance is thought to be associated with prepotency. There is no evidence for this. Since prepotency depends mostly on dominance and homozygosity, it is unlikely that an animal would be prepotent for all characters. For example, in a Hereford X Angus cross, the Hereford would appear to be prepotent for white face, whereas the Angus would appear to be prepotent for black body and polledness since offspring will have these characters.

It is sometimes implied that prepotency is transmitted and that prepotent sires have prepotent sons. This may well be true, but the only way it can occur is for the prepotent sire to have been mated to females carrying the same dominant genes he carried so that his sons will also be homozygous for the genes. Obviously, offspring of a cross to a homozygous recessive could not be prepotent for the same character.

Maternal Impressions

Belief in maternal impressions assumes that what a pregnant mother sees, hears, or experiences may influence her offspring. In general, this old belief can be repudiated, because, as Marshall points out, if it were true, all calves born in the spring up north would tend to be white because the mothers have viewed a white landscape all winter. Similarly, calves born in the fall would be green. There is no direct nerve connection between parent and offspring to afford means for transporting the effects of experiences. Animal experiments have all given negative results in this field. It is indeed a fortunate provision of nature that in the higher animals the embryo is well protected from all external influences. If maternal impressions were actually registered on the offspring, all types and races would have long ago become a hideous conglomerate.

Telegony

Telegony is the belief that after a female has borne young by a certain male, her subsequent offspring will show characteristics derived from the previous sire. The classic example is that of a mare which bore offspring by a quagga and later produced horse colts said to show some striping. Numerous attempts have been made to confirm this by crossing mares and zebras, but in all cases they have failed. The basic elements that determine the characters of any individual are the ovum and the spermatozoon that unite to produce it. Spermatozoa from one service could not possibly live through a gestation period in the female organs of any higher species to fertilize some future ovum, since a spermatozoon lives a very few days at the most. If this first offspring had any influence on other undeveloped ova in the ovaries, it would come under the heading of the inheritance of acquired characters, for which there is no conclusive proof. In the light of our present knowledge of inheritance, there is neither an experimental nor a theoretical basis for telegony.

Blood

In older animal husbandry literature and among breeders the term *blood* is often used more or less interchangeably with the word *inheritance*. Saying that an animal has three-fourths of the blood of a noted sire or, in the case of animals of mixed breeds, saying that a given animal is a half-blood means in each case that these fractions of their hereditary material came from the source indicated. The use of this term is based in antiquity when it was believed that actual blood passed from mother to offspring during fetal life. We now know that this is not the case—rather the embryo is nourished by nutrients from the mother's blood, and the only thing she transmits directly to it is a sample half of her hereditary material. Since the term is based on a misconception, its use should be avoided.

SUMMARY

Lethals and other hereditary abnormalities occur with enough frequency in many cases to make them economically important problems for breeders. Most are recessive in inheritance. In the heterozygous condition some have effects which are considered desirable by breeders and thus constitute a special problem. Genetic defects are responsible for a wide range of abnormal morphological and physiological variants. In many cases it is difficult to distinguish between abnormalities which are genetic and those which are caused by the environment. Each abnormality is a unique problem. Corrective measures to be taken in individual cases depend upon the nature of the defect and the kind of herd in which it occurs. If defects are genetic, outcrossing, culling affected individuals and their relatives, and progeny testing are all useful tools in specific cases for reducing the frequencies of undesirable genes. Farm animals have a large number of hereditary blood types and biochemical polymorphisms in the blood and other body fluids and tissues. These are useful in parentage determination and as tools in estimating phylogenetic relations among breeds and types of animals; some may also have relationships to characters of importance in economical production. A number of outdated animal breeding terms, most of which are based on false or unproven assumptions and are related to qualitative inheritance, are defined and discussed. Even though these terms are primarily of historical interest, students may encounter them and should understand their limitations.

SUGGESTIONS FOR FURTHER READING
Books

Adams, T. E., and M. R. Brandon. 1986. The Bovine Major Histocompatibility Complex and Disease Resistance. Pp. 178–203 in: "The Ruminant Immune System in Health and Disease," Cambridge University Press, Cambridge.

Blunt, M. H. (ed). 1975. "The Blood of Sheep. Composition and Function," Springer-Verlag New York Inc., New York. Sections of special interest relative to this chapter:

Blunt, M. H., and T. H. J. Huisman. The Haemoglobins of Sheep. Pp. 155–183.
Curtain, C. C. The Ovine Immune System. Pp. 185–195.
Tucker, E. M. Genetic Markers in the Plasma and Red Blood Cells. pp. 123–153.
Fries, R., and F. H. Ruddle. 1986. Gene Mapping in Domestic Animals. Pp. 19–37 in:
"Biotechnology for Solving Agricultural Problems," Martinus Nijhoff Pub.,
Dordrecht, Netherlands.
Hutt, F. B., and B. A. Rasmusen. 1982. "Animal Genetics," 2d ed., John Wiley &
Sons, Inc., New York.
Leipold, H. W., S. M. Dennis, and K. Huston. 1972. Congenital Defects of Cattle: Na-
ture, Cause, and Effect. Pp. 103–150 in: "Advances in Veterinary Science and Com-
parative Medicine," vol. 16, Academic Press, New York.
"Proceedings, First World Congress on Genetics Applied to Livestock Production,"
Madrid, October 7–22, 1974. Graficas Orbe, S. L., Padilla 82, Madrid (3 vols.). Vol-
ume 1 contains many review articles. Those written in English of interest relative to
this chapter include:
Adalsteinsson, S. Colour Inheritance in Farm Animals and Its Application to Selec-
tion. Pp. 29–37.
Kidd, K. K. Biochemical Polymorphisms, Breed Relationships and Germ Plasm Re-
sources in Domestic Cattle. Pp. 321–328.
Oosterlee, C. O., and J. Bouw. Structure of "Loci" in Animals. Pp. 243–252.
Rapacz, J. Immunogenic Polymorphism and Genetic Control of Low Density Beta-
Lipoproteins in Swine. Pp. 291–298.
Sandberg, K. Blood Typing of Horses: Current Status and Application to Identifi-
cation Problems. Pp. 253–265.
Spooner, R. L. The Relationships between Marker Genes and Production Charac-
ters in Cattle, Sheep and Pigs. Pp. 267–271.
Rasmusen, B. A. 1975. Blood-Group Alleles of Domesticated Animals. Pp. 447–457 in:
King, R. C. (ed.). "Handbook of Genetics," vol. 4, Plenum Press, New York.
Stormont, C. 1958. Genetics and Disease. Pp. 137–162 in: "Advances in Veterinary
Science," vol. 4, Academic Press, New York. (This article contains an extensive
listing of hereditary defects in farm animals.)
Stormont, C. 1973. A Survey of Blood Groups in Several Species of Large Animals
Used in Medical Research. Pp. 505–513 in: Harmison, L. T. (ed.). "Research Ani-
mals in Medicine." USDHEW (NIH) 72–333, Washington.
Thompson, M. P., and H. M. Farrell, Jr. 1974. Genetic Variants of the Milk Proteins.
Pp. 109–134 in: Smith, V. R., and B. L. Larson (eds.). "Lactation," vol. III, Aca-
demic Press, New York.

Articles

Anderson, D. E. 1963. Genetic Aspects of Cancer with Special Reference to Cancer of
the Eye in the Bovine, *Ann. N.Y. Acad. Sci.* 108:948–962.
Baker, C. M. A., and C. Manwell. 1980. Chemical Classification of Cattle. I. Breed
Groups. *Anim. Blood Groups Biochem. Genet.* 11:127–150.
Bovard, K. P. 1960. Hereditary Dwarfism in Beef Cattle, *Anim. Breed. Abs.* 28:223–237.
Briles, W. E. 1964. Current Status of Blood Groups in Domestic Birds, *Z. Tierzucht.
Zucht. Biol.* 9:371–391.
Gonyon, D. S., R. E. Mather, H. C. Hines, G. F. W., Haenlein, C. W. Arave, and S.

N. Gaunt. 1987. Association of Bovine Blood and Milk Polymorphisms with Lactation Traits: Holsteins. *J. Dairy Sci.* 70:2585–2598.

Haenlein, G. F. W., D. S. Gonyon, R. E. Mather, and H. C. Hines. 1987. Association of Bovine Blood and Milk Polymorphisms with Lactation Traits: Guernseys. *J. Dairy Sci.* 170:2599–2609.

Johansson, I. 1965. Hereditary Defects in Farm Animals, *World Rev. Anim. Prod.* 3:19–30.

Lin, C. Y., A. J. McAllister, N. F. Ng-Kwai-Hang, and J. F. Hayes. 1986. Effects of Milk Protein Loci on First Lactation Production in Dairy Cattle. *J. Dairy Sci.* 69:704–712.

Ng-Kwai-Hang, N. F., J. F. Hayes, J. E. Moxley, and H. G. Monardes. 1986. Relationship between Milk Protein Polymorphisms and Major Milk Constituents in Holstein-Fresian Cows. *J. Dairy Sci.* 69:22–26.

Nordskog, A. W. 1984. Immunogenetics as an Aid to Selection for Disease Resistance in the Fowl. *World's Poultry Sci. J.* 39:199–209.

Rasmusen, B. A. 1981. Blood Groups and Pork Production. *Bioscience* 31:512.

Warner, C. M., D. L. Meeker, and M. F. Rothschild. 1987. Genetic Control of Immune Responsiveness: A Review of Its Use as a Tool for Disease Resistance. *J. Anim. Sci.* 64:394–406.

Watson, T. O. 1986. Immunity to Gastrointestinal Parasites in Domestic Stock with Particular Reference to Sheep: A Review. *Proc. New Zealand Soc. Anim. Prod.* 46:15–22.

Whitney, R. McL., J. R. Brunner, K. E. Ebner, H. M. Farrell, R. V. Josephson, C. V. Morr, and H. H. Swaisgood. 1976. Nomenclature of the Proteins of Cow's Milk: 4th rev. *J. Dairy Sci.* 59:795–815.

CHAPTER **5**

VARIATION AND ITS STATISTICAL MEASURES

When we consider that probably no two blades of grass, no two calves, and no two humans have ever been exactly alike (though monozygotic twins are essentially so), we get a glimpse of the resourcefulness of nature. It would seem that new plans and specifications must at some point be depleted, yet life flows along in its thousands of varieties and millions of individuals, each different from all the rest of its kind. It is fortunate that this is so, because without this perpetuation of variability, populations would not be capable of adapting to changed conditions or demands. Variation is the raw material which the breeder has available for herd or flock improvement. An often-used quotation states that "variation is at once the hope and despair of the breeder"; the hope since when it is present, offspring better than their parents may be produced; the despair because after animals have been improved greatly, they may, and often do, vary again toward mediocrity.

Variation among animals in size, rate of growth, efficiency of feed utilization, carcass characteristics, disease resistance, milk production, speed, stamina, wool quality, and color has been observed and recorded numerous times. Of two steers, one gains 900 grams (g) per day during a 140-day feeding period, and the other gains 1,400 g; of two cows, one produces 8,000 kilograms (kg) of milk in a year, the other 12,000 kg milk; of two litters of two sows, one litter gains 100 kg from 500 kg of feed, the other litter requires only 360 kg of the same feed for 100 kg of gain; of two sheep, one shears a fleece of 3 kg, and the other shears a 5-kg fleece.

Both heredity and environment are important in producing differences among individual animals. In some instances the specific hereditary and environmental influences may produce variability which is not directly attributable to either but which is a result of their joint action or interaction. By studying variation alone, we are not able to determine which portion of the variation is certainly due to environment and which to heredity. Nevertheless, as will be evident in the succeeding chapters, the relative importance of hereditary and environmental influences on the variation for individual traits can be determined.

Environmental differences include the nongenetic variation resulting from managemental, nutritional, and climatic influences. Some animals may have been born in large litters and others in small litters; some may have had better care than others; some may have been born when the temperature was extremely hot, and others when it was extremely cold. Certain animals may be infected with parasites, whereas others are not.

NATURE OF VARIATION

Discontinuous and Continuous Variation

Traits generally are grouped into those which show qualitative differences and those which show quantitative differences. In the former the variations fall into a few clearly defined classes. This is usually due to the fact that these traits are under the control of one or a few pairs of genes, whose final expression is not greatly influenced by external environmental factors. The polled or horned condition in cattle is an example of this type of trait. In practically all instances an animal clearly displays one or the other of these two characteristics. Quantitative traits, on the other hand, show all manner of slight gradations from small to large as are found in milk production, wool clip, and rate of gain in the feed lot. The two sorts of variations associated with qualitative and quantitative traits are described as *discontinuous* and *continuous*. Qualitative suggests discontinuous variation (in sharply defined classes), and quantitative suggests continuous variation (many small gradations shading or intergrading almost imperceptibly into one another). Mendel dealt with one trait in peas which was quantitative (tall versus dwarf) but which behaved in a qualitative manner; i.e., the peas were either tall or dwarf.

Students and livestock breeders often develop a feeling of frustration over the seeming inability to find clear-cut evidence of the known simple principles of heredity in farm animals. The basic reason that heredity in livestock cannot be resolved into simple terms is, of course, the fact that most of the economically important traits of our higher animals are highly complex. Few of them are determined by the action of only one or two pairs of genes, and very few are immune to environmental influences. With many genes involved and with expressions readily subject to environmental influences, we naturally would expect to observe a graded series rather than sharply defined or discrete classes in most traits.

The number of genes present in the hereditary complex of farm animals is not known, but, conservatively, the number probably runs at least into the thousands. A substantial number of genes in any species are present in homozygous form; hence, they do not contribute to the observed variation in the species. Most quantitative traits such as size, growth rate, milk production, egg production, and prolificacy depend for their expression upon the interaction of a large number of heritable factors and environmental influences.

We do not know how many genes influence body size, but we do know that every organism begins as a single-cell zygote. Each organism is destined by its heredity to have its cells continue to divide until a certain size is attained, provided a suitable environment is available. The pituitary, thyroid, and thymus glands, perhaps also the adrenals and gonads, through their individual secretions and probably through complex interactions among them have an influence on both the rate of growth and mature size. How many individual genes and gene interactions function in this complicated process? Even in a relatively simple organism like the fruit fly, it is known that as many as 30 genes distributed through the four pairs of chromosomes influence eye color.

Other factors besides size play major roles in the ultimate performance of an animal. They include efficiency of digestion, absorption, and elimination; rate of blood flow, or basal metabolism in general; number and functional activity of the secreting cells in the udder; the animal's disposition; proper levels of feeding, the presence or absence of good management practices; and freedom from various diseases. Enough has been said to indicate that the overall physiological functioning of any animal is highly complex, probably involving the actions and interactions of hundreds or thousands of gene pairs (loci).

Variation from Recombination

Recombination of genes takes place as the chromosomes from the uniting gametes come together. In an organism with two pairs of chromosomes there are four possible ways to recombine the intact chromosomes. In each resulting germ cell, one member of each pair of chromosomes is found. The possible number of different gametes with various numbers of pairs of chromosomes (n) where only a single gene pair is considered to be different for homologous chromosomes is 2^n, as shown in Table 5-1.

If an Aa male is mated to an Aa female, two different types of gametes (A or a) can be produced by each parent, and offspring can be of three different genotypes (AA, Aa, or aa).

If a male of the genetic constitution $AaBb$ (with genes A and B in different chromosomes) is mated to a female of the same genetic make-up, four different types of gametes can be produced by each parent, and the offspring can be of nine different genotypes.

If two animals were of the genetic constitution shown below (loci on different chromosomes):

Sire: *Aa bb CC Dd*
Dam: *Aa Bb cc DD*

TABLE 5-1
NUMBER OF DIFFERENT KINDS OF GERM CELLS POSSIBLE FROM VARIOUS SPECIES
WHEN EACH PAIR OF CHROMOSOMES IS HETEROZYGOUS AT ONLY ONE LOCUS

Species	Number of pairs of chromosomes	Number of different kinds of germ cells	
Ascaris megalocephala	2	2^2	4
Fruit fly (*Drosophila melanogaster*)	4	2^4	16
Corn (*Zea mays*)	10	2^{10}	1,024
Swine	19	2^{19}	524,288
Human	23	2^{23}	8,388,608
Sheep	27	2^{27}	134,217,728
Cattle, goat, horse	30	2^{30}	1,073,741,824

then it would be possible to have 12 different genotypes among the offspring:

$$3 \times 2 \times 1 \times 2 = 12$$

Thus from mating the above parents, we could have 12 different sorts of full brothers or sisters as to genotype. All these individuals would have identical pedigrees; but their genotypes might all be different, ranging from *AA Bb Cc DD* to *aa bb cc Dd*.

If the capital letters represented desirable traits and the small letters undesirable ones, there would be a vast difference between these full sisters or brothers. Identity of pedigree, then, does not mean identity of inheritance, except when the parents are homozygous for all pairs of genes. Such a degree of homozygosity is almost never realized practically, but it is approached in crosses among highly inbred lines.

Our example has dealt with four pairs of genes, or eight in all. We do not know how many genes are present in the chromosomes of our farm animals. There are 60 chromosomes in some of these animals, and if there were an average of only 100 genes in each chromosome, there would be a total of 6,000 genes. Each parent would then be transmitting not 4 genes, as in our simple example, but 3,000. If two animals were heterozygous for all these 3,000 pairs of genes, then there could be $3^{3,000}$ different genotypes, assuming that all combinations were possible by means of independent assortment and crossing over (see Table 5-2). Such a number is, of course, beyond human comprehension. Even if linkage is complete in all chromosomes, we could still get 2^{30}, or 1,073,741,824, different recombinations of complete chromosomes, whereas there are probably fewer than 1 billion cattle now in the world. It is evident that the simple recombination of intact chromosomes in most of our farm animals can alone provide a tremendous amount of variability when considerable heterozygosity exists. The evidence available indicates that most of our animals are rather heterozygous.

TABLE 5-2
INDEPENDENTLY SEGREGATING GENE RECOMMENDATIONS

Number of pairs of heterozygous genes	Number of different gametes	Number of different genotypes in F_2	Total number of combinations in F_2
1	2	3	4
2	4	9	16
3	8	27	64
n	2^n	3^n	4^n

Variation from Gene Mutations

Gene mutations represent changes in the individual loci. The nature of mutational changes in genes is, of course, unknown, but it might conceivably be comparable to a change in the spatial arrangement of atoms in complex organic molecules.

Observable mutation rates under laboratory conditions are generally in the range of one mutation for each 100,000 to 1 million loci. The rates of observable mutations vary for individual loci and can be increased under the influence of x-ray, chemicals, ultraviolet light, and other extreme conditions.

Even though mutation rates can be increased by special treatment, they are still so rare as to impose a distinct limitation on the role of mutation in providing new variability. Furthermore, most of the mutations in our domestic animals appear to be harmful. Many are recessive and deleterious or lethal when the gene is in homozygous condition. Some few mutations have not been harmful and have been preserved. The best-known mutation of this type probably is the one influencing polledness, or hornlessness, in cattle. It has been estimated that the mutation rate from horned to polled is about 1 in 20,000, much higher than the rate generally observed for most loci. The dominance of the polled gene was also of major importance in its detection and increase in frequency. The polled condition has been reported in practically all our breeds of cattle. This example also points up how the harmfulness or usefulness of a mutation may be determined by the environmental conditions under which it is expressed. While we may consider polledness to be desirable under domestication, horns were undoubtedly of much value for the protection of the animal in the wild state.

Gene mutation is a reversible process, and loci which have mutated from the normal type are subject to mutation back to their original form. This can be summarized by a simple diagram:

$$A \underset{v}{\overset{u}{\rightleftharpoons}} a$$

in which A and a are alleles and u represents the mutation rate of locus A to a. The reverse mutation rate from a to A is v. These two rates of mutation, u

and *v,* may differ greatly. With the generally recessive nature of most mutations, the mutation rate of a dominant to a recessive usually exceeds the reverse mutation rate.

The genetic variability with which a breeder works is the result of the accumulation of mutants which arose over the long span of evolutionary time.
However, the rarity, recessiveness, and generally harmful nature of most gene
mutations prevent them from being important in providing favorable variation
for animal improvement within the lifetimes of even several generations of
breeders. In fact, the breeder must exert at least some effort in attempting to
purge the undesired mutants from the flock or herd as they appear.

Chromosomal Aberrations

In addition to gene or point mutations, a considerable variety of chromosomal
aberrations involving various-sized blocks of genes, whole chromosomes, or
whole sets of chromosomes have been discovered in farm animals in recent
years. Although several of these are associated with phenotypic abnormalities,
they are rather quickly lost from the population. Thus, such chromosomal
changes in farm animals are chiefly of evolutionary significance; plant breeders in many instances have been able to utilize these chromosomal changes to
produce new varieties or species. Changes of this nature are generally brought
to light by discrepancies in breeding behavior resulting in disturbances of the
normally expected Mendelian ratios, in the creation of new linkage relations,
or in the appearance of variable offspring. When such things occur, cytological
study generally has revealed certain changes in the number or arrangement of
the loci within the whole chromosome complex or a change in the actual number of chromosomes. This close correspondence between the visible changes
in the chromatin and changes in ratios or in individuals is the strongest evidence supporting the chromosome theory of inheritance. Additional coverage
of this topic was given in Chapter 2.

Environmental Variation

The many changes in the genes and chromosomes have been pointed out as
being responsible for genetic variation in animals. This genetic variation is expressed, however, only as the environmental stimuli and conditions permit its
expression. The environment does not directly change the hereditary make-up
of an individual, but the environmental circumstances do determine the extent
to which the inherited tendency is expressed. Animals of a good strain may
themselves be stunted and not developed to the limit of their inherent capac-
ities, but these animals may still be capable of producing offspring of good genetic potential. The genes and chromosomes from the parents form the actual
"bridge of inheritance" from parent to offspring, and these are not affected
directly by the stunting.

Even though environment does not directly influence the germ plasm, it is almost certain that in the evolutionary process the environment must have wielded a marked influence on life and caused gradual change in germ plasm. Examples from both the plant and animal kingdoms can be cited. Capsella, a wayside weed, has gradually climbed to a mountainous habitat. As it climbed, it developed a dwarf character—although it grew luxuriantly in the valley, it became small, compact, and dwarfed in its mountain home. When taken back to the valley, it continues its dwarf characteristics. Presumably, those plants genetically tending to be dwarfed had a better chance of surviving under the more rigorous conditions and decreased food supply that prevail to an ever-increasing degree as the mountains are ascended. The force bringing about the change was not the inheritance of acquired characters but "natural selection," which no doubt always has and always will, provided humans do not interfere, weed out those types and individuals poorly fitted to their environment and preserve and proliferate those better-fitted to survive under the existing conditions.

Light, temperature, food, and moisture represent environmental stimuli which modify the development and expression of inherent qualities. Low temperatures stimulate the development of red flowers in the Chinese primrose, but high temperatures stimulate the formation of white flowers. The kind of food supplied to the larvae of bees determines whether the females shall be fertile (queen bees) or infertile (workers). The moisture supply influences the yearly growth in trees, as can be depicted from the width of the annual growth rings.

Rabbits of the genotype c^h/c^h show the Himalayan pattern and are white with black at the tips of the feet, ears, nose, and tail. The black pigment at the extremities results because the enzyme responsible for this pigment is inactivated at normal body temperatures. Only those parts normally cool develop color. This has been demonstrated experimentally by placing an ice pack on another portion of the rabbit with the resulting development of dark hair in this area.

We should always keep in mind the constant interplay between heredity and environment. Even though environmental variation is not transmitted, this should not deter breeders' efforts to provide as favorable an environment as is economically feasible to permit the attainment of the animals' inherent potentialities. In many cases our animals already have the inherent potential for a much higher level of performance than their environment permits them to express. A striking example of this comes from the report of a Danish dairy demonstration. Six cows which in the previous 2 years had produced an average of 168 kg of fat each in 365 days were taken from an average dairy herd. During the experimental demonstration the cows were fed 4 times a day and induced to eat as much as possible. They were given the best care that could be provided and were milked 4 times per day. At the end of the 365-day demonstration, these cows had produced an average of 412 kg of fat per cow, an increase of 244 kg of fat over their average performance of the previous 2 years.

For most of our farm animals it is difficult to single out special environmental circumstances. The environment encompasses a complex of many measurable items, as well as numerous seemingly intangible ones working in cooperation and competition. For most of our larger animals the herds supervisor, or manager, provides an important but often indeterminate part of the environment. With two groups of animals of comparable genetic worth and with comparable housing and feed available for each group, but with two different supervisors, wide differences in performance may be expected. The impossibility of providing managemental recommendations on a recipe or formula basis makes it difficult to standardize environmental conditions with large animals. Nevertheless, it is in this special area that the art and experience of the skilled breeder pay handsome dividends to increased performance.

The breeder's success depends ultimately on the ability to detect and breed from those animals in the herd and flock which show favorable variation of a genetic nature. To do this, the breeder must keep records of performance on all animals and make as reasonable an allowance as possible for the contribution that either good or bad environment has made to the merits or deficiencies which the animals exhibit. Only in this way can the breeder build a firm genetic foundation under any attempts to breed better livestock.

MEASURING VARIATION

Probability

Variation within a population for qualitative traits can be summarized by assigning each individual to its appropriate discrete class. A frequency diagram showing the proportion of the population which possesses each of the qualities being studied provides a vivid picture of the population for the characteristic cataloged. In many situations the observations may be summarized as a ratio (such as 1:1 or 3:1) which may be expected on the basis of knowledge of the inheritance of the trait. Small samples may deviate widely from the expected ratios, but the true ratio will be closely approximated if the numbers are large.

An example of coin tossing provides an excellent analogy to certain genetic situations. Except for sex-linked genes, the genes in most farm animals are found in pairs. The composition of the gametes appears to be largely a matter of chance at segregation.

If we toss a coin, it has an even chance of falling heads or tails; the probability or likelihood of occurrence of either is 1 in 2 ($\frac{1}{2}$). If we toss two coins, the probability that the first will fall heads is $\frac{1}{2}$, and the independent probability that the second will fall heads is $\frac{1}{2}$. The probability that both will fall heads is the product of the likelihood of each separate event ($\frac{1}{2} \cdot \frac{1}{2} = \frac{1}{4}$). The same situation prevails for the likelihood that both coins will display tails. In addition, the two other possibilities, that the first coin displays heads and the second tails and that the first coin displays tails and the second heads, each has a probability of $\frac{1}{2} \cdot \frac{1}{2}$, or 1 chance in 4. If we toss three coins, the probability that all of them will fall heads is $\frac{1}{2} \cdot \frac{1}{2} \cdot \frac{1}{2}$, or 1 chance in 8.

Referring again to the coin-tossing example, if we let p equal the probability of heads (H) and q equal the probability of tails (T), the sum of these two probabilities for the toss of a single coin $(p + q)$ equals 1.0. The principle can be stated that the sum of the probabilities of all possibilities of occurrence for a single or a series of independent events is 1.0. For the single coin toss,

$$(p + q) = (\tfrac{1}{2} + \tfrac{1}{2}) = 1.0$$

We can extend the use of this principle to cover the possibilities for the two independent coin tosses discussed above:

$$\tfrac{1}{4}(\text{HH}) + \tfrac{1}{4}(\text{HT}) + \tfrac{1}{4}(\text{TH}) + \tfrac{1}{4}(\text{TT}) = p^2 + pq + qp + q^2 = 1.0$$

A second important principle is that the likelihood of two or more of a series of independent events occurring together is the product of the probabilities of occurrence of the independent events. Thus the probability that three tosses of a coin would give three heads is:

$$\tfrac{1}{2} \cdot \tfrac{1}{2} \cdot \tfrac{1}{2} = \tfrac{1}{8}$$

Binomial Distribution

Each of the probabilities mentioned above can be derived from the expansion of the binomial $(p + q)^n$, where p equals q equals $\tfrac{1}{2}$ and n is the number of observations in the series. In tossing the coin 3 times, the probability of any one series of events can be determined from the expansion of the binomial

$$(p + q)^3 = p^3 + 3p^2q + 3pq^2 + q^3$$

The coefficient of each succeeding term in the binomial expansion can be obtained by multiplying the coefficient of the present term by the exponent of p and dividing by the number the present term represents in the series. In the second term, the coefficient is 3 and the exponent of p is 2. Thus the coefficient of the third term is $(2 \times 3)/2 = 3$. The q^3 term represents the occurrence of three tails, and this is $\tfrac{1}{8}$. Of the eight possibilities that could arise from three tosses of a coin, the likelihood of getting three heads or three tails would be $\tfrac{1}{8}$. The probabilities of two heads and one tail or one head and two tails are each $\tfrac{3}{8}$. Again the sum of these probabilities adds to 1.0 ($\tfrac{1}{8} + \tfrac{3}{8} + \tfrac{3}{8} + \tfrac{1}{8}$).

This procedure can be followed for more complex problems where the probabilities of the events may not each be $\tfrac{1}{2}$, such as $\tfrac{3}{4}$ or $\tfrac{1}{4}$, or where more than two alternatives may be possible as with dice or multiple alleles at a specific locus. First find the probability of each separate event. These values are inserted in their proper place in the binomial (polynomial) expansion. In the case of three alternatives, one would have to expand the trinomial $(p + q + r)^n$ to examine the individual probabilities.

The binomial can be drawn upon most extensively in animal breeding. Where n represents the number of observations, the probability of a particular combination of two qualities can be obtained without needing to expand the complete binomial by using the following formula:

$$\frac{n!}{r!s!}p^r q^s$$

where factorial n = Total number involved
  factorial r = Number of one alternative
  factorial s = Number of the other alternative (and $r + s = n$)
    p = Probability or expectation of obtaining the first alternative
    q = Probability or expectation of obtaining the other alternative (and $p + q = 1$)

If we assume that the probability of a male or female from a particular birth was ½ for most practical purposes, we can determine the probability that 4 females and 4 males would be born in 8 births. Following the formula above, we have

$$\frac{8!}{4!4!}\left(\frac{1}{2}\right)^4\left(\frac{1}{2}\right)^4 = \frac{8 \times 7 \times 6 \times 5 \times 4 \times 3 \times 2 \times 1}{(4 \times 3 \times 2 \times 1) \times (4 \times 3 \times 2 \times 1)}\left(\frac{1}{2}\right)^4\left(\frac{1}{2}\right)^4 = \frac{70}{256} = .273$$

or about 1 chance in 4.

Chi-Square Values

How well a given set of results fits those to be expected on the basis of a particular genetic hypothesis can be tested by chi-square (χ^2). This test is often termed the test of goodness of fit. First the expected ratio is determined on the basis of the hypothesis to be tested. Then the contribution which each of the classes in the ratio makes to the chi-square value is computed by squaring the difference between the observed number (O) and the expected number (E) of observations in each class and dividing it by its expected number. The results for all classes are summed to give the chi-square value.

$$\frac{(O - E)^2}{E} = \chi^2$$

Suppose there were 83 black animals and 37 red animals arising from crosses among heterozygous Bb parents. Since black is dominant, we expect a ratio of 3 black animals to 1 red animal among the F_2 offspring on the basis of a monohybrid-dominant hypothesis. We now want to test whether the observations fit this hypothesis for this character or whether we must search for a more plausible theory. In short, we want to know what is the probability of getting as large a deviation or even larger deviations from the expected 3:1 ra-

TABLE 5-3
COMPUTATION OF χ^2 FOR A 3:1 EXPECTED RATIO

Class	Observed frequency, O	Expected frequency, E	Deviation, $O - E$	$\dfrac{(O - E)^2}{E}$
Black	83	90	−7	.54
Red	37	30	+7	1.63
			χ^2	2.17

tio in the F_2. The computations required to provide the chi-square value are shown in Table 5-3.

We now consult the table of chi-square (Table 5-4) and read across the line for one degree of freedom. We find that our χ^2 value of 2.17 is less than but near the value for .10. We interpret this to mean that the observed variation from the expected 3:1 ratio would be anticipated to occur by chance in slightly more than 10 percent of such cases. This degree of departure would not usually be considered statistically significant, and so we say that the monohybrid-dominance hypothesis can be accepted.

It is evident that the larger the deviation, the larger χ^2 becomes. If it gets to the size of the figures in the .05 column of the chi-square table, we say the difference is significant (theory probably not true), because that large a deviation from expectation would occur only once in 20 times by chance alone. If it gets as large as or larger than the figure in the table under the probability column headed .01, we say the deviation from expectancy is highly significant;

TABLE 5-4
TABLE OF CHI-SQUARE*

Degrees of freedom	Probabilities†						
	.99	.95	.75	.50	.10	.05	.01
1	.00	.00	.10	.45	2.71	3.84	6.63
2	.02	.10	.58	1.39	4.61	5.99	9.21
3	.12	.35	1.21	2.37	6.25	7.82	11.34
4	.30	.71	1.92	3.36	7.78	9.49	13.28
5	.55	1.15	2.67	4.35	9.24	11.07	15.09
6	.87	1.64	3.45	5.35	10.64	12.59	16.81
7	1.24	2.17	4.25	6.35	12.02	14.07	18.48
8	1.65	2.73	5.07	7.35	13.36	15.51	20.09
9	2.09	3.33	5.90	8.34	14.68	16.92	21.67
10	2.56	3.94	6.74	9.34	15.99	18.31	23.21
20	8.26	10.85	15.45	19.34	28.41	31.41	37.57

*Reprinted by permission from *Statistical Methods, Seventh Edition*, by G. Snedecor and W. Cochran, © 1980 by Iowa State University Press, Ames, Iowa 50010.
†This is the probability that a χ^2 value equal to or larger than the tabular value would be expected to occur by chance alone.

i.e., it would occur by chance only once in 100 or more times. In using χ^2, always consult the line corresponding to one less than the number of F_2 classes involved in the theory. In our examples we expected two F_2 classes, and so we used the values in line 1 in interpreting χ^2. If we had been testing a dihybrid theory with its expected four F_2 classes, we would have used line 3, and so on.

Normal Curve and Mean

We have noted that the majority of the traits which are economically important probably are influenced by genes at a large number of loci and by many environmental factors. As a consequence most of these traits are measured quantitatively, and they exhibit what we already have described as continuous variation. A number of special procedures and techniques have been devised to assist in characterizing and summarizing such continuous variation.

A single measurement describes the weight of a cow. A group of cows will vary in their weights, and a complete description of the weight of the entire group of cows requires an enumeration of the weight of each individual. Such a mass of data for a large herd of cows is unmanageable, and a statistical description of the population is desired. This description in its simplest form consists of a measure of the central tendency and a measure of the variability of the population. Such a description of the central tendency and the variation for the entire population of measurements is called a *parameter*. Ordinarily we do not have measurements on the entire population. By studying a sample of observations from the population (see Figure 5-1), statistics can be derived which are descriptive of the sample and which provide estimates of the corresponding population parameters. Statistics derived from a small sample may not be representative of the population parameters. A more reliable statistical description of the population usually can be obtained by increasing the size of the sample studied.

Our ability to describe a population of biological measurements in a simple yet meaningful manner is greatly aided by the fact that most of our biological measurements can be considered to be normally distributed. A distinctive feature of this normal distribution is that the values are clustered at a midpoint, thinning out symmetrically toward both extremes. Figure 5-1 shows a theoretical normal distribution. The height of the curve at a particular point represents the frequency of individuals having that particular value.

There are at least three common measures of central tendency: the median, representing the class value halfway between the two extreme values; the mode, representing the class with the highest frequency; and the mean, the average of all measurements in the population. For a true normal distribution, the median, the mode, and the mean will coincide. However, the mean is our most useful statistic in estimating the central tendency of most populations, since samples from a normally distributed population may show departures from normality.

The population mean is merely the arithmetic average of all the values in-

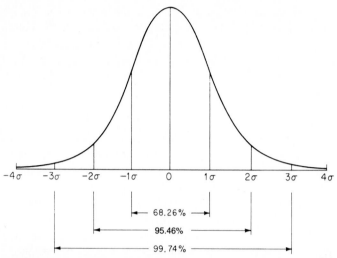

FIGURE 5-1
A theoretical normal curve showing the area bracketed by one, two, three, and four standard deviations from the mean. Percentages of total area bracketed by one or more standard deviations above or below the mean are indicated.

cluded in the population. It is conventionally (see Figure 5-2) represented by the symbol u. Obviously it is usually impossible to measure all individuals in the population; hence a sample chosen so as to be representative of the population is ordinarily used to estimate the population mean. The symbol $\bar{x}$ is conventionally used to represent the sample mean. Individual measurements in the sample can be symbolized by $X_1, X_2, X_3, \ldots, X_n$, and the mean can be computed according to the formula

$$\bar{x} = \frac{(X_1 + X_2 + X_3 + \cdots X_n)}{n}$$

where n represents the number of observations included in the sample.

Table 5-5 illustrates the use of the above formula in showing the sums and the means for heart-girth measurement of 190 cm and the mean weight of 570 kg to provide a base to which the individual cows in the sample can be compared.

Variance

The degree of dispersion or variation exhibited by a population can be expressed as the average deviation or difference from the mean, ignoring signs. The range between the extremes of the population also provides some indication of variability. However, each of these measures lacks flexibility, and com-

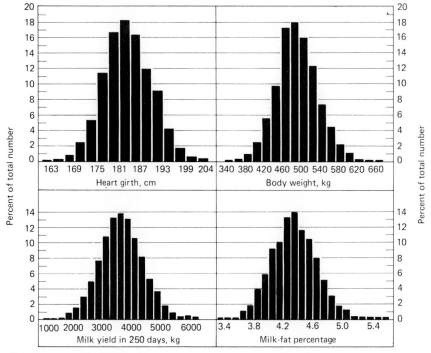

FIGURE 5-2
Frequency distribution of heart girth, body weight, milk yield in 250 days, and milk-fat percentage for 2,407 cows showing trend toward normal curve. (*From Johansson, I., Zeitschrift fur tierzuchtung and Zuchtungsbiologie 63:110, 1954.*)

plete reliance upon them soon blocks the detailed analysis of the observed variation. Variance (σ^2), which is the average squared deviation of the individual measurements from the population mean, is the most useful measure of variation for studying the variability of populations. Actually we deal with a sample from the population and compute estimates of the population variance according to the formula

$$s^2 = \frac{(X_1 - \bar{x})^2 + (X_2 - \bar{x})^2 + \cdots + (X_n - \bar{x})^2}{n - 1}$$

Since the deviations are squared, variance is a positive value with zero as a lower limit. The use of this formula is shown in Table 5-5, where the sample variance for heart girth is 118.8 and that for weight is 4,067. The sum of the squared deviations from the mean is divided by $n - 1$ rather than n to compute the average squared deviation, since a limited sample ordinarily does not encompass the entire range of the population. As a consequence, division by n consistently provides an underestimate of the population variance, but division by $n - 1$ provides an unbiased estimate. Unbiased estimates are ones which show no consistent tendency to be either above or below the population parameter.

TABLE 5-5
HEART-GIRTH MEASUREMENTS AND WEIGHTS OF 25 HOLSTEIN COWS TAKEN
APPROXIMATELY 1 MONTH AFTER CALVING, SHOWING CALCULATIONS FOR
DETERMINING LINEAR CORRELATION AND REGRESSION COEFFICIENTS

Cow no.	Heart girth, cm	Weight, kg	Deviations from means*		Squares and products of deviations		
	X	Y	x	y	x^2	xy	y^2
1	210	680	20	110	400	2,200	12,100
2	195	600	5	30	25	150	900
3	195	630	5	60	25	300	3,600
4	190	580	0	10	0	0	100
5	215	720	25	150	625	3,750	22,500
6	185	560	−5	−10	25	50	100
7	195	610	5	40	25	200	1,600
8	210	670	20	100	400	2,000	10,000
9	190	540	0	−30	0	0	900
10	195	570	5	0	25	0	0
11	195	600	5	30	25	150	900
12	185	560	−5	−10	25	50	100
13	190	550	0	−20	0	0	400
14	180	520	−10	−50	100	500	2,500
15	190	590	0	20	0	20	400
16	185	510	−5	−60	25	300	3,600
17	180	520	−10	−50	100	500	2,500
18	175	460	−15	−110	225	1,650	12,100
19	180	500	−10	−70	100	700	4,900
20	180	520	−10	−50	100	500	2,500
21	175	510	−15	−60	225	900	3,600
22	185	540	−5	−30	25	150	900
23	205	650	15	80	225	1,200	6,400
24	185	560	−5	−10	25	50	100
25	180	500	−10	−70	100	700	4,900
Sums	4,750	14,250	0	0	2,850	16,020	97,600

$$\bar{x} = 190 \qquad \bar{y} = 570 \qquad s^2_x = 118.8 \qquad s^2_y = 4,067$$
$$s_x = 10.9 \qquad s_y = 63.8$$

*x = deviation of X's from their $\bar{x}$.
y = deviation of Y's from their $\bar{x}$.
xy = product of a paired set of deviations.

Table 5-6 shows how the variance due to different environmental and genetic causes can be subdivided. This classic study by Plum (1935) of Iowa dairy records represents one of the first comprehensive analyses of the total variance in a particular trait. Variance is a most useful mathematical concept. Its mastery requires much thought and experience. Note that the total variance can be divided into component parts to assess the relative importance of the several identified sources.

TABLE 5-6
RELATIVE IMPORTANCE OF CAUSES OF VARIATION IN FAT PRODUCTION*

Source of variation	Percent of total variance	
Breed		2
Herd:		
Feeding policy	12	
Other factors (genetic or environmental)	21	
		33
Cow (mostly genetic)		26
Residual (year-to-year variations):		
Feeding variation within the herd	6	
Length of dry period	1	
Season of calving	3	
Other year-to-year differences	1	
Other factors	28	39
Total		100

*Plum, M., *J. Dairy Sci.* 38:824, 1935.

Standard Deviation

The population standard deviation which is the square root of the variance is symbolized by s. The estimate of the standard deviation from a sample is the square root of the sample variance s^2. While variance is expressed in kilograms or centimeters squared, the standard deviation is expressed in kilograms or centimeters, just as the original items were measured. Since the standard deviation is the square root of the variance, it can be computed as follows:

$$s = \sqrt{\frac{(X_1 - \bar{x})^2 + (X_2 - \bar{x})^2 + (X_3 - \bar{x})^2 + \cdots + (X_n - \bar{x})^2}{n - 1}}$$

This is illustrated in Table 5-5, where the standard deviation for heart girth is 10.9 cm and the standard deviation for weight is 63.8 kg.

When reliable estimates of the mean and standard deviation are available for a normally distributed population, the expected proportion of the population which will fall within a designated area of the normal distribution can be computed from specially prepared tables. It has been possible to develop such tables because a consistent relationship between the mean and the standard deviation exists for a normal distribution. Figure 5-1 shows this relationship, with 68.26 percent of the observations being included in the area bound by one standard deviation on either side of the mean. Similarly, 95.46 percent of the population is included in the area bracketed by two standard deviations from the mean, and 99.74 percent of the population is included in the area bracketed by three standard deviations on either side of the mean.

As an example of the application of this concept, studies show that the standard deviation of average breeding (genetic) values for sires, as appraised from a study of their daughters' lactation milk records, is about 500 kg. In a population of 1,000 bulls whose breeding values are normally distributed, 161 would be expected to have breeding values which are at least one standard deviation (500 kg) above the population mean. Only 23 would be expected to be at least two standard deviations (1,000 kg) above the mean, and only one would be expected to be at least three standard deviations above the mean (1,500 kg). When a sire's daughters are compared with their dams, increases of 2,000 kg of milk are sometimes found. Since the bull transmits only a sample half of his genetic make-up, a bull that is genetically 2,000 kg above the population average would be expected to have daughters that were 1,000 kg above average dams of the population. Hence, when the daughters of a bull average 2,000 kg of milk more than their dams, an environmental contribution to this increase should be suspected. Bulls capable of genetically increasing their daughters to this extent are extremely rare.

Coefficient of Variation

It is sometimes desirable to compare the variability of traits measured in different units. Generally, large things vary much and small things little, making it convenient to express the standard deviation as a percentage of the mean to make such comparisons. The standard deviation as a percentage of the mean is termed the *coefficient of variation (C)*, and the formula for computing it from a sample of the population is

$$C = \frac{s \times 100}{\bar{x}}$$

For example, the variance of heart girth in the example in Table 5-5 is 118.8 and that for weight is 4,067. However, the coefficients of variability are 6 and 11 percent, respectively. Certain traits for a class of animals have characteristic coefficients of variation, a knowledge of which is valuable for planning and evaluation experiments.

Correlation

We are often interested in learning whether one trait in an animal is associated with another trait. For example, is the height at the withers in beef cattle associated with body length? Is the weight of a dairy cow related to the amount of milk she will produce? Is there an association between back-fat thickness and the percentage of lean cuts in swine carcasses?

The *correlation coefficient r* measures the degree of association between two traits or variables in a sample from the population. It ranges in value from −1.0 to +1.0. A correlation of +1.0 indicates that for each standard unit increase in one variable there is a standard unit increase in the correlated trait. The correlation coefficient may be anywhere between these two extremes, with a value of zero indicating no association between the two variables.

Heart-girth measurements and weights for 25 Holstein cows were given in Table 5-5. The high correlation between heart girth and body size has long been recognized and has provided the basis for measuring tapes used to estimate body weight. Table 5-5 also shows the calculations necessary to compute the correlation coefficient.

The correlation coefficient r is then computed according to the following formula:

$$r = \frac{\Sigma xy}{\sqrt{(\Sigma x^2)\,(\Sigma y^2)}} = \frac{16,020}{\sqrt{(2,850)\,(97,600)}} = \frac{16,020}{16,678} = .96$$

Correlation coefficients are subject to sampling fluctuations, and the value (.96) obtained from this sample of 25 cows will be limited in its representation of the association between these two traits, say, in the population of Holstein cows in the United States. Methods have been developed to ascertain the probability that a correlation of the magnitude found in the sample actually exists in the population. A significant correlation, from the statistical standpoint, means that there is a high probability that there is a real association between the traits examined of the magnitude indicated by the sample value. With a large volume of data, statistical significance may be realized, but logical judgment is necessary to determine whether the correlation is large enough to be practically useful in prediction and culling for a particular situation.

Correlation and variance are related and in many respects represent two different ways of viewing variation. The square of the correlation coefficient r^2 measures the portion of the variance in one variable, say y, that can be accounted for by variation in a related variable x. The square of the correlation (.96) shows that 92 percent of the variance in body weight is associated with variation in heart-girth measurements. The remaining 8 percent $(1 - r^2)$ is associated with variation in things other than heart girth.

Too often when a correlation is found between two variables a cause-and-effect relationship is assumed. One must be most cautious in arriving at a conclusion regarding cause and effect. The correlation coefficient provides no evidence of itself as to which variable is the cause and which is the effect. Such evidence must come from a specific investigation of the biological relations between the traits.

Regression

Whereas the correlation coefficient measures the degree of association between two variables, the regression coefficient b measures the amount of change in one variable associated with a unit change in the second variable. For example, from the data in Table 5-5 we might ask how many kilograms change in body weight on the average are associated with each centimeter change in heart girth.

The information in Table 5-5 has been plotted in Figure 5-3 to demonstrate visually the association between heart girth and body weight. The regression line has been drawn to provide the best linear fit to the series of paired observations so as to minimize the sum of the squared deviations from this line. Again referring to Table 5-5, the regression coefficient which represents the slope of the line is computed as follows:

$$b_{xy} = \frac{\Sigma xy}{\Sigma x^2} = \frac{16,020}{2,850} = 5.6$$

In this example the regression coefficient indicates that for each change of 1 cm in heart girth an average change of 5.6 kg in body weight is expected.

FIGURE 5-3
Diagrammatic representation of linear regression of body weight on heart-girth measurements for the data in Table 5-5.

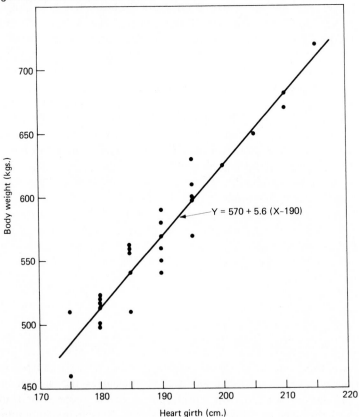

The regression coefficient and the correlation coefficient are related. If the correlation between heart girth and body weight were perfect, all the plotted points in Figure 5-3 would be on the regression line. Under these circumstances we would be able to predict the weight of a cow exactly by knowing the heart-girth measurement. The relationship between the correlation coefficient and the regression coefficient is pointed out further, since, if r is known, b can be obtained as follows:

$$b_{xy} = r\frac{s_y}{s_x} = (.96)\frac{63.8}{10.9} = 5.6$$

The regression finds most use in predicting or estimating one variable, provided the other variable is known. Presume that we know the heart-girth measurement of an animal and that on the basis of the analysis of the sample of data in Table 5-5 we wish to predict her weight. Symbolically the equation for predicting a value of $Y(\hat{Y})$ *is*

$$\hat{Y} = \overline{Y} + \frac{\Sigma xy}{\Sigma x^2}(X - \overline{X}) = \overline{Y} + b_{xy}(X - \overline{X})$$

If a cow's heart girth measured 180 cm, the predicted body weight for the cow would be

$$\hat{Y} = 570 + 5.6(180 - 190) = (570 - 56) = 514$$

Note that this and other predicted values will lie on the regression line.

Regression as originally introduced by Sir Francis Galton had a slightly different, but related, meaning. In studying human stature, Galton found that the progeny of tall parents usually are not as tall as their parents and that progeny of short parents are not as short as their parents. In each case the progeny were nearer the population mean, or, as Galton expressed it, the progeny regressed toward the population average.

Average parents tend to produce average offspring; parents above average tend to produce progeny above average; parents below average tend to produce progeny below average. However, the progeny of extreme parents, above or below the average, exhibit the parental characteristics to a less marked degree than these traits were expressed in the parents themselves.

This regression in the Galtonian sense is universal in material where the correlation between the two variables is not perfect. In the case of stature, in Galton's studies the correlation between the average of the two parents and their offspring approached 1.0, and though regression was evident, the progeny retained a large proportion of the parental superiority. The high parent-offspring correlation indicated that most of the variation in stature was genetically conditioned, hence transmissible from parent to offspring. For traits where the parent-offspring correlation is low, nongenetic variation is comparatively more influential. Only the parental superiority resulting from superior genetic potential is transmissible, and a more marked regression, in the Galtonian sense, toward the population average is evidenced. It should be obvious that this law of regression can be applied with confidence only to the average of large numbers of observations.

Analysis of Variance

A more comprehensive assessment of the variation in a trait can be obtained from the statistical technique commonly described as the *analysis of variance*. In our illustration of the computation of the variance, standard deviation, and coefficient of variation, only the total variance was considered. However, in Table 5-6 the concept that variance can be subdivided according to its various sources was suggested. Such a subdivision of the total variance into its genetic and environmental components is most useful in analyzing problems in animal breeding. Detailed procedures involved in the analysis of variance and the estimation of components of variance are described in statistical texts listed at the end of this chapter. The example to be discussed here is presented to illustrate the biological interpretation of components of variance.

We shall consider data from litters of swine analyzed so that farrowing season and breed differences are not included in the measure of variation. The thickness of the back fat of each of the pigs was obtained from back fat probes at 154 days of age. An analysis of v~~ari~~ mputed for the measurements as shown in T ~~edom,~~ first introduced in connection with the (an be taken to represent the number of indepe ing, and generally it will be one less than the t group $(n - 1)$.

The sums of squar ed deviations of the respective values from t of squares is the sum of the squared deviations a (2.79 cm). Likewise, the sum of squares am nted as the sum of the

$$\sigma_p^2 = .091927$$
$$\sigma_f^2 = .01918$$
$$(N \bar{z} w \, \sigma^2)$$

TABLE 5-7
ANALYSIS OF VARIANCE FOR BACK-FAT THICKNESS (IN CENTIMETERS) IN SWINE AT 154 DAYS OF AGE*

Source of variation	Degrees of freedom	Sums of squares	Mean squares	Components of mean squares
Total	7,590	1,906.2	0.2512	
Among sire families	342	395.4	1.1561	$\sigma_w^2 + 7.07\,\sigma_f^2 + 18.16\,\sigma_s^2$
Among dam families in sire	734	552.6	0.7529	$\sigma_w^2 + 6.59\,\sigma_f^2$
Among full sibs	6,514	958.2	0.1471	σ_w^2

*Data adapted from Cox, D. F. 1964. *J. Anim. Sci.* 23:447–450.

squared deviations of the sire family means (paternal half-sib averages) from the overall mean.[1]

When the sums of squares are divided by the degrees of freedom, we obtain the mean squares. These represent variances on a per-item basis. A further subdivision of these mean squares into components of variance can be made to assign the variance to its identified source.

The value of the components of the total variance included in the mean squares for the lower three lines of the analysis of variance in Table 5-7 are given in Table 5-8. We shall begin with the simplest variance component, σ_w^2, and then consider the added sources of variance. Numerical values for the components are obtained by setting the mean squares equal to their combination of components. The value for σ_w^2 is obtained directly as 0.1471 from Table 5-7.

The variance component σ_w^2 arises from differences among pigs from the same litter, i.e., full sibs. Average differences in the measurements due to the sex of the pigs were adjusted before analysis. Although the pigs of the same litter have the same parents, they do not necessarily have the same genes since each received independent samples of genes from each parent. They also occupied different positions in their dam's uterus and undoubtedly experienced slightly different conditions after birth. One might have contracted a disease while nursing, but the other remained free. The magnitude of these effects is reflected in σ_w^2.

Measurements of two pigs each by a different dam, but by the same sire, may vary owing to the differing genetic contributions of the two mothers. Also prenatal and postnatal influences which were alike for littermates may differ from one litter to another and the variance related to these influences is often designated as σ_c^2. Maternal environment, including the milk available from the mother, may be most influential on early growth and development; whereas, in later life these effects diminish in importance. The contribution of genetic and maternal influences to the differences among pigs is included in the component

---

[1] This is precisely true only when there are the same number of offspring per sire. However, the concept holds even when the family size varies.

TABLE 5-8
VALUES FOR COMPONENTS OF VARIANCE AND PERCENTAGE OF TOTAL VARIANCE REPRESENTED BY EACH COMPONENT

Component	Magnitude	Percent
σ_s^2	0.0204	8
σ_f^2	0.0919	36
σ_w^2	0.1471	56
Total	0.2594	100

of variance, σ_f^2. Additional variation is usually evident when two pigs have different sires. This would arise because of the different samples of genes contributed by the gametes from their sires. If there is much additively genetic variation in the trait, σ_s^2 should be relatively large, since it is associated with the genetic contribution of the sire. The value for σ_f^2 is

$$\frac{0.7529 - 0.147}{6.59} = 0.0919$$

and σ_s^2 is obtained by further substitution after σ_w^2 and σ_f^2 are available:

$$\frac{1.1561 - 7.07(0.0919) - 0.1471}{18.16} = 0.0204$$

The coefficients (6.59 and 18.16) in the denominator come directly from Cox,[2] and those interested in specific details as regards their derivation should consult this reference and the statistical texts listed at the end of this chapter.

The relative importance of the components of variance for back-fat thickness is shown by the percentage values in Table 5-8. Such percentages vary considerably for different traits depending upon the genetic and environmental influences.

SUMMARY

Variation is the raw material the breeder has available for herd and flock improvement. The variation which is observed arises from both genetic and environmental influences. Genetic variation results from differences in the genes provided by genetic recombination and less frequently by gene mutations and chromosomal changes. Traits are usually grouped into those which are qualitative or show discontinuous variation and those which are quantitative or show continuous variation. Specific ratios can be expected for qualitative traits, depending on the number of loci influencing them, the degree of linkage, and the degree of dominance. The observed ratio can be tested against the expected theoretical ratio using chi square. Quantitative traits cannot be characterized so easily. The mean or average of the observations and a measure of the variability about the mean are needed to describe the distribution of values for a particular trait. Correlation and regression coefficients are useful to express relationships between two quantitative traits. The correlation coefficient expresses the degree of association between two variables, being +1.0 or −1.0 when there is complete association. It is zero, when the two traits are independent, and they are not associated. The regression coefficient expresses the amount of change in one trait that is expected for each unit of change in the

[2] Cox, D. F. 1964. *J. Anim. Sci.* 23:447–450.

second trait. Variance, which is the average squared deviation of individual measurements from the population mean, is a much used statistical concept. Components of the total variance can be determined to assess the impact of identified factors as contributors to the variation observed.

SUGGESTIONS FOR FURTHER READING

Books

Bailey, N. T. J. 1981. "Statistical Methods in Biology," 2d ed., Hodder & Stoughton, Ltd., London.

Fisher, R. A. 1970. "Statistical Methods for Research Workers," 14th ed., Oliver and Boyd, Ltd., Edinburgh.

Snedecor, G. W. and W. G. Cochrane. 1967. "Statistical Methods," 6th ed., The Iowa State University Press, Ames, Iowa.

Srb, A. M., R. D. Owen, and R. S. Edgar. 1965. "General Genetics," 2d ed., W. H. Freeman and Co., San Francisco, California.

Steel, R. G. D. and J. H. Torrie. 1980. "Principles and Procedures of Statistics," 2d ed., McGraw-Hill Book Co., New York.

Strickberger, M. W. 1976. "Genetics," 2d ed., The Macmillan Co., New York.

CHAPTER 6

POPULATION GENETICS AND ANIMAL BREEDING

Thus far we have been concerned primarily with a specific mating and with the genetics of the individual or individuals resulting from this mating. The discovery of many important genetic principles has resulted from such studies. However, a different complexion is given to several of these principles when the heredity of a population or a group of interbreeding individuals is considered. *Population genetics* is a field of inquiry in which genetics as it is related to a group or population is considered in contrast to the genetics of individuals. This is sometimes termed statistical, or quantitative, genetics, but the term population genetics is preferred here. Statistics has entered the picture to characterize and to aid in providing a rational explanation of the dynamics of the population. The need to examine populations from the statistical viewpoint is also necessitated because most of the economically important traits are influenced by a large number of segregating loci. As was pointed out in the previous discussion of quantitative inheritance, the action of this large number of genes in the expression of individual traits precludes the separation of the population into discretely distinct genotypes.

The desire to unravel the mystery of evolution has provided a major stimulus to the growth of the field of population genetics, especially the work of Fisher (1918) and Wright (1921, 1931). For most of the economically important traits in animals, our efforts toward improvement really represent attempts to accelerate the action of evolutionary forces. Before humans realized that plants and animals could be modified to fulfill certain purposes, natural selection was actively developing populations better adapted to existing environmental conditions. The underlying biological circumstances which produced inherent change from generation to generation are the same for natural selection as for artificial selection. Where evolutionary change is measured against a base of centuries, breeders try to operate in animal improvement against a time base of a few years or generations.

GENERAL CONSIDERATIONS

Genetic improvement of a population results from underlying changes in gene frequency and/or changes in the way in which the mating system in the population permits the genes to unite as the zygotes are produced. Although the frequencies of most of the genes influencing quantitative traits are not known, an understanding of gene frequency is essential in order to appreciate and understand the genetic dynamics of populations.

Gene Frequency

The frequency of a gene, in its most general sense, means the degree to which it is rare or abundant. More precisely, the frequency of a gene is the proportion of the loci of a given allelic series occupied by a particular gene. When only two members of an allelic series, say A and a, are involved, and there are N individuals in the population, the gene frequency can assume any one of $2N + 1$ values. The frequency of a gene is conventionally designed by the letter p and that of its allele by q. For example,

$$p = \frac{A \text{ loci}}{A \text{ loci} + a \text{ loci}} \quad \text{and} \quad q = \frac{a \text{ loci}}{A \text{ loci} + a \text{ loci}}$$

When several members of an allelic series are involved

$$p_{A_1} = \frac{A_1}{A_1, A_2 + \cdots + a}$$

Gene frequency as defined takes on a range of values between 0 and 1.0.

For a numerical example, consider the case of coat color in Shorthorn cattle. The genetic situation is reasonably well explained on the basis of a single locus where an RR animal is red, an Rr animal is roan, and an rr animal is white. In a herd of 100 Shorthorns there are 47 red, 44 roan, and 9 white animals. There are 200 genes at this locus in the population, and 138 of the 200 loci are occupied by the R gene (two in each of the 47 red animals plus one in each of the 44 roan animals), giving it a frequency of .69. The r gene occupies 62 of the 200 loci in the population, giving it a frequency of .31:

$$p_R = \frac{(2 \times 47 + 44)}{200} = .69 \qquad q_r = \frac{(2 \times 9 + 44)}{200} = .31$$

For loci where dominance is exhibited, it is not possible to detect the heterozygotes directly and determine the gene frequency. However, the

Hardy-Weinberg rule, proposed independently by Hardy, a British mathematician, and Weinberg, a German physician, in 1908, has provided a basis for deducing the gene frequency in many situations where all of the individual genotypes are not distinguishable. This underlying idea of gene distribution postulates that, in large populations and without selection, the relative frequencies of each allele in a population mating randomly tends to remain constant from generation to generation.

Random Mating

To explore this idea in more detail we must understand what is meant by random mating. The situation of random mating exists when the probability of the mating of a given female to a particular male is in direct proportion to the frequency of the different types of males in the population. Random mating with respect to a trait exists when an individual which possesses this trait is no more or no less likely to mate with another possessing the trait than would be expected from the frequency of individuals of the opposite sex possessing the trait in the population. Using the example of human blood groups, data that have been collected indicate that mating is essentially random with respect to the AB, A, B, and O blood types. For example, a person with blood group *AB* is not influenced in choosing a mate by the prospective mate's blood type. The same is true of individuals with the other blood types. This means that in a population the proportion of individuals of any blood type which will marry individuals of other blood types can be predicted from the proportion of individuals of each blood type of the opposite sex in that population.

Some researchers may prefer the term *panmixia* on the grounds that "random mating" may suggest carelessness. Such is not the case, however. It should also be pointed out that the method of selecting animals may in no way be random, although the manner in which these selected individuals are combined may be random. The method of choosing the individuals and the mating system employed generally will not be contingent on one another.

The concept of random mating may be further illustrated by again referring to information on the ABO blood group system. If we let d, a, b, and o represent the relative frequencies of the AB, A, B, and O phenotypes, respectively ($d + a + b + o = 1.0$), then the expected frequencies of the various types of matings can be determined by squaring the phenotypic frequencies $(d + a + b + o)^2$. The proportion of the various types of matings can then be cataloged as in Table 6-1. The relative frequencies of the various mating types are directly proportional to the relative frequencies of the various genotypes in the population.

Hardy-Weinberg Law

In a population satisfying the Hardy-Weinberg law, the proportion of the different types of gametes produced in a population is directly proportional to

TABLE 6-1
SUMMARY OF THE SYMBOLIC FREQUENCIES OF VARIOUS TYPES OF MATINGS FOR
THE ABO BLOOD GROUP SYSTEM, ASSUMING RANDOM MATING

Female blood types	Male blood types			
	AB	**A**	**B**	**O**
AB	d^2	da	db	do
A	da	a^2	ab	ao
B	db	ab	b^2	bo
O	do	ao	bo	o^2

their respective gene frequencies. Thus, in our previous example with Short-horns where p was .69 and q was .31, the gametic frequencies in that population would be $(0.69R + 0.31r)$. In a large population mating randomly with no selection, mutation, or migration, the frequency of each allele tends to remain constant, and the square of the population's gametic array gives the zygotic distribution. With only two alleles the gametic array would be $[pA + qa]$. The zygotic array under random mating would be the square of this expression or

$$[pA + qa]^2 = p^2A + 2pqAa + q^2 aa$$

This expression will permit the extension of the checkerboard method for determining the expected proportion of the different kinds of genotypes from a particular mating to include the entire range of gene frequencies. Most of the previous examples in basic genetics were special cases in which p was equal to .5, as in F_1's and F_2's (see Figure 6-1).

The Hardy-Weinberg formula permits an indirect estimation of gene frequencies, when we are working with populations in which (1) the reproductive and survival rates of individuals carrying *AA, Aa,* or *aa* genotypes are equal, (2) mating is random, (3) mutations infrequent, and (4) the population is large enough to make accidents of sampling inconsequential.

When complete dominance is exhibited, the heterozygotes are indistinguishable from the homozygous dominants. Nevertheless, in a population that satisfies reasonably well the requirements enumerated above, an estimate of gene frequency can be obtained from an accurate count of the proportion of recessives in the population. Although recent counts are not available, earlier information has indicated that in Holsteins about 1 of 200 calves born is red and white. Red is recessive to black, and from the Hardy-Weinburg rule the recessives should make up q^2 of the population. Hence,

$$q^2 = \frac{1}{200} = .005 \qquad q = (.005)^{1/2} = .07 \qquad p = .93$$

For sex-linked traits the zygotic and gene frequencies in the heterogametic sex are the same. If complete dominance is expressed, the gene frequency in the

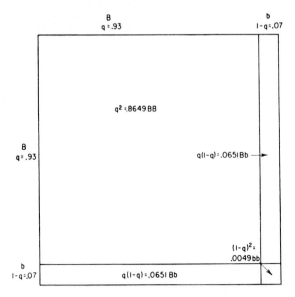

FIGURE 6-1
An example of the Hardy-Weinberg law where *BB* and *Bb* individuals have red coat color. Rather than a ratio of 3 black to 1 red which would be expected when gene frequency *q* was .5, the ratio here is approximately 199 black to 1 red.

homogametic sex could be determined in the same manner as with autosomic inheritance. Red-green color blindness in humans is sex-linked. If 5 percent of the males are color-blind, the frequency of the recessive gene would be $q = .05$. Assuming the same gene frequency in the females, the incidence of such color blindness would be $(.05)^2$, which equals .0025, or .25 percent. Red-green color blindness would be about 20 times as common in males as in females under random mating.

Extension to *n* Loci

We can extend the Hardy-Weinberg principle to include two, three, or more loci. The gametic distribution can be determined if the frequencies of the genes at the several loci are known. We will use polled-horned factor and black-red color to illustrate the situation with two independent loci. Assume a population in which the frequency of the gene for polledness *P* is .6 and the frequency for the gene for black color *B* is .9. Gametes derived from the population would be expected to be in the following proportions:

$$(.6\ P + .4\ p)(.9\ B + .1\ b) = .54\ PB + .06\ Pb + .36\ pB + .04\ pb$$

The product of the gene frequencies at each locus gives the expected gametic distribution. The above is comparable to the ratio of

$$(.25\ PB + .25\ Pb + .25\ pB + .25\ pb)$$

for an F_2 with a gene frequency of .5 for each allele.

As shown in the previous section relative to a single locus, the square of the gametic array gives the expected zygotic array. Thus $(.54\ PB + .06\ pb + .36\ pB + .04\ pB)^2$ gives an expectation of .8316 polled black + .0084 polled red + .1584 horned black + .0016 horned red. This is in contrast to the standard 9:3:3:1 ratio which is expected when the gene frequency is .5 at both loci. These operations can be summarized as follows:

$$[(.6\ P + .4\ p)\ (.9\ B + .1\ b)]^2$$

In the case of n independent loci, the procedure for deriving the zygotic array can be represented as

$$[p_A + q_a]^2\ [p_B + q_b]^2 \cdots [p_n + q_n]^2$$

As illustrated, the square of the gametic array or distribution gives the expected zygotic distribution. Extended further, the square of the zygotic array gives the distribution of the various types of matings when mates are randomly chosen. The frequency of the several mating types in relation to a single locus with two alleles is

$$[p^2\ AA + 2pq\ AA + (q^2)\ aa]^2$$

		Progeny		
Matings	**Frequency**	**AA**	**Aa**	**aa**
AA × AA	p^4	p^4		
AA × Aa	$4p^3q$	$2p^3q$	$2p^3q$	
AA × aa	$2p^2q^2$		$2p^2q^2$	
Aa × Aa	$4p^2q^2$	p^2q^2	$2p^2q^2$	p^2q^2
Aa × aa	$4pq^3$		$2pq^3$	$2pq^3$
aa × aa	p^4			q^4
Totals	1.0	p^2	$2pq$	q^2

The frequency of specific matings in the herd or population is dependent on the frequency of the genes involved. When two, three, or more loci are included, the same principles can be extended, although the expressions can become tedious.

CHANGES IN GENE FREQUENCY

Since changes in the mean of a population under random mating are dependent on gene-frequency change, the forces influencing gene frequency are found to be in various states of balance from time to time. The gene frequency at a particular instant is a result of the equilibrium between selection, migration, chance, and mutation. Wright has summarized the effects of these forces on evolutionary change and has provided a basis for interpreting their effects on gene frequencies in livestock populations.

Mutation

This source of genetic variability was discussed in the preceding chapter. Mutation proposes new genes, and selection dictates whether these genes are retained or rogued from the population. Mutations are generally much less important in changing gene frequency than De Vries originally presumed in 1901 after studying Oenothera. In contrast to Darwin's belief that small variations were responsible for eventual species formation, De Vries looked upon mutation as providing the means for drastic changes.

It has already been pointed out in Chapter 5 that several things limit the role of mutation in providing usable variation. These same conditions impose limitations to its influence in changing gene frequency. The quantitative change in gene frequency by mutation can be expressed by letting u be the mutation rate from A to a and by letting v be the reverse mutation rate from a to A. The change in p, symbolized as Δp, per generation can be expressed

$$\Delta p = -up + vq$$

The balance between the two mutation rates in the absence of other disturbing factors makes for a decrease, an increase, or a stable gene frequency in the population. Under circumstances where the mutation rates are permitted to seek their own equilibrium $\Delta p = 0$. At this point

$$p = \frac{v}{u + v}$$

Mutation must be recognized for the important part it plays in providing new variation over long periods of time, permitting the population to adapt to changing environmental stress. It is not a potent force in changing gene frequency in most economically important species, because most mutation rates are low and most mutations are harmful. Their general recessive nature makes it difficult to increase their frequency in initial selections.

Migration

The introduction of new genes into a population can change gene frequency. Widespread introductions are utilized in crop breeding, but in pure-bred livestock populations it is not usually of major importance. The pedigree barrier imposed by the closed registration in most of the pure breeds restricts the introduction of new genes, although some relaxation of the restrictions has been made in recent years. Any change which migration invokes is dependent on the frequency of the gene in the immigrant population and the extent to which the immigrants are allowed to propagate, i.e., the proportion of gametes they are permitted to furnish.

Several examples are available in which animals have been imported to introduce genes for specific traits. Zebu cattle were introduced early in this century into the beef populations in Texas and other Gulf states because of their resistance to tick fever and high environmental temperatures and because of their ability to gain on the poor-quality native roughage. Several new breeds have combined Zebu and European breeds in their formative stages i.e., Santa Gertrudis, Brangus, Charbray, and Beefmaster. The introduction of the Danish Landrace hog in the 1930s to provide genes to improve bacon quality and the litter size of our American breeds of swine is another interesting example.

In a more limited sense, certain seed-stock herds or flocks may serve as centers of gene radiation. The purchase of animals from these herds for use in commercial herds and flocks might be looked upon as selection for the breed as a whole. However, for the individual herd or flock, it is comparable to migration, with possible new genes being introduced from the seed-stock herds or flocks.

Chance

Changes in gene frequency can also result just from chance due to Mendelian sampling which takes place and which provides the gametes which represent the gene pool for the next generation. To the extent that these sampling changes do occur, they should be random and cancel one another in a large population. However, with a small population the continuous chain of sampling to provide the gametes for each succeeding generation may become a potent force in changing gene frequency.

If the sampling from each parental population were entirely random, the knowledge of the characteristics of the binomial distribution permits the examination of the chance variation expected in a given-sized population. We shall consider two alleles, A and a, where $p = .5$ in two populations, one consisting of 100,000 individuals and the other consisting of 50 individuals. The larger population's 200,000 gametes arise from a pool consisting of equal numbers of A and a gametes. Exactly equal numbers of the two gametes will not necessarily enter into the formation of the next generation. By chance there might be more or less than 100,000 of each kind of gamete. According to the sampling of a binomial distribution the expected number of A or a gametes in the large population would be

$$100,000 \pm \left(\frac{100,000 \times 100,000}{200,000}\right)^{1/2} = 100,000 \pm 223$$

Likewise in the small population of 50 individuals the expected number of A or a gametes would be

$$50 \pm \left(\frac{50 \times 50}{100}\right)^{1/2} = 50 \pm 5$$

The sampling or standard deviation of 223 for the large population is numerically larger than the 5 for the small population. Comparatively, the standard deviation for the large population is only 0.223 percent of the expected number of gametes; whereas in the small population it is 10 percent of the expected number. From this, one can see that in the absence of selection, mutation, and migration, the proportion of gametes, and hence the gene frequency, can remain seemingly unchanged from generation to generation in a large population. With small populations much variation in gametic proportions is expected, and the extinction or fixation of a given locus is not uncommon. Intermediate between extinction and fixation, the gene frequencies may drift in one direction or another, providing for a change which has been referred to by Wright as *genetic drift*.

The decrease in heterozygosity accompanying inbreeding is a manifestation of the above property. Under inbreeding the population number is restricted since matings take place within a closed herd or flock. The smaller the population, the more drastic may be the genetic drift or possible change in gene frequency due to chance, just as the smaller the intermating population, the more intense the rate of inbreeding.

Selection

Selection occurs whenever forces, either natural or imposed by humans, permit some individuals to produce more offspring than others. Individuals which leave no offspring are genetically culled from the population. Selection creates no new genes, but it permits the possessors of some genes and gene combinations to have more offspring than individuals lacking those genes or gene combinations. The underlying influence of selection is to change gene frequency, and the consequences of selection depend on the magnitude of the change in p. The apparent creativeness of selection in producing—after continuous unidirectional selection—types not present in the original population results from the accumulation of genes and gene combinations over many generations of selection.

Although actual selection involves saving or rejecting the individual and its full complex of genes, for illustrative purposes we shall simplify the situation to involve a single pair of two allelic genes. Even in this simplified situation, selection cannot be made directly for individual genes; rather individuals (zygotes) must be saved or rejected.

The amount of change in gene frequency that is to be expected from a generation of selection can be determined if we know the relative number of offspring that parents with each genotype will produce. Wright has developed the background for measuring this change in gene frequency Δp from one generation to the next when constant selection pressure is applied. We shall assume that the *BB*, *Bb*, and *bb* genotypes arising from segregation at a single locus reproduce in the ratio $1:1 - hs:1 - s$ (see Table 6-2).

TABLE 6-2
RELATIVE FREQUENCIES OF THREE ZYGOTES ARISING FROM ONE LOCUS UNDER
RANDOM MATING WITH THEIR RELATIVE REPRODUCTIVE RATES

Parental genotypes	Relative frequencies	Reproductive rates
BB	p^2	1.0
Bb	$2pq$	$1-hs$
bb	q^2	$1-s$

The symbol *s* is the selection coefficient representing the intensity of selection against the *bb* zygotes, assuming the *BB* zygotes are most desirable. The value of *s* represents the proportion of the *bb* types which are rejected. Thus in the case of the red genotypes for registered Angus and Holsteins, *s* would be 1.0, since red animals are not accepted in the regular registry.[1] Likewise for a recessive lethal condition, *s* equals 1.0.

The symbol *h* represents the intensity of selection against *Bb* types in terms of selection against the *bb* types. With zygotic selection and complete dominance ($h = 0$), the heterozygote is indistinguishable from the desired homozygote (*BB*); consequently, no selection pressure can be applied to the heterozygote. With no dominance ($h = \frac{1}{2}$), the heterozygote is intermediate in breeding value between *BB* and *bb*. The value of $h = \frac{1}{2}$ means that selection against *Bb* types is one-half as intense as it is against *bb* types. Where the recessive type (*bb*) is desired ($h = 1$), the *BB* and *Bb* types would be equally restricted in their reproductive rates.

On the basis of the above definitions Wright[2] has shown that the expected change in gene frequency Δp for one generation of selection with complete dominance ($h = 0$) is

$$\Delta p = \frac{spq^2}{1 - sq^2}$$

For no dominance where $h = \frac{1}{2}$,

$$\Delta q = \frac{spq}{2[1 - sq]}$$

When the recessive gene is the desired allele, $h = 1$, the reproductive rates in Table 6-2 become $1 - s$ for *BB*, $1 - hs$ for *Bb*, and 1.0 for *bb*:

$$\Delta p = \frac{sp^2q}{1 - s(1 - p^2)}$$

[1] Red-and-white Holsteins and red Angus are now recorded for registry.
[2] See appendices to this chapter for development of these expressions.

The effectiveness of selection for genes exhibiting these degrees of dominance can be seen in Figure 6-2, which was originally presented by Wright. The heights of the curved lines represent the rate of change in gene frequency to be expected from selection when the original gene frequency in the population ranges from 0 to 1.0. Fluctuation in the height of the curves makes it clearly evident that Δp is dependent on the existing gene frequency in the population when selection is being practiced.

For the gene showing no dominance Δp is largest when $p = .5$. Beginning on the left of the figure, where $p = .1$, progress will become more rapid with each successive generation of selection until p reaches .5. From this point the rate at which p approaches 1.0 will slow down more and more as the actual value of p in the selected population increases.

Dominance is an aid to increasing the frequency of a dominant gene when p is low. So long as the undesired recessive is more abundant than the desired dominant, even the retention of heterozygous individuals serves to increase the frequency of the dominant gene. Nevertheless, after the desired dominant gene has reached a frequency of .5, saving the heterozygotes does not contrib-

FIGURE 6-2
Rate of change of gene frequency under selection and mutation only. Equilibrium frequencies are shown at the junction of the mutation line and the curve for the respective type of desired gene. Mutation rate (u) assumed equal to $s/32$. (*From Wright, 1931.*)

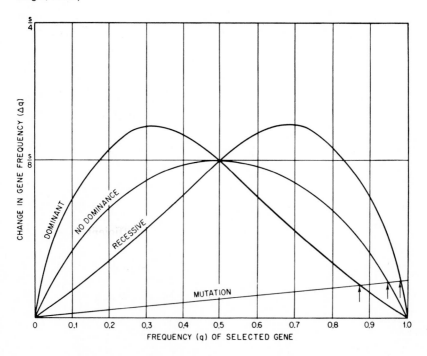

ute to increasing p. At this stage the complete dominance which assisted the change in p when the frequency of the desired gene was low now actually hinders progress.

In Figure 6-2 the curve shows that efforts to increase the frequency of a rare gene are not so rewarding when the gene is recessive. This is another way of pointing up the difficulty experienced in establishing a recessive mutation. During the early stages, even the saving of the heterozygotes would aid in increasing the frequency of the desired gene. However, it cannot be recognized and differentiated from the even less desired homozygote.

Mutation and Selection

As has been pointed out, the gene frequency at a given time is the result of the equilibrium between selection, mutation, chance, and migration. Through the search for the explanation of evolutionary changes, considerable progress has been made in expressing the relationship among these forces and in predicting the ultimate equilibrium values of gene frequency for a specific population.

Illustratively, we shall consider the special case in which only selection and mutation have an important influence on gene frequency. We shall further assume that reverse mutation is negligible; hence, the net change in gene frequency due to mutation would be uq. At equilibrium between mutation and selection, Δp would equal zero. When selection is for a dominant gene, the equilibrium point between mutation and selection is obtained by solving the following expression for q.[3]

$$up = spq^2$$

The equilibrium value for gene frequency (p_e) becomes

$$p_e = 1 - \left(\frac{u}{s}\right)^{1/2}$$

For the other two cases discussed in the preceding section the equilibrium value of p is $1 - 2u/s$ when selection is for a gene showing no dominance, and $1 - u/s$ when selection is for a recessive. The equilibrium points for each of these special cases are indicated by an arrow in Figure 6-2.

For an illustration of the equilibrium frequency, consider the problem of eliminating the red gene from either the black-and-white Holstein or the black Angus cattle. In these situations the double recessives can be eliminated, but the heterozygotes are allowed to breed. The selection coefficient s equals 1 and h is 0, since the desired gene is dominant. Information on the actual mutation rate u for this locus is not readily available. Assuming a mutation rate of 1 for each 20,000 loci (0.00005), which is a value suggested for the polled locus in cattle, an equilibrium gene frequency would be obtained when

---

[3] Approximately, since denominator for spq^2 is slightly less than 1.0.

$$p_e = 1 - \left(\frac{u}{s}\right)^{1/2} = 1 - \left(\frac{.00005}{1}\right)^{1/2} = .993$$

Earlier in this chapter it was pointed out that the frequency of the black gene was probably near .93. The above equilibrium value indicates that mass selection can still increase the frequency of the black gene, but it will be a time-consuming process, as illustrated in the curve of Δp for a dominant gene in Figure 6-2.

Progeny Testing for Recessive Genetic Defects

The equilibrium values for gene frequency shown in Figure 6-2 are based on mass or individual selection. If in the above example, breeders considered eliminating the red gene extremely important, and they were willing to go to the trouble and expense to test all males by breeding each of them to five or six red cows, progress toward reducing the frequency of the red gene would be accelerated. Under such a plan, selection against the red gene would be more intense, since most of the heterozygous males would be detected. If such a plan were kept active for many generations, a new equilibrium value for gene frequency would be reached somewhat nearer 1.0 than .993.

In Chapter 4, it was suggested that progeny testing of males for the presence of an undesired recessive gene might be used in certain breeder or seed-stock herds. Suspected carrier males most likely would be progeny-tested before being used extensively in artificial insemination.

The principle of progeny testing for deleterious recessives is based on the mating of a male that is suspected to be a carrier or to be heterozygous for the recessive gene to females whose genotype for the recessive is known or can be predicted. Let the deleterious recessive gene be designated by r and its dominant normal allele by R. A most common situation is to progeny-test the male by mating him to a number of heterozygous carrier females (Rr). If the male is heterozygous, $\frac{1}{4}$ of his progeny would be expected to exhibit the abnormality. On occasion, affected females (rr) may be available for test matings. Then, if the male is heterozygous, $\frac{1}{2}$ of his progeny would be expected to exhibit the defect.

The mating of a sire to his daughters may be useful where neither homozygous nor known heterozygous carriers are available. If the sire is a carrier (Rr), $\frac{1}{2}$ of his daughters would have received the undesired gene. Thus the expected frequency of the gene in a random group of the sire's daughters whose dams were not carriers would be $\frac{1}{4}$. If the sire were Rr, then $\frac{1}{8}$ of his inbred daughters would be expected to exhibit the recessive condition, assuming a negligible frequency in the population.

The above situations can be summarized by representing the frequency or proportion of r genes in the test population of females by q. Then

$q = 1.0$ for rr or homozygous testers
$q = .5$ for Rr or heterozygous testers (carriers)

$q = .25$ for daughters of a suspected male

$q =$ gene frequency in the population when the suspected carrier is mated to a random sample of females

We assume that the suspected male is heterozygous for the undesired gene (Rr); hence q is taken to be .5. The probability that a single offspring is defective is $q/2$. Then the probability that one or more defective progeny would be observed among n progeny would be $1 - (1 - q/2)^n$. In contrast, the probability that the animal being tested was in fact a carrier, but no defective individuals were detected among n progeny, would be $1 - [1 - (1 - q/2)^n]$. Table 6-3 summarizes these probabilities for three types of test matings.

Note in Table 6-3 that 10 offspring from matings to homozygous-affected females are as equivalent for detection as 20 offspring from heterozygous carriers and 50 from sire-daughter matings.

Any desired level of detection of an undesired gene can be chosen, e.g., a probability of .99. Then

$$1 - (1 - q/2)^n = .99$$

The appropriate value of q for the test population is inserted, and the term $(1 - q/2)^n$ is solved by use of logarithms.

Sire-daughter test matings provide a test for all deleterious recessives, not just the specific one in question. In this regard it is more effective than the other two methods. When only a single recessive is in question, however, approximately twice as many offspring are required for a desired probability of detection as when heterozygous carrier females are used. The principal disadvantage of this test is that the sire will necessarily be well-advanced in age be-

TABLE 6-3
SUMMARY OF PROBABILITIES OF DETECTION OF A HETEROZYGOUS CARRIER AND PROBABILITIES OF NONDETECTION FOR THREE TYPES OF TEST MATINGS FOR A SUSPECTED MALE

Number of offspring	Probability of detection			Probability of nondetection		
	rr mate	Rr mate	Daus* mate	rr mate	Rr mate	Daus mate
1	.500	.250	.125	.500	.750	.875
2	.750	.437	.234	.250	.563	.766
3	.875	.578	.330	.125	.422	.670
4	.937	.684	.414	.063	.313	.586
5	.969	.763	.487	.031	.237	.513
10	.999	.944	.737	.001	.056	.263
15		.987	.866		.013	.134
20		.997	.931		.003	.069
50			.999			.001

*q is taken as .5, assuming negligible frequency in the population.

fore offspring of his daughters can be observed. In a few cases (the mule-foot condition in cattle is an example) defects are expressed early, and it has been possible to speed up the progeny test by removing embryos surgically and classifying them. It can be a valuable procedure for suspected sires being considered for widespread use in artificial insemination. Depending upon the frequency of the various undesired genes in the general population to which the male is mated artificially, a less rigorous test is obtained by general sampling of the male.

Progeny testing procedures emphasize the probability of *failing to detect* a heterozygous sire. Since the birth of a single defective offspring identifies a sire as a carrier, the probability of *detection* of a carrier equals one minus the probabilities of nondetection. Thus, the mating of a sire to even a few known carrier females or to a few of his daughters is of considerable help in finding most carriers.

The same principles outlined for males will apply to progeny testing of females. Cows and ewes produce so few offspring that they cannot be adequately progeny-tested in a normal lifetime. The practice may sometimes be useful with litter-bearing animals.

HEREDITARY VARIANCE

Both hereditary and environmental influences are responsible for the observed variation in records of performance. The hereditary variation results from the action of the genes and gene combinations in response to the environmental conditions provided the individual members of the population. Only genes and chromosomes conditioning the hereditary variability can be transmitted from parent to offspring or selected for in breeding programs.

The most important quantitative traits are usually influenced by an unknown number of genes. Moreover, their mode of action, either chemically or descriptively, such as additive, multiplicative, etc., is also rarely known. Fortunately this, in most cases, is not an insurmountable handicap to animal breeders. While somewhat more accurate choices of animals might be possible if all the details as to the genes present and their mode of action were known for the more simply inherited traits, what the breeder would do corresponds closely to what can be done when the individual performance, the mean, and the relative importance of the several sources of variability are known.

Subdivision of Hereditary Variance

The fundamental genetic theory which provides the basis for the logical direction of applied breeding programs as developed largely by Fisher and Wright has been interpreted in the light of animal breeding applications by Lush and his associates. Its main thesis relies on the partitioning of the total hereditary variance into three components based on a statistically convenient description of gene effects. These three component parts of the hereditary variance ($\sigma^2 h$)

are the additively genetic variance ($\sigma^2 g$), the variance due to dominance deviations from the additive description ($\sigma^2 d$), and the residual genotypic variance ($\sigma^2 i$) termed *epistatic variance*.

The simplest way for genes to be combined is for a gene to produce a certain plus or minus change in the trait or quality being observed (Figure 6-3). Some gene effects presumably do combine in this way, and the effects on the phenotypic variability of other genes can be explained statistically by an additive description or model. The variance associated with the additive or average effects summed up gene by gene for all loci influencing the trait is termed the *additively genetic variance* ($\sigma^2 g$). The sum of the additive or average effects for all loci influencing a trait represents what is commonly referred to as the *average breeding value* for an individual.

In some instances the full effect of a gene on the phenotype cannot be adequately represented by the additive description. For many of such instances the effect of a gene may be dependent on the other genes present in the genotype. In the case of dominance, the substitution of A for a in an aa genotype has a more marked effect on the phenotype than when A is substituted for a in an Aa genotype. Dominance represents a condition of nonadditivity where the phenotype of the heterozygote is not exactly midway between the two homozygotes. The effect of the substitution of A for a is not additive for all genotypes in the population.

Even with complete dominance, much of the relationship between pheno-

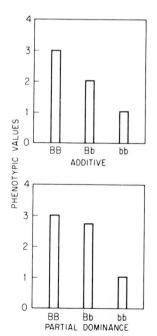

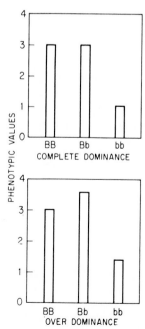

FIGURE 6-3
Phenotypic values for a single locus where additive effects and varying degrees of dominance are expressed.

types and genotypes in the population can be accounted for by an additive relationship. The portion of variability which cannot be thus accounted for is termed *dominance deviations* from the additive description, and the variance contributed by these deviations is the *dominance variance* (σ^2d). (See Appendix: "Partitioning the Heredity Variance" for further development.)

Joint influences (interactions) of nonallelic genes also may influence a trait. A classical example is that of purple and white flower color in sweet peas reported by Bateson. Two independent pairs of genes (*Cc* and *Pp*), each showing dominance, were found to be responsible. White flowers result when either one or both of the dominant loci are absent. At least one of each of the dominant genes at the two loci, however, are required to produce the alternative purple color. Hence the presence of these genes, each at different loci, is required for the expression of purple color. Though dominance may be regarded as interaction among allelic genes, this example represents the interaction also among nonallelic genes and is referred to as *epistasis*. Hereditary variance not included in the additive or dominance portions is termed *epistatic variance* (σ^2i).

The three types of genetic effects are shown in Table 6-4 for two loci. Phenotypic values have been assigned the genotypes to illustrate additive effects, dominance, and epistatic deviations. The values given in the table represent deviations from the mean, and the relative frequencies are those of a simple dihybrid F_2 ratio. Phenotypic values have been arrived at by assuming complete dominance at the two loci with *BB* = *Bb* = *CC* = *Cc* = 2 and *bb* = *cc* = −6, expressed as deviations from the mean. Then an interallelic effect was introduced by allowing *B−C−* combinations to add 2 to the phenotypic values, *B−cc* and *bbC−* to add −3 to the phenotypic values, and *bbcc* to add zero.

These phenotypic values were then analyzed to partition the variance into that attributable to the additive effects at the two loci plus the dominance and

TABLE 6-4
ILLUSTRATION OF PHENOTYPIC VALUES AS COMPOSED OF ADDITIVE, DOMINANCE, AND EPISTATIC EFFECTS

Genotypes	Relative frequency	Phenotypic value	Sum of additive effects	Sum of dominance deviations	Sum of epistatic deviations
BBCC	1/16	6.0	11.0	−5.5	0.5
BBCc	2/16	6.0	5.5	0.0	0.5
BBcc	1/16	−7.0	0.0	−5.5	−1.5
BbCC		6.0	5.5	0.0	0.5
BbCc	4/16	6.0	0.0	5.5	0.5
Bbcc	2/16	−7.0	−5.5	0.0	−1.5
bbCC	1/16	−7.0	0.0	−5.5	−1.5
bbCc	2/16	−7.0	−5.5	0.0	−1.5
bbcc	1/16	−12.0	−11.0	−5.5	+4.5

TABLE 6-5
SUMMARY OF THE COMPONENTS OF HEREDITARY VARIANCES FOR THE GENETIC
MODEL OF TABLE 6-4

Component	Variance	Percentage
$\sigma^2 g$	30.250	63.5
$\sigma^2 d$	15.125	31.8
$\sigma^2 i$	2.250	4.7
$\sigma^2 h$	47.625	100.0

epistatic deviations.[4] The sum of the additive effects represents the average breeding value of the genotypes as deviations from the mean. No environmental influences have been added to the phenotypic values; hence, all of the variance will be genotypic in the broad sense ($\sigma^2 h$). As was suggested earlier, the total hereditary variance can be subdivided into its additive $\sigma^2 g$, dominance $\sigma^2 d$, and epistatic $\sigma^2 i$ components: ($\sigma^2 h = \sigma^2 g + \sigma^2 d + \sigma^2 i$).

The components of variance have been computed for this example, but the procedures used here are too complex to be given in detail. The magnitudes and the relative contribution which each type of variance makes to the total variance are given in Table 6-5.

Note that the assignment of numerical values to the various genotypes and effects provides no explanation of the physiological basis of the gene effects. It has permitted a statistical description of the variation, which can be used to predict genetic changes in large populations. It should be noted that the gene frequency for both alleles in the above example was .5. If the gene frequencies changed, the relative frequencies of the various genotypes would change, as well as the relative importance of the three types of hereditary variance.

The usefulness of this concept of partitioning hereditary variance will be pointed out numerous times. In partial explanation, the gene effects which can be described in an additive manner and which are responsible for the additively genetic variance represent those influences which are subject to molding and change by selection. Those effects expressed as dominance and epistatic deviations are not directly amenable to mass selection pressure. They represent specific effects which provide the basis for heterosis in inbred-line and breed crosses.

Heritability and Repeatability

Two important concepts which tie together the statistical methodology and the principles of population genetics are repeatability and heritability. Their introduction into animal breeding literature can be credited to Lush.

---

[4] Additive effects, dominance, and epistatic deviations and components of variance computed according to Kempthorne, O., "An Introduction to Genetic Statistics." 1957. John Wiley & Sons, Inc., New York, p. 428.

Heritability is the fraction of the observed phenotypic variance which results from differences in heredity—among the genes and gene combinations of the individual genotypes as a unit. This is the broad concept of heritability in which the hereditary variance is considered as the sum of the additively genetic, the dominance, and the epistatic variances.

A more restrictive or narrow definition of heritability is more useful in most aspects of animal improvement. The narrow definition of heritability represents the fraction of the observed phenotypic variance which is additively genetic or which is associated with differences in average breeding values. This is expressed as

$$h^2 = \frac{\sigma^2 g}{\sigma p^2} = \frac{\sigma^2 g}{\sigma^2 g + \sigma^2 d + \sigma^2 i + \sigma^2 e}$$

where $\sigma^2 e$ represents all other nongenetic variance.

Theoretically, heritability can range from 0 to 1.0, but these extreme values are rarely encountered in practice. A particular heritability value is descriptive of a trait in a particular population at a given time. Since it is a fraction, its value can be varied by changes in the additively genetic variance of the numerator or by changes in any one or all the components of variance in the denominator. The additively genetic variance is closely associated with the gene frequency of the genes influencing the trait. For most situations $\sigma^2 g$ is largest when the frequency of the genes influencing the trait is near .5.

Repeatability, a concept closely allied to heritability, is the tendency for successive records of the same animal to be more alike than those of different animals. It is useful for those traits which are expressed several times during an animal's lifetime, such as lactation milk yield for dairy cows, number farrowed and litter weight for swine, and weaned weight of lamb or calf for sheep or beef cattle. Since the same genes or gene combinations influence the successive expressions of a trait, repeatability should be at least as large as heritability in the broad sense. It may be larger, since certain permanent environmental influences may be included in the numerator of the repeatability fraction, but they, of course, would be nongenetic.

Repeatability can be computed as the regression of future performance on past performance. In addition to being computed as a regression, it may be derived from an analysis of variance as an intraclass correlation among records or observations of traits on the same individual. Repeatability r can be expressed as

$$r = \frac{\sigma^2 g + \sigma^2 d + \sigma^2 i + \sigma^2 pe}{\sigma^2 g + \sigma^2 d + \sigma^2 i + \sigma^2 pe + \sigma^2 e}$$

where $\sigma^2 pe$ represents variance associated with possible permanent environmental influences which makes for differences in the expression of a particular trait for the several individuals in the population. For example, a cow may accidentally, but permanently, damage one-quarter of her udder. This would influence milk yield in all future lactations. The feeding and care of young animals may be such as to stunt them, and the influence of this poor feeding and care would appear again and again in the subsequent expression of a trait.

Some data of Lush et al. illustrate both the concepts of heritability and repeatability. They studied 676 daughter-dam comparisons used to prove 103 sires in Iowa dairy herd improvement associations. The lactation milk and fat yields for each female were expressed on a standardized mature basis. When the mates of a bull had only one record, the data for her and her daughter were discarded. The mates of each sire were then divided into a low and a high half on the basis of the first available record on each dam. All available later records were then averaged to provide a single value which was used in representing the later records of these cows. Each dam (mate) was then represented by only one value in the average of first and later records. The summary of these data is given in Figure 6-4.

If a bull had an even number of daughter-dam comparisons, all were used. Where a bull had an odd number of such comparisons, the mate having a median first record was discarded along with the information on her daughter. In this way each of the 103 sires had exactly the same number of mates in the high group as in the low group. Thus, differences in herd averages or in the genetic merit of the sires would not effect the differences between the high and low groups.

As shown in Figure 6-4, the differences between the mates' first records was 46.1 kg of fat (X). This would be representative of the total variability between these two groups of cows. The difference (Y) between the later records of these groups of cows was 19.8 kg. This is representative of the real difference in producing ability (repeatable difference) among these cows if successive records of these cows in the same herds were obtained. When the mates for each sire were divided into a low and a high half, the low half, for example, was represented by cows with lower inherent ability. In addition, by selecting

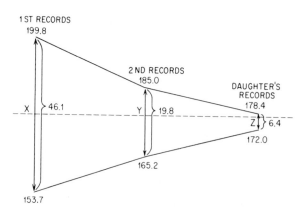

FIGURE 6-4
Regression of later records and daughter's records on the dam's selected record. Dotted line represents midpoint between selected high and selected low records. (*From Lush, J. L., et al. J. Dairy Sci. 24:699, 1941.*)

on the basis of the size of the record, poorer-than-average environmental circumstances were represented also. In the successive records a new sample of herd environment, now presumably representative for both low and high groups, was included. Hence, only the repeatable or real difference between these groups of cows is exhibited in the contrast of the later records. The above procedure provides an approximation to the regression of future performance on present performance, and as such is an estimate of repeatability. The repeatability of differences in single lactation fat yields in these data is 19.8/46.1 = .43. A range of values for the repeatability of several traits estimated from many studies is given in Table 6-6.

The averages for the daughter lactations for these two groups of cows are shown in Figure 6-5. The daughters from the high groups of dams averaged 178.4 kg and those from the low group averaged 172.0 kg. Each sire was represented by an equal number of daughters in the low and the high groups; hence, sire differences do not contribute to the 6.4-kg difference between the daughters of these two groups of dams. However, since the daughters received only a sample half of the dam's genotype, the 6.4-kg difference represents only one-half of what might have been expected if the sires as well as their mates (dams) could have been divided into comparable low and high groups.

TABLE 6-6
REPEATABILITY VALUES FOR SELECTED TRAITS*

Trait	Range of probable values
Cattle	
Milk yield	.40–.55
Fat yield	.40–.55
Fat percentage	.50–.65
Protein percentage	.45–.55
Days open	0–.10
Services per conception	0–.10
Calving interval	0–.15
Gestation length	.20–.30
Birth weight	.20–.30
Weaning weight	.40–.45
Sheep	
Greasy fleece weight	.50–.65
Clean fleece weight	.45–.55
Number of lambs born	.10–.15
Length of wool fiber	.60–.65
Swine	
Litter size	.05–.15
Litter weaning weight	.15–.20

*Based on estimates from a large number of studies.

On the basis of the first available record, the difference between the low and the high group of cows was 46.1 kg. However, on the basis of future records, only 19.8 kg represent real or repeatable differences among cows. The daughters of these two groups of cows are evidence that of the 46.1-kg initial difference in the cows, one-half of the average difference in breeding value among these cows was 6.4 kg. Hence the heritability differences in the lactation fat records is estimated to be 2(6.4/46.1), or .28. Multiplication by two is necessary since the effect of the sires is canceled out in this example. Each sire had the same number of daughters in each of the two groups.

The concept of regression in the Galtonian sense as referred to in the preceding chapter is aptly illustrated in the present example (Figure 6-4) with the regression of the average of the first records and the average of the daughters toward the mean of X. With perfect repeatability, the second records also should have averaged 199.8 kg and no regression would have been evident. With complete heritability the difference between the daughters of these two groups of dams should have been 19.8 rather than the 6.4 actually observed.

A study by Kincaid and Carter provides data which illustrate the concept of heritability utilizing information about the sire's transmission to his progeny. Thirty-eight bulls were placed on a standardized 168-day gain test prior to selection for breeding. After testing, they were divided into high- and low-gaining groups with 19 bulls in each group. Each of the 38 bulls were mated to a random sample of cows, and their steer progeny were also placed on test to determine their rate of gain. These data are summarized in Figure 6.5.

The high-gaining bulls averaged 1,026 g per day, whereas the 19 low-gaining bulls averaged 748 g per day. The difference (D_1) represents the phenotypic difference between the two groups. Ninety-one steer progeny of the high-gaining bulls averaged 821 g per day and 86 steer progeny of the low-gaining bulls averaged 775 g per day. The difference (D_2) between the two progeny groups represents one-half the difference in average breeding values for the two groups of sires. This is the case because the individual sires transmit-

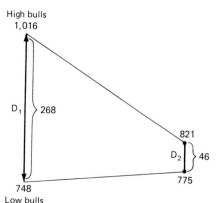

High bulls
1,016

D_1 } 268

821

D_2 } 46

775

748
Low bulls

FIGURE 6-5
Sire-offspring performance for growth rate on test ration as a basis for estimating heritability. (*From Kincaid, C. M., and R. C. Carter. 17: 675–683, 1958.*)

ted a sample half of their genes to their progeny and both groups of bulls were mated to an equivalent sample of cows.

Using these data, heritability of steer gain based on sires' gain is estimated as 2(46)/268, or .34. If heritability had been 1.0, D_2 would have been 134 g rather than 46.

Determining Heritability

Numerous determinations of heritability have been made for economically important traits in farm livestock. Due to statistical sampling and variation, rather large bodies of information are required to provide meaningful estimates. Average heritability values for individual traits are given in Table 6-7 for each specific class of livestock.

Numerous procedures for computing heritability are used in research studies. The underlying basis for determining heritability depends on evaluating how much more alike individuals with similar genotypes are than unrelated or less closely related individuals. The parent-offspring and the paternal half-sib relationships are most widely used to determine heritability in farm animals.[5]

Parent-Offspring Relationship The regression of offspring on parent is commonly used to estimate heritability, and the examples previously discussed with milk-fat yields and beef cattle gains approximate this approach. The more precise statistical procedure begins by obtaining measurements of the traits on both the parents and their offspring. These measurements are corrected for such influences as age or sex when these are likely to influence the measurements. The sum of products (Σxy as in Table 5-5) for the paired observations—x for parent and y for offspring—is obtained. When this is divided by the degrees of freedom ($n - 1$) an estimate of the *covariance* between parent and offspring is expected to include one-half of the additively genetic variance for the trait. The variance of the parents' measurements is obtained and the

$$b_{xy} = \frac{\text{cov } yx}{\sigma^2 x} = \frac{\sigma^2 g/2}{\sigma^2 x}$$

regression of offspring on parent represents one-half of heritability. Since the relationship to only one of the two parents is involved, the regression must be multiplied by 2 to compute heritability.

An example of the results from using this procedure to compute heritability for percentage of nonfat milk solids in dairy cattle is given in Table 6-8. The regression has been computed on a within-herd-year basis to avoid the influ-

---

[5] See Falconer, D. S. 1981. "Introduction to Quantitative Genetics," The Ronald Press Co., New York, for information on other methods of estimating heritability.

TABLE 6-7
HERITABILITY VALUES FOR SELECTED TRAITS*

Traits	Range of probable values
Cattle	
Milk yield (lactation)	.20 to .30
Fat yield (lactation)	.20 to .30
Protein yield (lactation)	.20 to .30
Fat percentage	.50 to .60
Protein percentage	.45 to .55
Calving interval	.05 to .10
Service per conception	0 to .05
Calving difficulty	.05 to .15
Birth weight	.25 to .35
Weaning weight	.30 to .40
Daily rate of gain (feed lot)	.40 to .50
Feed efficiency	.35 to .45
Cancer eye	.20 to .40
Dressing percentage	.35 to .40
Mastitis resistance	.05 to .30
Swine	
Number of pigs farrowed	.05 to .15
Litter weaning weight	.10 to .20
Daily growth after weaning	.35 to .45
Feed efficiency	.30 to .40
Back-fat thickness (probe)	.40 to .60
Loin eye area	.40 to .60
Sheep	
Greasy fleece weight	.30 to .40
Clean fleece weight	.30 to .40
Length of wool fiber	.40 to .50
Lambs born per ewe	.10 to .20
Birth weight	.10 to .30
Daily growth after weaning	.30 to .40
Poultry	
Sexual maturity	.20 to .50
Body weight	.30 to .70
Eggs per hen housed	.05 to .15
Egg weight	.40 to .70
Feed efficiency (layers)	.40 to .70

*Based on estimates from a large number of studies.

TABLE 6-8
HERITABILITY OF THE PERCENTAGE OF NONFAT MILK SOLIDS FROM OFFSPRING (y)
PARENT (x) REGRESSION

Source	Degrees of freedom	x	xy	Variance	Covariance
Total	813	55.54	17.87		
Among herd-years	257	22.16	7.34		
Within herd-years	556	33.38	10.53	.0600	.0189

$$b_{xy} = \frac{.0189}{.0600} = .315$$

$$h^2 = 2\, b_{xy} = (2)(.315) = .63$$

ence of herd or yearly effects on the regression. Note that the variance for the parents and the covariance of parents and offspring for the regression are on a within-herd-year basis. Often the intrasire regression of daughter on dam is used since the progeny of a sire are nearly contemporary and perform under similar managemental conditions. Furthermore, if specific mates are chosen for the different sires, the intrasire regression does not permit the peculiarities of the mating system to contribute to the daughter-dam regression. For the above example the regression of offspring or parent is .315, and heritability of the percentage of nonfat solids is 0.63. If maternal influences are known to affect a trait, the dam-offspring regression would provide an overestimate of heritability.

Paternal Half Sibs.   Half sibs have on the average only one-quarter of their genes in common rather than one-half as with parent and offspring. Usually paternal half sibs are used since there are few maternal half sibs in a set, and their similarity may be influenced by maternal effects. Heritability is determined from an analysis of variance, such as was illustrated in Table 5-8. The intraclass correlation t between paternal half sibs can be computed as a ratio of the components of variance given in Table 5-8. Since half sibs have only

$$t = \frac{\sigma^2 s}{\sigma^2 s + \sigma^2 f + \sigma^2 w}$$

one-fourth of their genes in common, the covariance among them ($\sigma^2 s$) represents $\sigma^2 g/4$. Interpretations for $\sigma^2 f$ and $\sigma^2 w$ are provided for paternal half sibs and are given on pages 119–120. Heritability is computed by multiplying the paternal half-sib correlation t by 4. The heritability value for back-fat thickness (Table 5-8) would be .32 on this basis since ($t = .0240/.0204 + .0919 + .1471 = .08$).

In cattle, monozygotic twins have been used to determine the relative influence of heredity and environment on various traits. These one-egg twins have the same genetic composition. Differences between them should be of an environmental nature. Heritability values obtained from identical twin studies, almost without exception, have been larger than those computed from field data. They usually approximate heritability in the broad sense, since all the dominance and epistatic effects contribute to the similarity among monozygotic twins. Possible maternal and contemporary environmental factors could also serve to inflate these heritability values. As a consequence, heritability estimates from identical twin data are usually high, and they do not provide a true indication of the progress that can be expected from individual selection for a particular trait.

The parent-offspring and paternal half-sib resemblances are intended to provide an estimate of the narrow definition of heritability. Nevertheless, practically all the estimates of heritability include some unavoidable environmental contribution to the resemblance among related animals. Dam-offspring relationships may include a contribution from maternal environment. In addition to possible environmental contributions to the resemblance, some epistatic influences may contribute to the observed likeness. Generally this contribution is believed to be small, and we shall consider the estimates of heritability using parent-offspring and half-sib relationships as being indicative of the importance of the additively genetic fraction of the phenotypic variance.

FIGURE 6-6
A pair of monozygotic twins. Such twins arise from the same fertilized ovum and carry identical inheritance. Approximately 5 percent of all twins in dairy cattle are monozygotic. (*Courtesy A. E. Freeman, Iowa State University.*)

Why Estimate Heritability?

The usefulness of the concept of heritability will be illustrated in the succeeding chapters. It would be beneficial to consider at this point some of the reasons why this is an important concept.

When the breeding program consists of individual selection with the most desirable animals being chosen as parents, the anticipated progress or change is the heritable fraction (heritability in the narrow sense) of the superiority of the parents over the average of the population from which they were chosen. The intensity of selection determines the superiority of the parents, but only the portion resulting from additively genetic differences is expected to be recovered on the average in the progeny. This can be represented as

$$\Delta M = h^2 \times SD$$

where ΔM = Expected change in the mean
SD = Average superiority of the parents

Thus heritability also can be estimated from the results of a planned selection experiment. Much more consideration of this point will be given in Chapters 7 and 8.

A reliable estimate of heritability is also needed to decide which breeding plan is likely to be most effective. Where heritability is high for the desired characteristic, individual selection with little emphasis on pedigree, family performance, or progeny testing may permit the most rapid improvement. If heritability is low, progress from selection on individual records may be slow enough to warrant emphasis on pedigree, family selection, or even the progeny test.

Where estimates of heritability in both the broad and narrow concepts are available, the relative importance of additively genetic variance in comparison to epistatic and dominance variance can be appraised. If there is much epistatic and dominance variance, breeding systems which will capitalize on the specific genetic combinations are to be recommended. This may call for inbred-line formation if progeny from crosses among already existing breeds do not express practically significant levels of hybrid vigor.

Genetic Correlations

The sum of the average effects of the genes influencing a trait represents the average breeding value for that trait. These average effects of genes (plus or minus) are responsible for the additively genetic variance $\sigma^2 g$ associated with our definition of heritability in the narrow sense. The additive or average breeding values for two traits also may be correlated in that a gene or more than one gene may be responsible for an enzyme or other product that influences both traits (pleiotrophy).

The observed or phenotypic correlation between two traits was discussed in Chapter 5 under the heading "Correlation." These phenotypic correlations include both genetic and environmental components or contributions.

$$rX_1X_2 = \frac{\text{cov } X_1X_2}{\sqrt{(\text{var } X_1)\,(\text{var } X_2)}} = \frac{\text{cov } g_1g_2 + \text{cov } e_1e_2}{\sqrt{(\text{var } g_1 + e_1)\,(\text{var } g_2 + e_2)}}$$

The genetic correlation is the correlation between the additive breeding values for two traits (g_1 and g_2) or between the sum of additive effects of the genes influencing these two traits. The genetic correlation coefficient can then be represented as

$$r_{g_1g_2} = \frac{\text{cov } g_1g_2}{\sqrt{(\text{var } g_1)(\text{var } g_2)}}$$

The numerator above is the covariance between the additive breeding values for traits X_1 and X_2. In the denominator we find the genetic variances for the two traits. These genetic variances are the same statistics required for the numerator of the heritability fraction, as given in the section on "Heritability and Repeatability."

Computation of the genetic covariances (cov g_1g_2) requires special procedures to ensure that the environmental effects do not inflate this covariance. In general the procedure involves the analysis of measurements on two related animals whereby the covariance is obtained between trait 1 in one relative and trait 2 in the second relative. Observations for a large number of such pairs of sets of relatives are necessary to provide statistically reliable values.

In the Appendix to Chapter 8 ("Computation of a Selection Index"), the genetic correlation between average daily gain X_1 and gain per unit of feed consumed X_3 is

$$r_{g_1g_2} = \frac{-0.001909}{\sqrt{(358.001)(0.01014)}} = .60$$

What this correlation reflects is that pigs with additive breeding values for more rapid gains require less feed per kilogram of gain. Thus the breeding values for gain per day g_1 and gain per unit of feed g_2 are negatively correlated; the rapid gainers require less feed per unit of gain. Genetic correlations theoretically can range in value from -1.0 to 1.0. Note that the genetic correlation between growth rate and back-fat thickness is positive (.35). Pigs which gain most rapidly also have genes which tend to increase back-fat thickness. Some probable values for genetic correlations are given in Table 6-9.

Genetic correlations result from pleiotrophy, a situation where a gene influences both traits. Among the progeny of recent crosses of highly inbred lines, genes which are linked may contribute to the observed genetic association.

TABLE 6-9
GENETIC CORRELATIONS BETWEEN SELECTED TRAITS*

Traits	Range of probable values
All species	
Growth rates at different life stages	.30 to .50
Rate of gain/feed efficiency	−.40 to −.70
Cattle	
Milk yield/fat percent	−.10 to −.60
Milk yield/protein percent	−.10 to −.50
Fat percent/protein percent	.50 to .60
Milk yield/fat yield	.60 to .90
Birth weight/weaning weight	.20 to .40
Birth weight/yearling weight	.20 to .40
Birth weight/mature weight	.30 to .50
Carcass lean yield/yearling weight	.20 to .40
Carcass lean yield/carcass quality grade	−.10 to −.20
Swine	
Weaning weight/rate of gain after weaning	.10 to .25
Rate of gain/back-fat thickness (probe)	−.10 to .10
Feed efficiency/back-fat thickness (probe)	.25 to .35
Yield lean cuts/back-fat thickness (probe)	−.60 to −.80
Sheep	
Greasy fleece weight/clean fleece weight	.65 to .75
Clean fleece weight/staple length	.30 to .40
Poultry	
Egg weight/body weight	.25 to .50
Egg number/body weight	−.20 to −.60
Egg number/egg weight	−.25 to −.50

*Based on estimates from a large number of studies.

SUMMARY

The field of population genetics seeks to relate genetics to a group or population rather than an individual. The genetics of a population is influenced by the frequencies of the various genes in the population. Gene frequencies are influenced by mutation, migration, chance, and selection. Selection is the most important force available to the breeder for changing gene frequencies. The expected genetic change for a trait in a population is closely related to the nature of the hereditary variation. The total hereditary variance ($\sigma^2 h$) has been partitioned into three portions: the additively genetic variance ($\sigma^2 g$) the variance due to dominance deviations ($\sigma^2 d$), and the variance due to epistatic deviations ($\sigma^2 i$). The expected progress from selection for improvement of a trait is closely related to the proportion or ratio of the total phenotypic variance which is additively genetic. This ratio is termed heritability, and it also can be defined as the fraction of the difference between the parents and their generation mean that is expected to be transmitted to their progeny. Heritability for a trait can be estimated by several methods. Methods which utilize the parent-offspring and paternal half-sib relationships are most widely used in animal populations. Variances due to dominance and epistatic deviations are much more difficult to estimate. When these kinds of hereditary variances are proportionately important, breeding systems requiring the crossing of various lines or strains are needed to obtain the specific genetic combinations which are most desirable. Genetic correlations are useful to describe the relationship between additive breeding values for two traits.

APPENDIXES:
Expressions for Gene Frequency Change Δp

The basis for the expression for Δp may be instructive to inquiring students. We shall use the model given in Table 6-2.

Parental genotypes	Relative frequencies	Reproductive rates
BB	p^2	1
Bb	$2pq$	$1 - hs*$
bb	q^2	$1 - s$

*s = Intensity of selection against bb (fraction rejected) in terms of desirability of BB (s = 1.0 for lethals)
h = Intensity of selection against Bb in terms of selection against aa
    $h = \frac{1}{2}$ No dominance
    $h = 0$ Complete dominance
    $h = 0$ Overdominance
    $h = 1.0$ Recessive desired

Further assumptions in the development include random mating, constant selection pressure, large population size, and no influence of mutation or migration. The expected gene frequency after one generation of selection ($p_1 - p$) is Δp. The selected gametes from one generation of selection are:

$$\begin{array}{cc} A & a \\ p^2 + pq(1 - hs) & pq(1 - hs) + q^2(1 - s) \end{array}$$

The new gene frequency for A is

$$p_1 = \frac{p^2 pq(1 - hs)}{p^2 + 2pq + q^2 - 2pqhs - sq^2} = \frac{p^2 + pq(1 - hs)}{1 - sq(q + 2h)}$$

and so therefore

$$\Delta p = p_1 - p = \frac{p^2 + pq(1 - hs) - p[1 - sq(q + 2ph)]}{1 - sq(q + 2ph)}$$

$$= \frac{spq[q + h(2p - 1)]}{1 - sq(q + 2ph)}$$

With the substitution of appropriate values for h and further simplification,

$$\Delta p = \frac{spq^2}{1 - sq^2} \quad \text{(Complete dominance)}$$

$$= \frac{spq}{2[1 - sq]} \quad \text{(No dominance)}$$

$$= \frac{sp^2q}{1 - s(1 - p^2)} \quad \text{(Recessive desired)}$$

Fisher[6] first pointed out that when there is overdominance at a locus, thereby favoring the heterozygote, an intermediate equilibrium value of p_e is attained as Δp approaches zero. At equilibrium,

$$p_e = \frac{1 - h}{1 - 2h}$$

This equation is satisfied only by values of h less than zero. Some examples of a preference for the heterozygote among some breeders are the roan color in Shorthorns, Blue Andalusian fowl, and the Kerry cattle.[7]

Partitioning the Hereditary Variance, $\sigma^2 g$ and $\sigma^2 d$

In the consideration of hereditary variance, it was mentioned that the deviation of the phenotypic values from the additive representation was responsible for dominance deviations and dominance variance. The following partitioning of the variance at a single locus may be of interest to those desiring a mathematical expression of the additive and dominance variances. The relationship between the average breeding value and the phenotypic value can be expressed as the regression of phenotypic values on the number of the most desired alleles in the genotype. Only additive and dominance effects and variances can be illustrated with this example involving a single locus.

The gene designated B is taken to be the desired one. The linear regression of phenotypic values Y on the number of desired genes X represents the average effect of substituting gene B for b. It can be derived as follows from the information in Table 6-10.

[6] Fisher, R. A. 1922. *Trans. R. Soc. Edinburgh* 42:321–341.
[7] Lush, J. L. 1945. "Animal Breeding Plans," The Collegiate Press, Ames, Iowa, pp. 129–130.

TABLE 6-10
GENOTYPIC FREQUENCIES AND PHENOTYPIC VALUES IN RELATION TO THE NUMBER
OF THE MOST DESIRED ALLELES IN THE GENOTYPE

Genotypes	Frequency	Phenotypic values, Y*	Number of desired genes, X
BB	p^2	a	2
Bb	$2pq$	d	1
bb	q^2	−a	0

*a=(BB−bb)/2 (one-half the difference between the two homozygotes)
d = phenotypic value for heterozygote (d = a for complete dominance, and d > a if overdominance is expressed)

$$b_{XY} = \frac{\Sigma XY - (\Sigma X)(\Sigma Y)/n}{\Sigma X^2 - (\Sigma X)^2/n} = \frac{\Sigma xy}{\Sigma x^2}$$

The frequencies of the genotypes are expressed as proportions, and n then is equal to 1.0.

$$\Sigma X = 2p^2 + 2pq = 2p$$
$$\Sigma Y = p^2a + 2pqd - q^2a = a(p - q) + 2pqd$$
$$\Sigma XY = 2p^2a + 2pqd$$
$$\Sigma X^2 = 4p^2 + 2pq$$
$$\Sigma xy = 2pa^2 + 2pqd - [(2p)\{a(p - q) + 2pqd\}]$$
$$\quad = 2pqa + 2pqd - 4p^2qd$$
$$\Sigma x^2 = 4p^2 + 2pq - [2p]^2 = 2pq$$
$$\Sigma y^2 = 2pqa + d - 2pd^2 + 4p^2q^2d^2$$

Then the regression of phenotype on the number of desired genes is

$$b_{YX} = \frac{2pqa + 2pqd - 4p^2qd}{2pq}$$
$$\quad = a + d(q - p) = \text{(average effect of substituting } B \text{ for } b)$$

The variance in phenotypic values Y accounted for by the linear regression of phenotypic values on the number of desired genes represents the additive genetic variance $\sigma^2 g$. The variance due to the linear regression of phenotypic values on the number of desired genes is the additively genetic variance:

$$V_{reg} = \frac{(\Sigma xy)^2}{\Sigma x^2} = \frac{\{2pq[a + d(q - p)]\}^2}{2pq} = \frac{(2pq\alpha)^2}{2pq}$$
$$\quad = 2pq\alpha^2 = \sigma^2 g$$

The additional variance is termed the dominance variance, and it results because the phenotypic value of the heterozygote does not lie exactly intermediate between the values for the two homozygotes. This dominance variance turns out to be

$$\sigma^2 d = (2pqd)^2$$

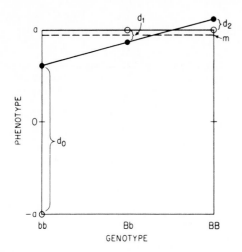

FIGURE 6-7
Diagrammatic representation of the regression
of phenotype or genotype for a single locus.

Since no environmental variance is included in this simple model,

$$\sigma^2 h = \sigma^2 g + \sigma^2 d = 2pq^2 + (2pqd)^2$$

Where many loci influence a trait, the contributions from the several individual loci are summed to give the total variance. This example involves only additive and dominance effects and no epistatic effects since only a single locus is involved.

Drawing upon the example for black and red pigmentation in Holsteins (Figure 6-7), a single locus representation can be given. Since dominance is complete, $a = d$ for the definition of phenotypic values in Table 6-10. Arbitrarily we will say that black-and-white individuals are 2 units superior to red-and-white individuals. Thus $a = d = 1.0$. The regression of Y on X, then, is

$$b_{YX} = a + d(q - p) = .14 = \alpha$$

since q is .93. The total variance for this locus can be computed from the above expressions for $\sigma^2 g$ and $\sigma^2 h$, remembering that a is .14 and q is .93.

$$\sigma^2 h = \sigma^2 g + \sigma^2 d = .0026 + .0173 = .0199$$

For this gene frequency, the dominance variance is larger than the additively genetic variance. The proportion of the total variance which the additively genetic and the dominance variances represent of the total hereditary variances differs as values of gene frequency change. In Table 6-11 this change in these proportions is illustrated for several values for gene frequency. When dominance is complete and the frequency of the desired allele is high, less of the hereditary variance is additively genetic. This situation reflects the difficulty that would be encountered in making further progress once a high frequency of the desired gene(s) had been attained (Figure 6-2).

TABLE 6-11
FRACTIONS OF THE HEREDITARY VARIANCE REPRESENTED BY THE ADDITIVELY
GENETIC AND DOMINANCE VARIANCES FOR VARYING VALUES OF GENE FREQUENCY
FOR COMPLETE DOMINANCE

Gene frequency, p	.10	.30	.50	.70	.90
$\sigma^2 g/\sigma^2 h$	.95	.82	.67	.46	.18
$\sigma^2 d/\sigma^2 h$	.05	.18	.33	.54	.82

SUGGESTIONS FOR FURTHER READING

Books

Crow, J. F., and M. Kimura. 1970. "An Introduction to Population Genetics Theory," Harper & Row, Publishers, Inc., New York.

Falconer, D. S. 1981. "Introduction to Quantitative Genetics," 2d ed., The Ronald Press Company, New York.

Hartzl, D. L. 1980. "Principles of Population Genetics," Sinauer Associates, Inc., Sunderland, Massachusetts.

Kempthorne, O. 1957. "An Introduction to Genetic Statistics," John Wiley & Sons, Inc., New York.

Li, C. C. 1976. "First Course in Population Genetics," The Boxwood Press, Pacific Grove, California.

Lush, J. L. 1945. "Animal Breeding Plans," 3d ed., The Iowa State University Press, Ames, Iowa.

Pirchner, F. 1983. "Population Genetics in Animal Breeding," 2d ed., Plenum Press, New York.

Van Vleck, L. D., E. J. Pollack, and E. A. B. Oltenacu. 1987. "Genetics for the Animal Sciences," W. H. Freeman and Co., New York.

Articles

Cockerham, C. C. 1954. An Extension of the Concept of Partitioning Hereditary Variance for Analysis of Covariance among Relatives When Epistasis Is Present. *Genetics* 39:859–882.

Fisher, R. A. 1918. The Correlation between Relatives on the Supposition of Mendelian Inheritance. *Trans. R. Soc. Edinburgh* 52:399–433.

Lush, J. L. 1949. Heritability of Quantitative Characters in Farm Animals. *Proc. Eighth Intern. Congr. Genet. (Hereditas, suppl. vol.)*, pp. 356–375.

Wright, S. 1921. Systems of Mating. *Genetics* 6:111–178.

———. 1931. Evolution in Mendelian Populations. *Genetics* 16:97–159.

CHAPTER 7

PRINCIPLES OF SELECTION

Selection in animal breeding refers to decisions made each generation by breeders to determine which animals will become parents and which will be culled and leave no offspring. Secondary decisions are required to determine the number of offspring each selected individual will be permitted to have.

In animal breeding enterprises, there are two broad categories of selection. The first, of major importance when an enterprise is initiated, is selection among the available breeds or types. Choice of the foundation breed or type that has a combination of traits which most nearly approaches the desired goal can save many generations of within-population selection. Such choices have been difficult to make due to the limited amounts of performance data and information for comparing different breeds or types. However, in recent years information for such comparison has been accumulating from research studies and organized performance testing programs.

The second category of selection is that carried on within populations. Its basic purpose is to increase the frequencies of desired genes. It is of such paramount importance that it has been aptly termed the *keystone in the arch of animal improvement*. Selection creates no new genes, but properly applied it can increase the frequency of genes with desired effects. Most of the material in this chapter relates to within-population selection.

Selection is not an invention of modern times. It has been going on in nature since life first existed on the earth. In nature, the animals best adapted to their environment survive and produce the largest number of offspring. This natural selection, or survival of the fittest, acting upon the variations produced by mutation and recombinations of genetic factors, eliminates the unsuccessful genetic combinations and allows the most successful to multiply. Selection by human intervention represents the addition of a new criterion, the animal's ability to serve human needs, to nature's criterion of ability to survive and reproduce.

Selection may be practiced at many stages in the life cycle of an organism. Some individuals may not be born because their potential parents were culled; others may experience birth but are culled before being allowed to reproduce. Certain individuals may be selected on the basis of their pedigree or their own performance, but they may be culled when information about their progeny test becomes available. Still other individuals may be retained to produce as many offspring as practicable. In farm animals the selection process also may be influenced by economic considerations. What might be most desirable for maximum genetic improvement may be unduly costly and may have to be modified by economic considerations in operational programs.

ROLE OF SELECTION IN CHANGING GENE FREQUENCIES

The change in traits due to selection stems directly from changes in the frequency of genes influencing the traits. Selection in practice can seldom be for genes at a single locus. Actually, selection cannot be based solely on the effects of the genes which may influence the expression of a single desired trait. Rather, the entire individual or zygote must be chosen.

Most of the traits of economic importance for farm animals are quantitative in nature, and they possess the following characteristics:[1]

1 They are influenced by many genes, most of which probably have small effects individually. Seldom, if ever, is it possible to identify individual gene effects on a quantitative or metric trait.

2 Although we are far from having complete knowledge of the nature of gene action, it appears that additive, dominance, and epistatic effects are all expressed, their relative importance varying from trait to trait.

3 The expression of the genes is conditioned by environmental effects. Consequently, these traits have a continuous distribution with no sharp demarcation between "good and bad."

In Chapter 6 the role of selection in changing gene frequencies for genes controlling a few qualitative characters was discussed. Formulae were given for predicting the rate of change in gene frequency for some specified situations for single loci.

The transition from qualitative to quantitative traits often evokes a feeling of frustration over the seeming inability to discretely identify the genes involved. Segregation of the genes at the individual loci cannot be observed as distinct phenotypic classes. Genes at the individual loci may have differing

---

[1] A few genes with identifiable major effects on quantitative traits have been found in farm animals. The most widely studied of these is the Boorola gene in Australian Merino sheep which when present apparently increases the percentage of multiple births appreciably. The same gene or one with similar effects has been found in some Indonesian sheep breeds. However, the frequency of such genes is not known to be high enough to invalidate the generalizations given here.

magnitudes of influence, and environmental effects may further obscure the difference between the phenotypic expressions of the different genotypes. In spite of these differences, however, the same principles apply to selection for quantitative traits that are influenced by many pairs of genes as to selection for qualitative traits controlled by only one or a few pairs of genes.

Looking at a much simplified example for a quantitative trait, let us assume the following for a population in sheep:

- Daily gain is controlled by four pairs of genes, *Aa, Bb, Cc,* and *Dd* that segregate independently.
- All inheritance is additive, no dominance or epistatic deviations.
- Each gene designated by a capital letter (plus gene) adds 5 g.
- Each gene designated by a lowercase letter (minus gene) subtracts 5 g.
- Mean rate of daily gain is 300 g.
- Environmental influences are considered to be zero, i.e., h^2 is 100 percent.
- Initial frequency of each plus gene is .50.
- Mating has been random.

With the above we have the following distribution of genotypes and phenotypes in the current generation:

Genotypes	Frequency	Number of plus genes	Number of minus genes	Daily gain, g
aabbccdd	1	0	8	240
Aabbccdd aaBbccdd, etc.	8	1	7	255
AAbbccdd AaBbccdd, etc.	28	2	6	270
AABbccdd AaBbCcdd, etc.	56	3	5	285
AABBccdd AaBbCcDc, etc.	70	4	4	300
AABBDcdd AaBBCcDc, etc.	56	5	3	315
AABBCcDc AaBBCcDD, etc.	28	6	2	330
AaBBCCDD AABbCCDD, etc.	8	7	1	345
AABBCCDD	1	8	0	360

If, as an example, only individuals in the top three classes of this generation were selected as parents of the next generation, the frequency of plus genes would be as follows:

Number of plus genes per individual	Number of selected individuals	Number of loci in selected individuals	Number of plus genes in selected individuals
6	28	224	168
7	8	64	56
8	1	8	8
		296	232

Then the frequency of plus genes equals 232/296 = .784. Thus, under these simplified assumed conditions, the frequency of plus genes could be increased from .50 to .784 in one generation. The average daily gain would increase from 300 to 334 g.

This example illustrates the expected change in the frequency of desired genes and the accompanying increase in the mean due to selection. It would not accurately describe most situations for several reasons. Probably many more than four pairs of genes influence most quantitative traits. Thus, the influence of any one gene is relatively smaller than in this example. There would be many more phenotypic classes, actually a continuous distribution, as environmental differences would tend to merge the discrete genetic classes. Some animals which have a high frequency of plus genes, but which have been exposed to poor environments, would have lower performance records than some animals which have fewer plus genes but which have enjoyed above-average environments. This also would result in inaccurate selection of genotypes in some cases.

Stated differently, heritability is always less than 100 percent, and some inaccuracies are inevitable. Nevertheless, if heritability is larger than zero, correct decisions will outnumber the inaccurate ones. Thus, consistent selection for a trait such as rate of gain would increase the frequency of desired genes and change the mean of the population.

FACTORS INFLUENCING PROGRESS FROM SELECTION

The initial task in estimating the progress which can be expected from selection is to obtain an accurate estimate of the average or additive breeding values of the animals available for selection. The breeding values of the individuals are predicted from available information, such as their own performance and their relatives' performance. Following the approach of Lush, much of our consideration of selection alternatives in this chapter will center around methods for enhancing the accuracy of selection, i.e., increasing the correlation between the breeding value G of the individual and the information I used for selection r_{GI}.

In general terms, the genetic progress per generation from selection depends upon the genetic superiority of animals selected to be parents of the next generation, as compared to the average of all animals in the population in the generation from which they are selected. In mathematical form,

$$\Delta G = \overline{G}_s - \overline{G}$$

where ΔG = Genetic superiority of selected individuals
$\overline{G}_s$ = Average breeding value of selected individuals
$\overline{G}$ = Average breeding value of the population from which the selected individuals came

We do not know exactly the breeding or genetic values of animals for quantitative traits, and must use their phenotypic measures in selection. Hence, only the heritable fraction (heritability) of the phenotypic superiority can be expected to be transmitted to the next generation. In mathematical form this becomes:

$$\Delta G = h^2 \times (\overline{P}_s - \overline{P})$$

where ΔG = Expected genetic change
h^2 = Heritability of the trait
$(\overline{P}_s - \overline{P})$ = Selection differential or difference between the mean of the selected parents $\overline{P}_s$ and the average of the population $\overline{P}$ for their generation

Figure 7-1 illustrates diagrammatically the relationships in the above formulae. The diagram illustrates "truncation" selection in which all animals above a given point on a scale of merit are selected and those below are culled. Actually, it is seldom that selection can be this precise. Some of the selected individuals may die or may be sterile or subfertile, etc., thus reducing the sharpness of selection.

The effectiveness or rate of improvement in performance from selection during a generation is dependent on three primary factors. These are the accuracy of selection, the intensity of selection, and the genetic variation of the trait elected. In certain circumstances the expected genetic improvement per year is an important consideration, since some testing methods may inordinately prolong the generation (selection) interval. Four factors determine genetic improvement per year from selection:

$$\text{Genetic improvement per year} = \frac{\text{Accuracy} \times \text{intensity} \times \text{genetic variation}}{\text{Years per generation or selection cycle}}$$

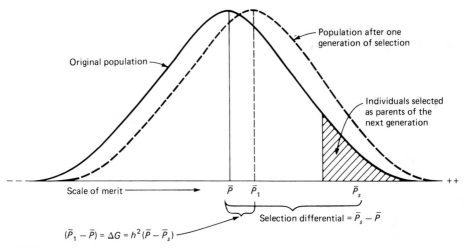

FIGURE 7-1
Diagrammatic illustration of truncation selection from a normally distributed population and the expected mean for the next generation with heritability of .25. Individuals represented by shaded area selected as parents.

In efforts to make the most rapid progress possible for a given period of time, the three factors in the numerator should be made as large as possible and the denominator reduced as much as practicable.

An alternative expression for the genetic change expected from selection enumerates more clearly the factors which influence the efficiency of the selection process. The prediction aspect of selection has been emphasized and the equation for ΔG can be expressed as a prediction equation. The predicted average breeding value $\hat{G}$ for the next generation is equal to the mean breeding value $\overline{G}$ of the herd

$$\hat{G} = \overline{G} + h^2 \times (\overline{P}_s - \overline{P})$$

or population, plus the average genetic superiority of the animals selected as parents. The average genetic superiority can also be expressed as

$$\Delta G = (\overline{G}_s - \overline{G}) = h^2 \times (\overline{P}_s - \overline{P})$$

In each of these expressions the value for heritability can be considered as a regression coefficient. For individual selection it represents the regression of breeding value on phenotypic value b_{GP}, or the change in average breeding value for a unit of phenotypic selection. Thus we may write

$$\Delta G = b_{GP} \times (\overline{P}_s - \overline{P})$$
$$= r_{GP} \times \frac{\sigma g}{\sigma p} \times (\overline{P}_s - \overline{P})$$

where the second expression follows from the relation between the regression coefficient and the correlation coefficient, page 117.

Most biological measurements are distributed normally. If the percentage of the population which is saved from a normally distributed population is specified, functional relations are available to express the selection differential in terms of the number of standard deviations by which the mean of the selected parents exceeds the population mean.[2] Thus if i is the selection differential in standard deviation, the selection differential can be written

$$(\overline{P}_s - \overline{P}) = i \times \sigma p$$

in terms of standard deviation units and σp, the phenotypic standard deviation of the trait.

The above expressions can now be put together to express ΔG in an alternative manner

$$\Delta G = r_{GP} \times \frac{\sigma g}{\sigma p} \times i \times \sigma p$$

which can be reduced to provide the following expression for ΔG.

$$\Delta G = r_{GP} \times i \times \sigma g$$

In the above form the specific factors which influence genetic change can be identified:

r_{GP} = Correlation between breeding value and phenotype or measurements on which selection is based
i = Intensity of selection in equivalent standard deviation units
σg = Standard deviation of breeding values

The significance of each of these factors will be considered in subsequent discussions.

One further point should be presented at this time. Humans are impatient in selection and seek to accomplish in a few years what nature may be content to wrestle with for centuries. Hence, a measure of genetic improvement per unit of time is often a useful practical concept:

$$\Delta G_t = S \frac{1}{L}(r_{GP} \times i \times \sigma g)$$

---

[2] A more thorough coverage of this point will be given below in the section on "Intensity of Selection."

where ΔG_t = Genetic change per unit of time
$\quad\quad\ L$ = Average generation interval

Accuracy of Selection

Accuracy in selection is evaluated by the magnitude of the correlation between the breeding value G of the individual and the variable(s) on which selection is based. For traits expressed only once by the individual, the correlation r_{GP} between breeding value and phenotype equals h.[3] This will be true whether the phenotype is the measurement of a single trait or whether it represents a combination of traits expressed in a single index value. The estimation of correlations between the breeding value of an individual and selection criteria based in whole or in part on repeated records or records of relatives becomes more complex, but the same basic principle applies. In general, higher heritabilities permit more accurate selection.

An important and effective means of increasing heritability and accuracy of selection is to manage potential breeding stock in as uniform an environment as possible and to make statistical adjustment for known variables such as age, sex, and age of dam. These techniques reduce the environmental component of variance in the denominator of the heritability fraction and thereby increase heritability.

Other methods for increasing accuracy of selection include:

1 Using additional measurements for the trait from the same individual
2 Using measurements of correlated traits
3 Using measurements on relatives (family selection) to predict the breeding values

These are often termed *aids to selection*. Their usefulness varies according to heritability of the trait(s) of interest and other factors. These aids to individual selection will be discussed in detail later in this chapter.

Intensity of Selection

The intensity of selection, from disease and other nongenetic factors, can provide an opportunity for increased intensity of selection and (usually) increased profitability. Selection intensity will also vary depending upon whether a population is expanding, remaining constant, or decreasing in size. Extending the use of males through artificial insemination procedures permits large increases in the selection intensity of males. Superovulation and embryo transfer have been accomplished with several species. To date it has been used principally with cattle for increasing the numbers in rare breeds and for increasing the

---

[3] This relationship can be demonstrated by several approaches. Perhaps the most direct at this point is to recall (Chapter 5) that $(r_{GP})^2$ would represent the fraction of the phenotypic variance associated with variation in average breeding values. Thus $(r_{GP})^2 = \sigma^2 g / \sigma^2 p = h^2 = b_{GP}$

TABLE 7-1
SELECTION DIFFERENTIALS* ATTAINABLE WITH TRUNCATION SELECTION WHEN
SPECIFIC FRACTIONS OF A POPULATION ARE RETAINED FOR BREEDING

Fraction saved (p)	Selection differential	Fraction saved (p)	Selection differential
.90	.20	.09	1.80
.80	.35	.08	1.86
.70	.50	.07	1.92
.60	.64	.06	1.99
.50	.80	.05	2.06
.40	.97	.04	2.15
.30	1.16	.03	2.27
.25	1.27	.02	2.42
.20	1.40	.01	2.67
.15	1.55	.005	2.89
.10	1.76	.001	3.37

*In units of phenotypic standard deviations.

number of offspring of outstanding females of established breeds, particularly
to provide outstanding males for progeny testing. These procedures permit increased selection intensity, but to date use has been limited due to high costs.

Selection differentials in standard deviation units for different selection intensities are shown in Table 7-1. They are equal to z/p or i (intensity of selection), where p is the proportion saved and z is the height of the normal curve
at the point where the selected individual with the lowest record falls. Selection of this type is often referred to as *truncation selection*. All individuals included in the normal distribution above the specific culling level would be selected, and those below this level are culled (see Figure 7-2).

When the value for z/p or i is multiplied by the standard deviation of the
trait being selected, this gives the maximum selection differential attainable
when a given percentage of a herd or population must be retained for breeding.
Selection differentials of this magnitude can be attained only if all the animals
above a certain point in the frequency distribution are selected and all those

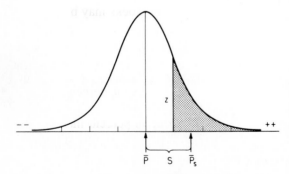

FIGURE 7-2
Diagrammatic representation of
selection differential in terms of
standard deviation units. Only
individuals whose phenotypes are
one or more standard deviations
(represented by shaded area) above
the mean are selected. This
represents about 16 percent of the
population (p). The selection
differential (z/p) is approximately
1.53 standard deviations above the
mean of the original population.

below it are culled. Failure to attain this intensity of selection for important characters or for an index can result from carelessness, inadequate evaluation, or from influences in selection for unimportant traits. Values of the selection differential when various percentages of the population are saved are given in Table 7-1.

A simple example will illustrate the importance of high reproductive rates to enhance selection intensity. Assume a beef cattle herd using natural service in which a breeding male is required for each 25 cows each breeding season. Assume that herd numbers remain constant, that all replacement bulls and cows are selected within the herd, and that each bull is used 2 years. Cows on average remain in the herd for 5 years. With varying percentages of calf crop raised, the percentages of males and females that would need to be selected for breeding and the selection differentials would be:

Percent calf crop raised*	Percent of animals raised that must be selected†		Selection differentials‡		
	Males	Females	Males	Females	Average
100	4.0	40	2.15	0.97	1.56
80	5.0	50	2.06	0.80	1.43
60	6.7	67	1.94	0.54	1.24
40	10.0	100	1.76	0	0.88

*Calves raised annually as a percent of cows exposed for breeding.
†Assumes half of calves are of each sex.
‡In terms of standard deviation.

Such large differences in reproductive rates within species are unusual and, in general, the potential sizes of selection differentials are largely determined by the typical reproductive rate of each species. Table 3-3 gives what are believed to be reasonable estimates for the percentages of males and females which must be retained to maintain constant herd size.

Genetic Variability

The genetic standard deviation (σg) is characteristic of the trait itself, and it is somewhat difficult to change. Crosses among divergent stocks may be used in certain instances to increase the genetic standard deviation and to provide a foundation population with a broader genetic base for selection.

An idea of the importance of the magnitude of the genetic standard deviation on the absolute response to selection can be gained from an example with milk composition in dairy cattle. The heritability of both fat percentage and protein percentage is .50 to .60. Yet the genetic standard deviation for fat percentage is about .30 and for protein about .20. Hence, 50 percent more absolute change in fat percentage would be anticipated with the same intensity of selection for both traits.

Concern is often expressed over the loss of initial genetic variability when selection is continued for many generations. Some increase in homozygosity eventually may result from selection. Also some unavoidable inbreeding will occur when particular individuals are selected and their influence is concentrated. In a program of selection and outbreeding, loss of original genetic variability should not be a major deterrent to selective improvement during the span of most breeding operations.

Generation Interval

Conventionally, the generation interval for a breeding population is defined as the average age of the parents when their offspring are born. This of course varies with species. Some practical values for generation length are given in Table 7-2.

Management practices that allow for early breeding of individuals can reduce generation interval. For example, when beef heifers are bred to calve first at 2 years of age rather than at 3 years, the female generation interval can be reduced approximately 1 year. On the other hand certain breeding practices such as progeny testing extend the generation interval, and this should be reckoned with in assessing the total merit of such a selection system.

INDIVIDUAL SELECTION

Individual or mass selection is selection based solely on the phenotype of individuals. The word phenotype as used here can be for a single trait or for a combination or index of several traits on the individual.

Individual selection is most useful for traits that can be measured in both sexes before sexual maturity or age of first breeding. Some such traits are growth rates, conformation scores, fleece weights, back-fat thickness, and others. For an effective selection program it is necessary that performance records be kept on the total population from which selections are to be made.

TABLE 7-2
APPROXIMATE GENERATION INTERVAL FOR VARIOUS CLASSES OF LIVESTOCK

	Average generation length, in years		
	Male		
Class of livestock	Mass selection*	Progeny testing†	Female
Beef cattle	3.0	8.0	4.5
Dairy cattle	3.0	8.0	4.5
Sheep	2.0	4.0	4.0
Swine	1.5	3.0	1.5

*Males used early and replaced after two breeding seasons.
†Males to be used three breeding seasons after progeny test is complete.

Individual selection has the very important advantage that all animals can be evaluated. This contrasts with progeny testing in which only a small population can usually be progeny-tested because of expense. Also, it is much easier to obtain unselected and interpretable performance records on all animals in a current population than to obtain such records for animals in pedigrees.

Individual selection does have certain limitations. Some of these are:

1 For those traits expressed only by females, such as milk and egg production or maternal traits in meat animals, males cannot be selected on their own performance.

2 Performance records for milk and egg production and maternal qualities become available only after sexual maturity. Thus, if initial selections must be made before sexual maturity, some other criterion must be used.

3 For traits of low heritability, individual performance may be a poor indicator of breeding value. Other aids to selection may need to be used.

4 The ease of evaluating individual characteristics, especially conformation or "type," sometimes tempts breeders to overemphasize it in selection as compared to its *optimum* use with other tools such as pedigree or progeny selection.

5 For traits which can be measured only after slaughter, frozen semen or embryos would be required for individual selection.

For traits to which it is applicable, however, individual selection should receive first consideration.

Repeated Observations

For many traits only one evaluation is possible during the lifetime of the individual. Weight at any given age is an example. In other cases, repeated records and observations are possible. Examples are several lactations of dairy cows, several wool clips from sheep, several litters from litter-bearing species, weaning weights of several calves from beef females, repeated evaluation of conformation at successive ages, etc. Successive records from the same animal tend to be more alike than those of different animals. This tendency is termed *repeatability* and is usually symbolized by r. It was defined in Chapter 6 as the ratio of the variance due to effects common to successive records of the same animal to the total variance. For convenience we can combine some of the sources of variance and define r as

$$r = \frac{\sigma^2 c}{\sigma^2 c + \sigma^2 e}$$

where $\sigma^2 c = \sigma^2 g + \sigma^2 d + \sigma^2 i + \sigma pe^2$, the variance due to effects similar or common to records of the same animal

$\sigma^2 e$ = Residual variance due to temporary environmental influences that differ from record to record

The temporary environmental effects are random and will tend to average out with repeated records. However, they never completely disappear and $\sigma^2 e/n$ will remain in the average of n records. This means that the $\sigma^2 e$ component of the denominator of the repeatability equation will be expected to be .5 $\sigma^2 e$ in the mean of two records, and .33 $\sigma^2 e$ in the mean of three records, etc. Obviously, most of the practically achievable reduction in the denominator will have been obtained with three or four records.

The estimate of real producing ability (often called most probable producing ability, or MPPA), expressed as a deviation from herd or population average, is

$$\frac{nr}{1 + (n - 1)r} \quad \text{(Individual average } - \text{ herd average)}$$

where n = Number of records
r = Repeatability

The expression $nr/ [1 + (n - 1)r]$ is a regression coefficient that estimates the fraction of the superiority or inferiority in the average of n records that will, on average, be found in subsequent records. We can symbolize it as r_n. It can be expressed as

$$r_n = \frac{\sigma^2 c}{\sigma^2 c + \sigma^2 e/n}$$

It can be shown that these two expressions for repeatability are equal by substituting $\sigma^2 c/(\sigma^2 c + \sigma^2 e)$ for r in the first and reducing. The same relationship holds for the expected heritability of averages of n records h^2_n, which is

$$h^2_n = \frac{nh^2}{1 + (n - 1)r}$$

Table 7-3 gives values of r_n for several values of r and n.

Several important traits in farm animals, including fleece weight in sheep, lactation records in dairy cows, and weaning weights of successive calves from the same beef cow, have been shown to have repeatabilities of approximately .40 to .60. For a repeatability of .50, the accuracy of evaluating an animal on the basis of two records should be increased by 15 percent as compared with evaluation on the basis of one record, and 22 percent if three records are available (Table 7-5).

The breeder is interested in properly evaluating animals with varying numbers of records, if some are to be culled and some retained. An example of a problem of this kind in beef cattle is given in Table 7-4. The cows listed had records (adjusted to a mature-cow equivalent) as indicated after their calves were weaned. What is the most probable producing ability of each cow for this trait?

TABLE 7-3
VALUES OF r_n FOR SEVERAL VALUES OF n AND r

Number of records, n	Repeatability, r				
	.10	.30	.50	.70	.90
2	.18	.46	.67	.82	.95
3	.36	.56	.75	.88	.96
4	.31	.63	.80	.90	.97
5	.36	.68	.83	.92	.98
6	.40	.72	.86	.93	.98
.	.	.	.	.	.
.	.	.	.	.	.
.	.	.	.	.	.
10	.53	.81	.91	.96	.99

TABLE 7-4
CALF PRODUCTION RECORDS OF SOME BEEF COWS

Cow number	Number of records	Average deviation of calf weaning weights from herd average, in kg	Probable producing ability as a deviation from herd average
18	1	+ 22	+ 8.8
1	3	+ 12	+ 8.0
9	2	+ 11	+ 6.3
12	1	+ 10	+ 4.0
2	1	− 6	− 2.4
6	3	− 8	− 5.4
25	1	− 15	− 6.0

Other studies indicate that repeatability of the weaning weight of calves of beef cattle is approximately .40. Using this value for repeatability and applying the formula given earlier, we arrive at the figures in the right-hand column as the estimates of the individual cow's real producing ability. Note that the estimate of real producing ability for cow 18, with one record of 22 kg above the herd average, is only slightly higher than that of cow 1, with three records averaging 12 kg above the herd average. This does not say that the real producing ability of cow 18 could not be 22 kg above the herd average. It merely says that in the absence of further records the best estimate is that her high record was due in part to a particularly favorable set of environmental circumstances. Future calves by this cow probably will be nearer the herd average. Similar reasoning will permit making comparisons among other animals in Table 7-4.

When the average of n records is used to evaluate the phenotype of an animal, more of the difference between the herd or population mean is expected to be repeated in future expressions of the trait. Likewise, more of this difference would be expected to be transmitted from parent to offspring. Partially offsetting the increased accuracy and higher heritability of the average of repeated records as compared to single records is the fact that between animals the standard deviation of n measurements is reduced to only $\{[1 + (n - 1)r]/ n\}^{1/2}$ of the variance of single measurements. Thus, selection differentials in absolute terms are somewhat lower for the average of n records as compared to selection based on single records.

Taking account of the fact that r and h^2 are expected to be increased by $n/ [1 + (n - 1)r]$ of that for single observations, the net gain in accuracy for both r and h^2 is $\{n/ [1 + (n - 1)r]\}^{1/2}$. Table 7-5 gives the relative accuracy of selection that can be expected if averages are used rather than single records as based on this formula.

Waiting for additional records before deciding whether to keep or cull an animal can be costly if inferior animals are kept for extended periods. It can lengthen the generation interval and possibly reduce total annual progress in selection for certain traits. This could happen if a dairy cattle breeder insisted on having five or six records from a cow before saving one of her sons. Thus, the advantages of obtaining additional records must be balanced against the costs and disadvantages of getting and using them.

There is evidence from the literature that all lactations of dairy cattle may not provide equal information. The first lactations have had higher heritabilities than other records in a number of studies. Permanent changes in the animal due to such things as udder damage and special treatment in later lactations for animals which perform well in their first record may prevent all records from being of equivalent value in predicting future performance. In any case, additional records for traits with high repeatability are not very useful simply because the first record has already accurately evaluated the animal. Thus, one record for a trait of high repeatability can give a more accurate indication of real producing ability than will several records for a trait with low repeatability.

TABLE 7-5
RELATIVE ACCURACY OF SELECTION EXPECTED WHEN USING AVERAGES AS COMPARED TO SELECTION ON SINGLE RECORDS

Number of records, n	Repeatability, r				
	.10	.30	.50	.70	.90
2	1.35	1.24	1.15	1.08	1.03
3	1.58	1.37	1.22	1.12	1.04
4	1.75	1.45	1.26	1.14	1.04
5	1.89	1.51	1.29	1.15	1.04
6	2.00	1.55	1.31	1.15	1.04
.	.	.	.	.	.
.	.	.	.	.	.
.	.	.	.	.	.
10	2.29	1.64	1.35	1.17	1.05

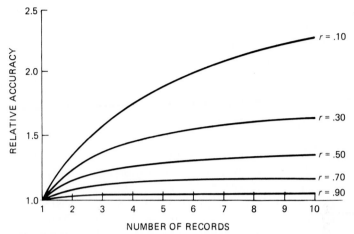

FIGURE 7-3
Relative accuracy of selection using the average of different numbers of
records (observations) per individual and assuming that each record in
the average provides equal information about the individual's breeding
value. The curves represent plots of values for $\{n/[1 + (n - 1)r]\}^{1/2}$

USE OF PEDIGREE INFORMATION IN SELECTION

Technically, a pedigree is a genealogical record of the ancestry of an individual. In animal breeding this definition is usually broadened to include collateral relatives in addition to ancestors. Further, for pedigrees to be useful in selection, information on the productivity of the animals in the pedigree is necessary. A mere listing of the genealogical records is meaningless unless we have performance information on the desired traits of the animals in the pedigree.

Production records have been generally available for dairy cattle for many years. The development and increasingly widespread participation of breeders in performance testing programs for beef cattle and swine during the past two or three decades are making useful pedigree records increasingly available in these classes of livestock. For situations in which production records are not available for broad cross sections of animals of a breed or industry, performance records of animals in specific herds can still be helpful for pedigree evaluation.

Pedigree selection is most useful for sex-limited traits such as milk yield in dairy cattle and egg production in poultry and to a lesser extent for maternal traits of meat animals. For milk yields the high costs of progeny testing dictate that only a few bulls can be tested, and it is important that the very best prospects be selected. Pedigree is the most useful aid for making these initial selections. In meat animals it is often necessary to make preliminary selections, particularly of males, before all growth and feed efficiency records are available. Pedigree information can be helpful in making these early preliminary selections.

Generally speaking, pedigree selection is of limited value for traits of medium to high heritability that can be measured early in the life of both sexes. If the animal's phenotype is known, relatively less can be gained in accuracy of estimating potential breeding value by attention to pedigree. However, a consideration of pedigree information in cases such as this can add slightly to accuracy of predicting breeding value (see Table 7-6). If records are available on performance of animals in pedigrees, its use as a supplement to individual performance data can sometimes be worthwhile. Its only cost will be that of computation.

Pedigree information is valuable because, with the exception of sex-linked genes, each individual receives half its genes from each parent. It must be remembered, however, that each parent transmits only a *sample* half of its genes to each offspring. At loci where heterozygosity exists, chance at segregation will determine which gene is transmitted from that locus. Even if heritability were 1.0, permitting us to know the exact breeding values of each parent, the correlation between a parent's breeding value and that of the progeny would only be .50. The same factor prevents our approaching a correlation of 1.0 between the average of the parents and the offspring. Its upper limit in a random mating population is .71, and this is one of the limitations of pedigree selection. Of course, with lower heritability values the expected correlations between pedigree information and the individual's breeding value would be much lower than the .50 or the .71 given above.

Probably the two major dangers of pedigree selection are (1) undue emphasis on relatives, particularly remote relatives, with the result that intensity of individual selection is reduced and (2) unwarranted favoritism toward the progeny of certain individuals.

The history of pure-bred breeding has been marred by numerous instances of unwise pedigree selection on the basis of remote relatives or meaningless family names. Fortunately, it appears that the possibility of such pedigree

TABLE 7-6
ACCURACY OF SELECTION r_{GI} ON INDIVIDUAL RECORDS OF SOME ANCESTORS AND RELATIVES WITH VARIOUS VALUES OF HERITABILITY

Records Used*	r_{GI}[†]	Heritability				
		.10	.30	.50	.70	1.0
Individual	h	.32	.55	.71	.84	.00
One parent	$.5h$	.16	.28	.36	.42	.50
Both parents only	$.71h$	.23	.39	.50	.60	.71
One grandparent	$.25h$	.08	.14	.18	.21	.25
One parent and one grandparent	$.56h$	.18	.31	.40	.47	.60
One half sib	$.125h$	.04	.07	.09	.10	.12
One full sib	$.25h$	.08	.14	.18	.21	.25

*One record assumed for each individual identified.
†General expression for correlation between breeding value G of the individual and the pedigree information I.

FIGURE 7-4
Pictorial representation in which the relative size of the ancestor is proportional
to its contribution to the inheritance of an individual. The folly of putting undue
emphasis on one or two remote ancestors is evident.

crazes becoming dominant factors in animal breeding is much less in recent
years than formerly.

Guides to Pedigree Use

The potential accuracy of pedigree records for estimating the average breeding
value of individuals can be visualized from the correlations as given in Table
7-6 and the calculated figures for records of some relatives.[4]

These correlations are based on the assumptions of random mating, no relationships among animals mated and therefore no inbreeding, the same heritability of successive records of the same animal, and uniform environment for
all animals. Thus, they represent *maximum usefulness* of pedigree evaluation.
In practice, related animals will quite likely have some common environmental
effects and there may be some degree of nonrandom mating. Also, the correlations ignore sex linkage, but this would be expected to affect their magnitude
little if at all.

Note that the estimates of accuracy in Table 7-6 are based on an assumption
of only one record for each individual in a pedigree. Additional records can
result in significant increases in accuracy.

It must be emphasized that the correlations and the calculated values in Table 7-6 represent idealized situations. It is almost inevitable that environments

---

[4] The student can use standard statistical procedures for combining correlations to estimate the
accuracy of other combinations. Skjervold, H. J., and A. K. Odegard, *Acta Agr. Scand.* 9:341–
354, 1959; and Young, S. S. V., *Heredity* 16:91–102, 1961.

will differ to at least some degree between animals and that, intentionally or by chance, some individuals or groups will have received better treatment than others. For some remote ancestors or relatives, especially if they have been in other herds, it becomes increasingly difficult for a breeder to be aware of and make allowances for environmental differences. For populations where performance testing is widespread, appropriate statistical procedures can be effective in adjusting for a portion of the environmental variation.

The correlations in Table 7-6 emphasize the fact that information on close relatives is much more valuable than that from more distant ones. The correlation between the breeding value of an individual and that of one parent ($R = .50$) is twice as great as that of one grandparent ($R = .25$). But, if parental records are available, grandparental records add relatively little to accuracy. For example, with a heritability of .30, a record of one parent has a correlation of .28 with the individual's breeding value. Adding a grandparent record increases this to only .31.

To summarize, pedigree selection properly used is a useful tool, but one that must be used with understanding and caution. The principal problems in its use relate to possible unknown environmental variables that may be confused with genetic differences and the ever-present temptation to give unwarranted emphasis to records of remote ancestors or other distant relatives.[5]

FAMILY AND SELECTION

Selection based wholly or in part on family is closely related to and could be represented as part of pedigree selection. However, because of some special aspects, and particularly because of attention that has historically been given to the concept of families in animal breeding, it is worthy of special treatment.

The term *family* is used in at least two senses in animal breeding. In several breeds the family has been defined as all animals descending in an unbroken line of females from a foundation female who was considered outstanding enough to have a family named for her. In horses such females are sometimes referred to as *tail mares*. In a few breeds, families have been named for outstanding males, and all animals descending in an unbroken line of males from the foundation male carry the family name. Obviously, if the foundation animal is many generations back in the pedigree, and there has been no inbreeding or linebreeding, her (or his) genetic superiority will have been halved so many times that she cannot now be the source of many genes that would be common to current members of the family. In such cases family names have about as much significance as human family names. Captain John Smith was probably a great man, but does that automatically make all the Smiths great?

Family names lend themselves readily to speculation and the history of several breeds has been marred by the unwarranted popularity of a few families in

---

[5] The student is referred to Appendix, "Example of Pedigree Evaluation," Chapter 11, for an additional example of pedigree evaluation.

certain periods of time. Fortunately, with the development of knowledge among breeders and the increased attention given to actual production records in recent years, family "fads" are becoming less frequent. Almost always, breeders are on sound ground in their selection programs if they pay attention to individual excellence and to the excellence of close relatives rather than the possible excellence of remote family foundations.

The other use of the term family is for groups of related individuals. Usually, in animal populations, these are groups of half sibs or full sibs. With random mating they have *inter se* genetic relationships of .25 and .50, respectively. With inbreeding, *inter se* relationships higher than this can be attained, but such lines are rare in farm animals, and their usefulness is very doubtful (see Chapter 9).

Individuals in groups of full or half sibs are only collaterally and not directly related to one another. An individual receives no genes from collateral relatives, but their average performance is an indication of the genes that may have been transmitted to each family member by one or more common ancestors. The information on the family can thus be used to estimate the breeding value of an ancestor and, indirectly, the breeding value of individual collateral relatives. The correlation between the breeding value of an individual and groups of n half or full sibs, respectively, is $.125h(N)^{1/2}$ and $.25h(N)^{1/2}$ where $N = n/[1 + (n - 1)t]$ and t is the intraclass correlation among sib groups. In the absence of nongenetic correlations among sib groups, $t = h^2/4$ or $h^2/2$, respectively, for half or full sibs.

In many large animal populations, several groups of half sibs may be available from which replacement animals may be selected. In swine there are litters of full sibs and in poultry large groups of full sibs may be available. Ordinarily, the two parents will have transmitted different samples of genes, and so groups of half or full sibs superior to the others should be available.

The question is: Should selection be (1) entirely on a family basis, saving only the animals from the families with the best averages; (2) entirely on individuality, saving only the best individuals regardless of the average performance of others in the family; or (3) on some basis which combines emphasis on the individual and the family? In the last case some plan of saving the better individuals from the better families would be followed.

The difference between these three alternatives can be illustrated in Figure 7-5. Four litters of pigs are represented. Presume that four pigs are needed for breeding and that a weight of 105 kg at an early age is the selection objective. If selection were on the basis of family merit alone, the four pigs in litter 3 would be chosen. If selection were based on individual performance, pigs A, E, I, and J would be chosen without regard for their family (littermates) average days to 105 kg. If information on the individual and the family were taken into consideration, some weighting of the individual's performance and that of the family average would be required. As contrast to the four pigs chosen on the basis of individuality, K might probably replace A. Pigs I and J would be selected on either of the three bases.

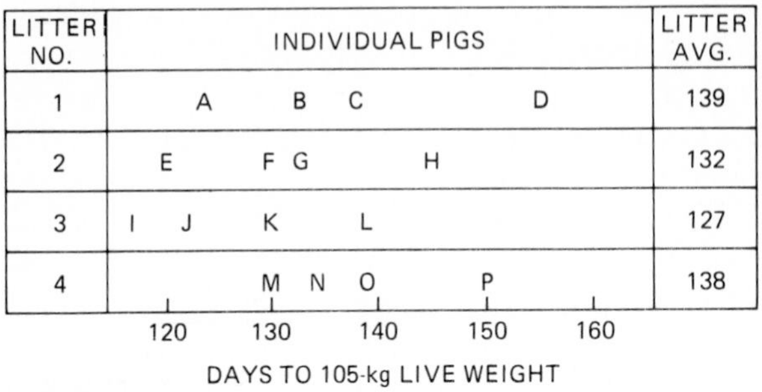

LITTER NO.	INDIVIDUAL PIGS				LITTER AVG.
1	A	B C		D	139
2	E	F G	H		132
3	I J	K	L		127
4		M N O	P		138
	120	130	140	150 160	

DAYS TO 105-kg LIVE WEIGHT

FIGURE 7-5
Example to show the options provided by individual selection, by family selection, and by the combination of individual and family selection. (*Adapted from Lush, J. L., Am. Naturalist 81:243. 1947.*)

Lush[6] studied this problem and found that the relative effectiveness of the three methods depends upon (1) R, the genetic correlation or relationship among family members, (2) t, the phenotypic correlation among family members, and (3) n, the number per family. A comparison of the accuracy of the above three alternatives was made by evaluating the correlation between the individual's breeding value, its own performance, its family average, and the combination of its own performance and the family average.

It was found that if selection was based on family average alone, its accuracy would be equal to $[1 + (n - 1)R]/ \{n[1 + (n - 1)t]\}^{1/2}$ times that of individual selection. This factor is less than 1.0 when R is small and t is large. Conversely, the factor becomes more than 1.0 when R is large and t is small.

If selection is on a combination of family and individual merit, an optimum combination would be $\{1 + [(R - t)^2/1 - t][(n - 1)/1(n - 1)t]\}^{1/2}$ times as effective as purely individual selection. This expression is equal to 1.0, and combination and individual selection are equally effective if $R = t$. Thus, combination selection is always at least as effective as purely individual selection. If R and t are unequal, a combination of family and individual selection will be more effective than individual selection alone.

Table 7-7 gives information on the comparative accuracy of combination selection as related to individual selection for several variations of n, R, and t. As can be seen, the advantages of combination selection for the usual combinations of R and t are small, and it may be questionable whether sufficient accuracy is gained to justify the effort. If t is smaller than R, the family average should be given positive emphasis in selection. If t is larger than R, we know that some environmental factor is producing at least part of the resemblance. This leads to the conclusion that a good individual in a better-than-average family may be good partly because of favorable environment and that we

[6]Lush, J. L. 1947. *Am. Naturalist 81:* 241–261.

TABLE 7-7
EFFECTIVENESS OF A COMBINATION OF FAMILY AND INDIVIDUAL SELECTION AS
COMPARED WITH PURELY INDIVIDUAL SELECTION UNDER CERTAIN ASSUMED
CONDITIONS

	Relationship, R								
	.25			.50			.60		
n	8	12	20	8	12	20	8	12	20
0.1	1.050	1.063	1.079	1.316	1.390	1.472	1.465	1.567	1.680
0.2	1.004	1.005	1.006	1.153	1.178	1.203	1.258	1.299	1.339
t 0.4	1.034	1.038	1.041	1.016	1.017	1.019	1.060	1.066	1.071
0.6	1.188	1.201	1.212	1.017	1.018	1.019	1.000	1.000	1.000
0.8	1.614	1.641	1.665	1.216	1.227	1.236	1.101	1.107	1.111

should discount the record. Although it may seem paradoxical, family averages should be given a negative emphasis in selection under such conditions.

PROGENY AND SELECTION

The evaluation of performance of an adequate sample of progeny in a properly
designed progeny test provides the most accurate possible estimate of breeding value. This has long been recognized in the old saying that "Individuality
tells us what an animal seems to be, his pedigree tells us what he ought to be,
but the performance of his progeny tells us what he is." The potential value of
progeny testing was recognized long before the basic laws of heredity were understood. Varro, a Roman, advocated progeny testing 2,000 years ago, and the
pioneer breeder, Bakewell, is said to have used it in the eighteenth century by
letting out bulls and rams on an annual basis. Those which proved to be outstanding transmitters could later be returned for use in his own herds.

The theoretical accuracy of the progeny test in estimating the breeding
value of an individual is equal to

$$r_{GO} = \frac{h}{2} \frac{n}{1 + (n - 1)t}$$

where h = Square root of heritability of individual records
n = Number of progeny records
$t(h^2/4)$ = Expected intraclass correlation among the progeny

With an h^2 of .30 the correlation is .677 for 10 progeny, .94 for 100, .988 for
500, and .994 for 1,000. These figures compare with an accuracy of .55 for estimating breeding value on the basis of individual performance and with the
accuracies given in Table 7-6 for various combinations of pedigree information.

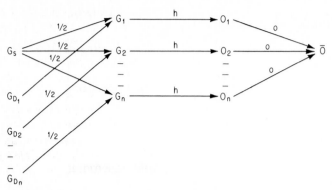

FIGURE 7-6
Diagram showing the genetic relations in a progeny test. The
genotypes of the progeny (G_1, G_2, G_n) represent independent
samples of genes from their sire (G_s). The square root of heritability
(h) represents the impact of the individual progeny's genotypes on
the phenotypic values $(O_1, O_2, \ldots O_3)$.

The basic reason for the potentially higher accuracy of the progeny test can
be readily envisioned from the biometrical relations between individuals involved in a progeny test, as shown in arrow form in Figure 7-6. Repeated samples of the genes of the individual being progeny-tested are figuratively observed in the progeny. Hence, under ideal conditions the repeated sampling
overcomes the limitations which chance at segregation imposes on pedigree
selection.

Because the other parents also contribute to the genetic composition of the
offspring of a parent being progeny-tested, it is essential that the mates represent a random sample of those available. Alternatively, procedures can sometimes be developed in which mates are classified into groups of similar productive potential and random samples of each group assigned to each test
animal, in order to equalize the average merit of mates of animals (usually
males) being progeny-tested.

In addition to ensuring that each animal being progeny-tested has a comparable group of mates, for valid test results it is necessary that (1) all the progeny, not just a selected sample, be included in the progeny test appraisal and
(2) different progeny groups have the same environmental treatment. If there
are differences in environments between progeny groups, the value of t in the
formula for estimating accuracy of the progeny test may be inflated owing to
nongenetic factors, and the accuracy of the test would be reduced.

Table 7-8 gives ratios of the accuracy of the progeny test to individual selection for varying numbers of progeny at three different levels of heritability,
with no nongenetic contribution to the intraprogeny group correlation, and
with an environmental contribution of .10 to the correlation. It can readily be
seen that the increased accuracy of the progeny test is largest for traits with
low heritability but that accuracy is much reduced with an environmental con-

TABLE 7-8
COMPARATIVE ACCURACY OF PROGENY TEST AND INDIVIDUAL PERFORMANCE*

	h^2 values: $t = h^2/4$			h^2 values: $t = h^2/4 + .10$		
n	.10	.25	.50	.10	.25	.50
1	.50	.50	.50	.50	.50	.50
2	.70	.69	.66	.66	.65	.63
3	.85	.82	.77	.77	.75	.72
4	.96	.92	.85	.85	.82	.77
5	1.07	1.00	.91	.91	.87	.81
10	1.43	1.26	1.08	1.08	1.01	.91
20	1.84	1.51	1.22	1.22	1.20	.97
50	2.37	1.75	1.32	1.32	1.16	1.02
100	2.68	1.87	1.37	1.37	1.21	1.06

*The tabular values are the ratio of the accuracy of selection on the progeny test to selection on individual merit with only one record for each animal $\{n/4[1 + (n - 1)t]\}^{1/2}$.

tribution to the intraprogeny group correlation. As an example, a situation with a nongenetic correlation of this kind could exist if several sires were being progeny-tested but each progeny group were maintained at a different location. Feeding and management as well as other environmental factors could contribute to progeny group differences or to similarities within progeny groups, and could mistakenly be considered to be caused by differences in the transmitting abilities of the sires.

The progeny test can provide an extremely accurate appraisal of an individual's breeding value. However, decisions as to whether to use it in practical breeding programs and to what extent to use it must depend upon many factors. Decisions must be made on a case-by-case basis. Disadvantages of the progeny test include the facts that it is costly and that the extra time consumed completing the test means that generation intervals are increased. Further, because of the excessive costs, only a few individuals can be progeny-tested, whereas all animals can be evaluated for individuality for traits expressed by both sexes. Thus, increased costs and generation intervals must be balanced against the accuracy of the progeny test.

For an effective large-scale progeny testing program, several requirements must be met. First many more animals must be progeny-tested than are required for eventual use. Unless such an initial provision is made, there will be little opportunity for selection after the progeny test is available. Effective progeny testing programs for males usually will require that at least four or five individuals be progeny-tested for each progeny-tested sire that is chosen. In most situations economic considerations may modify what is genetically optimum.

Secondly, a procedure must be available to accurately assess the progeny test information. Unless the progeny test is designed to provide a critical evaluation of the progeny-tested individual, there is little justification for undertaking the testing. Testing of progeny simultaneously at several locations, using artificial insemination, and the adoption of comparison of performance with contemporary animals can measurably increase the accuracy of progeny test evaluations.

Another extremely important requirement for effective progeny testing is that there must be a means of utilizing the superior animals extensively once they have been located. Artificial insemination and frozen semen have been most important in this respect for cattle. If a progeny-tested male can be used widely, then the industry can afford an extensive investment in progeny testing programs designed to locate such individuals.

The inability to obtain many progeny from females is one reason why planned progeny testing of females is not of major importance in livestock. In addition, the progeny test information accumulates so slowly that by the time cows and ewes have three or four tested progeny, many useful years of the dam's life have passed. This does not mean that progeny performance should not be used in female selection, but its value is definitely limited. Selection of females must rely quite heavily on individual merit.

Several studies have made it clear that effective progeny testing programs require large populations. In dairy cattle, the most efficiently designed progeny testing program that is practical for herds of 100 to 150 cows would only about equal a system in which sire selection is based only on records of their dams and other close relatives. But in larger populations with as many as 10,000 breeding females, properly designed progeny testing and artificial insemination programs should lead to an average annual genetic improvement in milk production of about 1.5 percent of the mean. In contrast, annual improvement without progeny testing would be expected to be only 0.7 to 1.0 percent of the mean.

For traits expressed by both sexes in meat animals, several studies have shown that progeny testing cannot be expected to increase progress as compared to individual selection when only natural breeding is used. With artificial insemination permitting more extensive use of outstanding sires, the prospects for progeny testing are somewhat more favorable. Further, breed programs utilizing progeny testing and artificial insemination provide genetic ties between herds and permit more valid breed-wide comparisons than were ever possible previously. (See Chapter 11 for information on some of these programs.)

SUMMARY

Selection is the process of determining which animals of a population are allowed to have progeny and contribute their genes to future generations. Most economically important traits are under the influence of many pairs of genes with individually small effects. They are also influenced in their expression by environmental circumstances. The purpose of selection is to increase the frequency of favorable genes. This can be accomplished by initial selection of populations most nearly meeting the goals of a specific breeding program, and then by within-population selection of individuals aimed at achieving further improvement. Selection may be based on individuality, on traits of ancestors and other relatives, or on performance of progeny. Optimum selection procedures are dependent upon many factors including reproductive rates of different animal classes, heritability and economic importance of traits, whether traits are expressed by only one sex, phenotypic and genetic correlations among traits, and costs of measurement.

APPENDIX: Comparative Accuracy of Sib and Individual-Performance Testing

The initial swine testing programs utilized family selection and were referred to as *sib testing*. Two females and two barrows from a litter were entered on test at central test stations. In Danish tests much emphasis was given to carcass merit and particularly bacon quality of the slaughtered sibs. The subsequent development of the back-fat probe and ultrasonic measurements for loin eye area and back-fat depth provided useful practical measures of carcass merit. This has contributed to the shift from sib testing of barrows and gilts to performance testing of boars.

The biometrical relations to examine the comparative accuracy of sib and individual performance testing for boar selection are shown in Figure 7-7. The relative accuracies of the two approaches can be compared by evaluating the correlations between the average breeding value of the boar in question G_B with its phenotype P and also the correlation of G_B with the average of the sibs ($\overline{O}$). The value of $r_{G_B P}$ is the square root of heritability h. The correlation $r_{G_B \overline{O}}$ can be derived using the path coefficient analysis giving

$$r_{G_B \overline{O}} = \frac{h}{2}\sqrt{\frac{n}{1 + (n-1)t}}$$

for the correlation where the sibs are full sibs. The relative value of the two correlations

FIGURE 7-7
Diagram showing genetic relations in a sib test using the performance of four sibs (O_1, O_2, O_3, O_4) to estimate the breeding value of a boar (G_B) in contrast to estimating G_B from the boar's performance (P). The square root of heritability (h) represents the impacts of the individuals' genotypes. The correlation between littermates is represented by t.

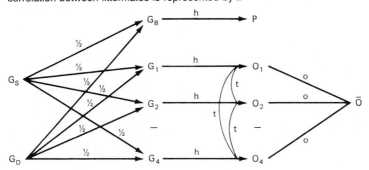

$$\frac{r_{G_B\bar{O}}}{r_{G_BP}} = \frac{\dfrac{h}{2}\sqrt{\dfrac{n}{1 + (n - 1)t}}}{h} = \sqrt{\frac{n}{4[1 + (n - 1)t]}}$$

For full sibs the value of t would be $h^2/2$ if there were no common environmental factors contributing to the similarities among full sibs (c^2). Since these sibs would have a common prenatal and postnatal environment prior to weaning, this would make t larger than $h^2/4$.

Table 7-8 shows the relative accuracy of sib selection as compared to performance testing in selecting boars with these three heritability values where e^2 is zero and when c^2 is assumed to be .10. Since all the values in Table 7-9 are less than one, selection on the basis of individual performance is more accurate than selection on the average of four sibs for all the situations identified. Furthermore, with boar testing each of the four animals tested is a potential candidate for breeding use.

Apart from the special care which must be taken to avoid spreading contagious diseases when boars are returned to herds as breeding boars, that is when artificial insemination is not used, the advantages to individual performance testing over sib testing are:

1 The generation interval can be reduced to a minimum.
2 The boar's breeding value is assessed more accurately than when as many as four sibs are tested.
3 The selection of boars can be intensified.

TABLE 7-9
COMPARATIVE ACCURACY OF SIB TESTING AND INDIVIDUAL PERFORMANCE*

	h^2 values: $t = h^2/2$			h^2 values: $t = h^2/2 + .10$		
n	.25	.35	.50	.25	.35	.50
4	.85	.81	.76	.77	.74	.70

*Tabular values are the ratio of the accuracy of selection on the basis of four full sibs to selection on individual performance when common litter environmental contributions (c^2) to t are assumed to be 0 and .10.

SUGGESTIONS FOR FURTHER READING

Books

Johansson, I., and J. Rendel. 1968. "Genetics and Animal Breeding," W. H. Freeman and Company, San Francisco.

Lerner, I. M. 1951. "Population Genetics and Animal Improvement," Cambridge University Press, Cambridge.

Lerner, I. M. 1958. "The Genetic Basis of Selection," John Wiley & Sons, Inc., New York.

Lush, J. L. 1945. "Animal Breeding Plans," 3d ed., Collegiate Press, Inc., Iowa State University, Ames, Iowa.

Pirchner, F. 1983. "Population Genetics in Animal Breeding," Plenum Press, New York.

Van Vleck, L. D., E. J., Pollak, and E. A. B. Oltenacu. 1987. "Genetics for the Animal Sciences," W. H. Freeman and Company, San Francisco.

Articles

Dickerson, G. E., and L. N. Hazel. 1944. Effectiveness of Selection on Progeny Performance as a Supplement to Earlier Culling in Livestock. *J. Agr. Res.* 69:459–476.

Lush, J. L. 1935. Progeny Test and Individual Performance as Indicators of an Animal's Breeding Value. *J. Dairy Sci.* 18:1–19.

Lush, J. L. 1947. Family Merit and Individual Merit as Bases for Selection. *Am. Nat.* 81:241–261; 362–379.

Minkema, D., M. Bekedam, D. Kroeske, and T. Stegena. 1964. A Selection Experiment with Pigs: Comparison of the Effectiveness of Individual Selection and Sib Selection. *Neth. J. Agr. Sci.* 12:318–329.

Robertson, A., and J. M. Rendel. 1950. The Use of Progeny Testing with Artificial Insemination in Dairy Cattle. *J. Genet.* 50:21–31.

8

SELECTION: MULTIPLE TRAITS: ENVIRONMENT AND RESULTS

Discussions in the previous chapter focused primarily on selection for one major trait. Such a situation will be experienced only rarely in practice. Even if one trait is of paramount importance, other qualities, such as fertility, need to be emphasized. Beyond this if selection is for only one trait, some change in other traits may be expected due to possible genetic correlations between them and the primary trait. Genetic correlations were defined in Chapter 5, and we will discuss their importance in other situations in this chapter.

MULTIPLE-TRAIT SELECTION

Selection for more than one trait reduces the selection pressure on any single trait. The reason the selection differential for a specific trait is reduced by selecting simultaneously for several traits can be understood from the following illustration. When 10 percent of a population is saved for breeding purposes, the selection differential could be 1.76 standard deviations if selection emphasized only one trait. However, to get 10 percent of the individuals which rank as high as possible in each of two traits, it is necessary that the product equal .10, or $(.10)^{1/2} \times (.10)^{1/2} = (.316)(.316)$. Hence, individuals would have to be drawn from the upper 31.6 percent for each of the two traits. The expected selection differential for the best 31.6 percent from a normal curve is about 1.13 standard deviations rather than 1.76, or only about 64 percent as large as can be attained for each trait singly.

The reduction in intensity of selection is not so severe when an index or total score is used. Deficiencies in one trait then can be overcome by superiority in another. When selection is directed toward improving n traits which

TABLE 8-1
RELATIVE INTENSITY OF SELECTION FOR INDIVIDUAL TRAITS WHEN n EQUALLY
IMPORTANT INDEPENDENT TRAITS ARE SELECTED

n	Relative intensity
1	1.00
2	.71
3	.58
4	.50
n	$1/(n)^{1/2}$

are equally important and independent, the intensity of selection will be $1/(n)^{1/2}$ as intense as if selection were for only one trait (Table 8-1).

The foregoing is not to argue that selection should be for only one trait. On the contrary, it is essential, as was pointed out earlier, to emphasize more than one trait in most breeding enterprises. However, the potential for selection should not be dissipated on unimportant traits. If all traits being selected are really desired, index or total score selection on all of them will give a larger increase in total merit than emphasizing only one or two.

The most important consideration is that the traits fully merit consideration. Selection for traits which are economically unimportant or extremely low in heritability, or which are based on details of color marking or conformation is often excused on the basis that it does no harm, since they do not detract from the economic value of the animal. This is fallacious reasoning, since such selection is indirectly harmful by reducing the selection intensity for more important traits which will respond to selection.

Alternative Methods

Several different procedures could be used to improve more than one trait in a selection program. Hazel and Lush[1] examined the theoretical accuracy of three methods of selection to improve several traits. These were the tandem method, the method of independent culling levels, and the total score or index method. In the tandem method, selection is practiced for only one trait at a time until a satisfactory level is reached; then a second trait is considered, etc. The efficiency of this method is dependent on the genetic relationships among the traits. Should there be negative relationships, the subsequent selection for a second trait could undo the progress previously made in earlier selections. Of course, strong positive genetic correlations among the traits would mean that selection for one trait would improve the correlated traits. Overall, the tandem method was found to be the least efficient of the three methods compared.

[1]Hazel, L. N., and J. L. Lush. 1942. *J. Hered.* 33:393–399.

With independent culling levels, selection may be practiced for two or more traits simultaneously. A minimum standard is set for each of the n traits, and all individuals below that level for any one trait are culled without regard to their merit for other traits. The major disadvantage of this method is that the superiority for one trait does not have an opportunity to offset lack of merit in another trait.

Using an index or total score for selection allows one to add up the merits and demerits for each trait and arrive at a total score. Such an approach was found to be more efficient than either of the other two methods. In practice, both of the other methods will probably have to be used, to some extent, since in most herds at least some culling must be done before all the information needed for the optimum use of the total-score method is available. For example, in beef cattle, weaning weight and rate of gain after weaning are both economically important traits. In using the total-score method of selection, we would want to give points for performance in each trait. However, if our facilities for testing the rate of gain after weaning would not permit testing all animals, we would usually cull at weaning those below a certain weaning weight.

Selection Indexes

The approach to selection using an index or total score is derived from the work of Smith[2] and Hazel.[3] Index procedures utilize scores or index values for each individual, so that the index values are as closely correlated to the individual's composite average breeding value as is possible to obtain with a linear combination of traits. The composite average breeding value g_w is the combination of the average breeding values for the traits to be included in the index, weighted by their relative net economic values (see Appendix, p. 206).

The computation of selection indexes involves the technique of multiple regression (several X variables), whereas we have considered only the simple linear regression (Chapter 5). The index (I) computed is of the form:

$$I = b_1 X_1 + b_2 X_2 + \cdots + b_n X_n$$

where X's = Phenotypic values for the different traits
$\quad\quad b$'s = Weights given to each of the traits

In statistical terms the b's are multiple regression coefficients which maximize the correlation between index values and composite breeding values $r_{g_w I}$. An example showing the computation of an index is given in the Appendix to this chapter.

Why is an index essential for efficient simultaneous selection for several traits? First, the traits considered usually are not all of equal economic impor-

[2] Smith, H. F. 1936. *Ann. Eugenics* 7:240–250.
[3] Hazel, L. N. 1943. *Genetics* 28:476–490.

tance. Some differential weighting in accord with the net economic return expected from a unit of improvement of each trait is needed. Second, all traits do not have the same heritability, and the same intensity of selection will not be expected to give the same proportionate improvement for each trait. Third, there may be phenotypic and genetic interrelationships among the traits. Emphasis on one trait may affect change in another, and these interrelationships must be appropriately considered.

In the computation of a selection index for an animal enterprise, certain information is needed to determine the appropriate weightings (b's):

1 The relative net economic importance of a unit of change in each of the traits independent of change in the other traits—This information effectively defines the goal of the selection program. A composite of the several traits weighted by their relative net economic importance is the goal of improvement rather than a single trait.

2 The heritability and the magnitude of the genetic variance and the phenotypic variance for each trait—In regard to economic returns to the breeding program, 50 percent of something worth 20 cents is more valuable than 20 percent of something worth 40 cents. Economic merit and heritability or genetic variances must both be considered.

3 The phenotypic and genetic covariances among each trait in the index—To the extent that the phenotypic covariances are environmental, they serve to adjust for the effects of common environment. The role of the genetic covariances as copredictors of the breeding value for other traits was discussed in Chapter 5.

The selection index approach has been used widely by research workers to obtain a clearer picture of the influence of changes in the parameters that are important when selection involves many traits. Many selection indexes have been suggested for specific purposes, and some have been and are being used in industry. However, for a variety of reasons the approach has not been universally adopted.

The reasons for the failure to use indexes include the complexity of computations necessary to develop an index. These are difficult for breeders to understand, and there is always reluctance to use something which is not well understood. The relative economic importance of traits may vary for different enterprises and particularly for different breeds. Heritability values and the genetic and phenotypic interrelationships may differ among populations making it difficult to recommend generalized selection indexes. Large bodies of data are necessary to obtain accurate genetic correlations, and the use of inaccurate estimates can reduce the efficiency of an index.

Most of the above arguments point up the possible limitations to the use of the same or general indexes by many breeders in an industry. To the extent that objectives differ from breeder to breeder or from breed to breed, such indexes should not be used. However, the authors are firmly of the opinion that indexes developed for specific populations are a tremendous aid to breeders in maintaining consistent goals and procedures. To the extent that goals across herds or breeds are similar, generalized indexes can be very useful.

If all information needed for the development of an optimum index is not available (e.g., good estimates of genetic correlations may be lacking), indexes based on available information often can still be very useful. The net economic weights and the heritabilities of the several traits are usually the most important parameters. Lush reported an early index for swine based on market score, litter weight, and sow productivity computed on the basis of the economic weights and the heritabilities of the traits. This index was later recomputed considering the genetic and phenotypic interrelationships among the traits. The expected accuracy of the first index was 90 percent of the second. In fact, if the traits in the index are independent, the b values become the heritability figure multiplied by the economic weight for a trait.

The National Swine Improvement Federation[4] (NSIF) developed indexes for boars tested at central boar testing stations. These are presented in Chapter 11, page 301. The indexes combine average daily gain, feed per unit of gain, and back-fat thickness, so that the choice of boars can be made on the basis of the index values rather than the individual measurements. Since the test station environments tend to be standardized, these indexes should be useful for the several test stations.

Indexes are also provided by the NSIF for data from on-farm testing programs. A general index has been developed, plus one for emphasizing maternal merit for female selection and one emphasizing paternal merit. The general index for overall merit is:

$$\text{General index} = 100 + 6.6(L - \overline{L}) + .4(W - \overline{W})$$
$$- 1.5(D - \overline{D}) - 63(B - \overline{B})$$

where L = Number of pigs born alive
$\quad\;\; W$ = Adjusted 21-day litter weight
$\quad\;\; B$ = Adjusted back fat at 230 lb
$\quad\;\; D$ = Number of days to 230 lb

The maternal index designed to give more emphasis to litter size is

$$\text{Maternal index} = 100 + 7.0(L - \overline{L}) + .4(W - \overline{W})$$
$$- 1.4(D - \overline{D}) - 53(B - \overline{B})$$

---

[4] National Swine Improvement Federation. 1987. "Guidelines for Uniform Swine Improvement Programs." U.S. Department of Agriculture Science and Education Administration, Washington.

The paternal index is designed to emphasize growth, feed efficiency, and back fat to develop paternal lines for terminal crosses.

$$\text{Paternal index} = 100 + 2.0(L - \overline{L}) - 1.9(D - \overline{D}) - 110(B - \overline{B})$$

Several selection indexes have been proposed for beef cattle, but the wide range of managemental conditions and breeding goals makes it difficult to develop and recommend a general index for industry use. Special indexes for beef cattle have been developed by Swiger et al.[5] and Wilson et al.[6] These illustrate the effect of variation in the phenotypic and genetic correlations, as well as economic values on the weights (b's) to be given the traits.

Correlated Response to Selection

Since traits are interrelated, selection for one could bring about change in others, if the interrelationship is due to a genetic correlation among the traits. The genetic correlations or covariances used to develop selection indexes take account of such relationships in arriving at the relative weights (b's) in the selection index. Correlated traits on occasion are referred to as *indicator traits* for the trait of primary interest. For decades there has been interest in simply identified or simply evaluated traits as indicators of merit for more complex or important traits.

Unfortunately, the history of animal breeding has been marred by many instances of misguided beliefs that some minor conformation item, color, shape of horn, patterns of hair worls on the rear quarters, etc., were related to economically important production traits. Needless to say, most of these beliefs have proved to be false when subjected to critical research. Even if relationships are real, they are unlikely to be strong enough to make selection for them useful when actual records on traits of interest can be obtained.

However, for some traits it is difficult or prohibitively expensive to obtain actual data on a large-enough sample of a population to be useful for genetic improvement. This is especially true for carcass traits. Obviously, an animal can be killed only once, and thereafter it cannot be used for breeding purposes regardless of the carcass superiority found.[7] Use of actual slaughter data means that selection will be on the basis of sib or progeny information. This almost always reduces selection intensity, and also therefore the annual rate of genetic progress as compared to direct selection on individual records.

In such situations indicator traits can be very useful. Perhaps the best example is the widespread use of determinations of back-fat thickness by probe

[5] Swiger, L. A., et al. 1965. *J. Anim. Sci.* 24:418–424.

[6] Wilson, L. L., et al. 1963. *J. Anim. Sci.* 22:1086–1090.

[7] An exception to this statement is that it may be possible to collect and store semen from a group of males before slaughter and subsequently use the semen only from the males with superior carcasses.

or ultrasonics in living swine as an indicator of carcass content,of preferred lean cuts. Back-fat thickness is a less-than-perfect indicator of carcass yield of preferred cuts. Nonetheless, selection can be so much more intense for it than for procedures involving actual slaughter and use of sib and progeny test procedures that genetic progress per year is considerably greater.

If the appropriate parameters are known, potential progress from selection for correlated traits versus direct selection for the trait of interest can be estimated. If we let 1 represent a trait of primary interest and 2 a correlated trait, the regression of breeding value of 1 on the breeding value of 2 is

$$b_{g_1g_2} = \frac{\text{cov } g_1g_2}{\sigma^2 g_2} = r_{g_1g_2} \frac{\sigma g_1}{\sigma g_2}$$

The direct response in 1 from selection for 1 itself is,

$$\Delta G_d = h^2{}_1 i\sigma p_1 = h_1 i\sigma g_1$$

and the correlated response in 1 from selection for 2 can be expressed as

$$\Delta G_c = r_{g_1g_2}h_2 \frac{\sigma g_1}{\sigma p_2} i\sigma p_2 = r_{g_1g_2}h_2 i\sigma g_1$$

where d = Direct response to selection
$\quad\quad c$ = Correlated response to selection

Then the ratio of the correlated to the direct response is

$$\frac{\Delta G_c}{\Delta G_d} = \frac{r_{g_1g_2}h_2 i\sigma g_1}{h_1 i\sigma g_1} = \frac{r_{g_1g_2}h_2}{h_1}$$

From the foregoing expression it is apparent that if $r_{g_1g_2}$ is 1.0 and $h_1 = h_2$, the direct and correlated responses would be equal. The correlated response would be less than the direct response if $r_{g_1g_2}$ is less than 1.0 or $h_1 > h_2$. The product of $r_{g_1g_2}h_2$ must be larger than h_1 for the correlated response to exceed the direct response. This would only rarely occur.

The general requirements of (1) a high genetic correlation between the primary trait of interest and a correlated or indicator trait and (2) a high heritability of the indicator trait usually limit the practical usefulness of indicator traits. As mentioned earlier, the prediction of carcass lean content from probe or ultrasonic estimates is probably the most widely used and effective procedure of this nature.

Partial Records

In some situations it is desirable to make selections before a record has been completed. For example, egg production over a period of a year is usually con-

sidered to constitute a full or complete record. However, to keep the generation interval as short as possible, it is desirable to make decisions about saving
hatching eggs from a female after only a short laying period. Similarly, annual
wool production is usually considered a complete record. In some breeds in
certain environments, young males can be used for breeding by 7 or 8 months
of age. This is desirable in order to shorten generation intervals. However, if
wool yield is an important trait in selection, selections must be based upon
only part of a year's wool production.

In diary cattle, 305-day records are the standard, but sometimes it is necessary to do some culling among first-lactation heifers before all records are
complete. Also sire generation interval can be reduced if partial records are
used initially rather than waiting for all daughters to complete 305-day records.

The correlation between phenotype for part record and the average breeding value for full record is

$$r_{p_1 g_2} = (r_{g_1 g_2} h_1)$$

where $r_{g_1 g_2}$ = Genetic correlation between part record (1) and complete record
(2)

h_1 = Square root of heritability of the part record

As related in the previous section, if $r_{g_1 g_2} h_1$ is equal to h_2, selection on the
part record would be as accurate as selection on the complete record. Usually
selection on the part record will be somewhat less accurate. In each case, the
loss in accuracy from selection on the basis of part records must be balanced
against the reduction in generation interval to estimate whether annual genetic
progress will be increased or decreased.

There is a possibility that continued selection for part records over a long
period will result in changing the genetic correlation, so that selection for the
part records will result in less progress for the residual portion of the record.
This has been reported in egg-laying poultry. Selection for high initial rates of
egg production apparently resulted in the creation of populations with ability
to make good partial records but with a decrease in the residual portion.

ENVIRONMENT AND SELECTION

The question of what environment a breeder should provide for his animals is
a broad one, and no specific answer is available. It has sometimes been assumed that animals in breeder herds should be maintained in the best practical
environment, and the commercial producer using the descendants of these animals should strive to improve the environment toward the level of the breeder
herds. From a practical economic point of view, excellent management and
care will probably pay well in improving the level of performance and appearance of breeding stock that will be available for sale.

The opinion is also expressed by some researchers that a poor or harsh environment will put a virtual ceiling on the expression of inherent differences in ability to gain, in type or conformation, or in ability to reproduce, making selection ineffective by reducing genetic variation. Hammond[8] recommended providing the animals with an environment which would permit maximum expression of the traits. Selections would be made from the animals with maximum performance, even though later descendants were to perform commercially under poorer conditions.

Genotype-Environmental Interactions

Hammond's proposition would be satisfactory if genotype-environmental interactions were not important. This would mean that a specific change in the environment would result in a proportionate change in the phenotype for the entire array of genotypes. If such were the situation, the productivity of individuals or strains would rank in the same order in different environments. Selection, though not necessarily equally effective in all environments, would be for the same goals, and changes in the environment would not influence which animals would be selected. Conversely, if individuals or strains changed their rank in different environments, a genotype-environmental interaction would be suggested. Selection for improvement in one environment would not necessarily result in improved performance in another. Hammond's proposal would be preferred if there were no genotype-environmental interactions, and if the heritability were higher in the optimal environment.

The representation of genotype-interactions as a trait measured in two different environments being considered as two different traits was developed by Falconer.[9] As an example, the growth rate of animals reared on a high plane of nutrition could be considered one trait, and the growth rate of these animals reared on a low plane of nutrition a second trait. If the genetic correlation between growth rate in the two environments was high, very little interaction of genotypes and environment would be suggested. The change in growth rate after selection in either environment can be examined in terms of correlated responses to selection. A correlated response has been defined as one in which selection is primarily for one trait, but due to a strong genetic correlation a change occurs in a second trait.

When the alternatives of direct or correlated response to selection are considered for a specific population, intensity of selection i will not be affected by the trait selected. For example, whether we select for increasing growth rate using the gains to 1 year of age or the conformation of the animals at 1 year as a correlated trait, we would still have to save the same percentage of the herd. However, when the response to selection for a trait in two different environ-

---

[8] Hammond, J. 1947. *J. Biol. Rev.* 22:195–213.
[9] Falconer, D. S. 1952. *Am. Naturalist* 86:293–298.

ments is considered, the selection intensities of selection could be different and should be taken into consideration. Then the ratio of the correlated to direct response to selection becomes:

$$\frac{\Delta G_c}{\Delta G_d} = \frac{r_{g_1 g_2} h_2 i_2}{h_1 i_1}$$

Experimental Evidence

Fowler and Ensminger[10] selected swine using an index which gave major emphasis to rate of gain in determining whether selection for growth rate was as effective when swine were limit-fed as when they were liberally fed. Foundation stock of a Danish Landrace × Chester White cross was divided into two equivalent lines. One line was full-fed from weaning to 150 days, limit-fed from 150 days to parturition, and full-fed during lactation. The second line was given 70 percent of the feed consumed by the liberally fed line during each of the above periods. Progress from six generations of selection for rate of gain was very close to that expected from the selection differentials and the heritability in each line.

Samples of the pigs from the two lines were tested on both feeding regimes after six generations of selection. The previously limited-fed line gained faster on the liberal ration than did animals of the line selected for fast gains on full feed. Likewise when animals of both lines were tested on the limited ration, pigs from the limited-fed line achieved the faster gains. Although the differences between the lines were not great, a genotype-environmental interaction was indicated. The results did not support Hammond's contention. Actually the line selected under the more rigorous condition performed better in both environments. The authors interpret the differential response as being the result of selection for appetite alone in the full-fed line and for both appetite and efficient feed utilization in the limited-fed line.

Selection experiments with mice for growth on unrestricted and restricted diets have been reported by Falconer.[11] In these experiments selection on the restricted plane of nutrition appeared to improve the genetic potential for growth under either feeding regime. Selection on the unrestricted diet or full-feeding regimen was not effective for improving growth rate on the restricted diet.

An interesting experiment with Hereford cattle has been reported in which animals from Montana were moved to Florida and vice versa.[12] Calves from Florida dams performing in Florida had heavier weaning weights than did calves from Montana dams performing in Florida. Conversely, the Montana calves performing in Montana had heavier weaning

---

[10] Fowler, S. H., and M. E. Ensminger. 1960. *J. Anim. Sci.* 19:434–449.
[11] Falconer, D. S. 1960. *Genet. Res.* 1:91–113.
[12] See articles by Koger, Burns, Pahnish et al. in "Suggestions for Further Reading" at end of chapter.

weights than the Florida calves performing in Montana. For yearling weight and gain after weaning, the heifers from Montana breeding ranked above those from Florida breeding at both locations. For reproduction, the Montana cows ranked higher in percent pregnant and percent of calves weaned in Montana; whereas, the cows of Florida breeding ranked higher for these traits in Florida. Statistically significant within-breed interactions were evident; whereas, in experimental studies of sire by location and ration interaction, generally statistically significant interactions have not been reported. The technique of comparing sire progeny groups is a less sensitive test of genotype-environmental interactions. Sire progenies, on average, have only one-fourth of their genes in common, and progeny differences are expressed under environmental conditions under which the dams have likely been selected.

Some other analyses of population data have suggested that genotype-environmental interactions may be present, but their magnitude is such that few are of broad practical significance. Our interpretation of available evidence on this important question in temperate regions is: (1) Adaptation tends to be fairly general for most traits. Only in exceptional circumstances would selection in one environment be totally ineffective for performance in another. (2) Genetic differences tend to be more fully expressed under more "favorable" environments. (3) The evidence indicates the desirability of selection under conditions reasonably like those of commercial animals of the area.

In spite of available evidence, however, the most common procedures for rearing animals in the best-known sheep and beef cattle breeder herds to date in the United States have been very nearly in line with Hammond's viewpoint. Seed-stock herds and flocks have for the most part been maintained in areas where feed conditions are good and other environmental conditions are not extreme. Animals from these herds have then provided foundation stock for commercial production in all areas of the country. This situation, together with the nationwide promotional activities of many breed associations, has sometimes tended to result in animals of the same general type being raised in virtually all parts of the country regardless of their adaptation. A critical look at this situation by the breeders would appear desirable.

With dairy cattle, the breeding operation has centered about the programs developed by artificial insemination studs. Sire selections are based on progeny performance in the tested herds throughout the area. Thus progeny perform in a broad cross section of environments in commercial herds.

Maternal Effects

Maternal effects are the phenotypic expressions arising from those influences which the mother may have on the expression of a trait in her offspring, apart from the direct influence of the genes she transmits. Maternal influences, which condition the maternal effects are the result of the dam's genotype and the environmental circumstances which affect its expression and influence on the offspring. From the standpoint of the offspring, maternal influences are environmental in nature, conditioning the expression of the offspring's genes for the trait being measured. Yet, from the standpoint of the mother there are both distinct genetic and environment components.

An example of a trait that is maternally influenced is weaning weight of beef calves. The calf's weight at weaning is the result of the influence of the genes of the calf—half of which were received from the mother—and of the environmental circumstances influencing the expression of the calf's genes. These environmental circumstances include the maternal influences plus other environmental influences on the calf, not directly related to the maternal influences. Such traits are often referred to as composite traits, since the phenotypic value results from a composite of the genetic influence of the offspring and the dam.

The following expression may help to clarify these relationships:

$$P_x = u + g_{ox} + e_{ox} + g_{my} + e_{my}$$

where P_x = Phenotypic measurement on the offspring for trait X, e.g., weaning weight

u = Population mean for the trait

g_{ox} = Effect of genotype of offspring for the trait

e_{ox} = Environmental effect on g_{ox}, apart from the maternal effect

g_{my} = Genotype for the maternal effect of the dam y

e_{my} = Environmental effect on the genotype for maternal effects in the dam y

It is apparent that any representation of weaning weight must recognize that it is a complex trait. When selection is based on the weaning weight of a calf, lamb, or a litter of pigs, genes for growth during the preweaning period are selected. In addition, the unexpressed genes for maternal effects are also included in the total genotype of the young, but they are expressed only in females at a later date.

Selection for weaning weight should result in some change in the calf's ability to grow as well as the dam's capability to provide a maternal environment for that growth. In the above simple model we have ignored a possible correlation between the genes for growth and those for maternal effects. Taking into account only the additively genetic effects or the average breeding values, the regression of the breeding value of the calf on the phenotype for weaning weight is

$$\frac{\sigma^2 g_o + \tfrac{3}{2}\, \sigma g_o g_m + \tfrac{1}{2}\, \sigma^2 g_m}{\sigma_p^2}$$

where $\sigma^2 g_o$ = Variance of average breeding value of calf for weaning weight

σg^2_m = Variance of average breeding value for maternal effects of mother

$\sigma g_o g_m$ = Covariance (corelation) between average breeding values for weaning weight and maternal effects[13]

The regression above can be viewed as an expression of heritability for this composite trait, since heritability is defined as the regression of average breeding values on phenotypic values. A prime concern is the magnitude and sign of $\sigma g_o g_m$. If the genetic correlation between g_o and g_m is zero, $\sigma g_o g_m$ is zero also. Selection for weaning weight would tend to be more for the genes for growth in the offspring g_o than for maternal performance g_m. If it is positive, selection improves both g_m and g_o. However, if it is strongly negative, optimum procedures would be to select for maternal genetic effects in females and genetic values for growth in males.

Many studies have sought to determine the sign and magnitude of the genetic correlation between the genes for growth in the individual and those for maternal performance of the dam. Most of these studies are inconclusive, but the preponderance of evidence suggests that the genetic correlation is slightly negative for many such situations. One well-confirmed finding is the negative relationship between the size of litter to which a female is born and the size of the litters produced by that female.

Maternal environmental effects are frequently classed as prenatal and postnatal. Prenatal effects are those which are primarily influenced by the uterine environment. Postnatal effects are largely lactational in nature for most farm mammals. While maternal effects are most evident for juvenile traits, there are numerous examples of carryover effects at least to preadult stages of development.

RESULTS OF SELECTION

Selection experiments with livestock are expensive to conduct and require several years to obtain definitive results. Yet evidence is accruing to attest to the effectiveness of selection in changing animal populations. The interest in selection, even in theoretical studies, has appeared to gain a new impetus. Considerable selection results with *Drosophila* have been reported during recent years. In addition, research with laboratory mammals has increased in tempo.

Undoubtedly the classical demonstration of the potential for selection has come from the Illinois selection experiment for high and low oil content. In the last published report[14] the oil content of the high line had increased to 16.64

[13] Willham, R. L. 1972. *J. Anim. Sci.* 34:864–869.

[14] Dudley, J. W., et al. 1974. "Seventy Generations of Selection for Oil and Protein in Maize," Crop Science Society of America, Madison, Wisconsin, pp. 181–212.

percent, and the content in the low line was less than 1 percent following 70 generations of selection. Further decline in the low line has been hampered by poor germination as the oil content became too low. From an initial oil content of 4.68 percent, the above changes are most striking (Figure 8-1). Even in the more recent years there has been a continuing response to selection.

Selection with Laboratory Mammals

The dramatic results of selection with laboratory mammals serve to remind us that directed selection can bring about change. Castle[15] pioneered work with quantitative traits in his selection for variation in the hooded (spotting) pattern in rats. Races with the extremes in hooding, as shown in Figure 8-2, were produced by selection. At the time that Castle conducted this research there was a scientific controversy over the fundamental effect of selection. Some researchers maintained that the genes themselves were changed by the selection process. Castle's comment on this point is illuminating:

> The question now arose whether the observed changes had occurred as a result of change in the single unit-character or gene clearly concerned in the case, or whether this was due to other agencies. To test the matter the selected races, now modified genetically in opposite directions, were crossed repeatedly with a nonhooded (wild) race. The recessive hooded character disappeared in F_1 but was recovered again in F_2 in the expected 25 per cent of this generation. These extracted hooded individuals, following each cross, were less divergent than their hooded grandparents from the ordinary hooded pattern. After three successive crosses (six generations) the whitest individuals extracted from the dark hooded race were no darker than the darkest individuals extracted from the white hooded race. In other words, repeated crossing with the non-hooded (wild) race had caused the changes in the hooded character, which had been secured by selection, largely to disappear. The conclusion was drawn that the hooded allelomorph itself had remained unchanged throughout the selection experiment, and that the phenotype had been altered by associating with the hooded gene a different assortment of other genes in each of the selected races, these serving as genetic modifiers. In the course of the selection experiment a mutation was observed to occur in the plus series to practically the Irish stage. This is not included in the summary but is mentioned to show how it is possible for selection to be aided in its progress by the occurrence of contemporaneous genetic changes, no less than by the sorting out of variations originally present in the foundation stock. Apart from the mutation mentioned the results of selection in this case show conclusively that the changes obtained had not occurred in the gene for the hooded pattern, but in the residual heredity. Other cases of apparent gradual change in unit-characters under the action of selection may be explained in a similar way. Accordingly, we are led to conclude that unit-characters or genes are remarkably constant and that when they seem to change as a result of hybridization or of selection unattended by hybridization, the changes are rather in the total complex of factors concerned in heredity than in single genes.

[15] Castle, W. E. 1930. "Genetics and Eugenics," Harvard University Press, Cambridge, Massachusetts, pp. 237–240.

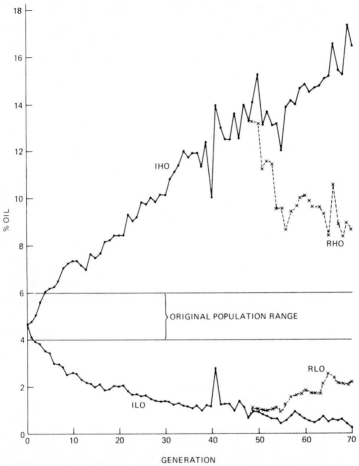

FIGURE 8-1
Response to 70 generations of selection for oil content of corn. Actual data
are indicated by solid lines. Reverse selections are shown as broken lines.
Note that the selection progress is symmetrical until a plateau is reached in
the low line. (*Dudley, J. W., R. J. Lambert, and D. E. Alexander. "Seventy
Generations of Selection for Oil and Protein in Maize," Crop Science
Society of America, Madison, Wisconsin, 1974, p. 188.*)

MacArthur[16] conducted two-way selection from a heterogeneous base population. After 21 generations of selection, mice from the large line at 60 days
were 3.3 times as heavy as those from the small line (Figure 8-3). More recent
selection experiments confirm the general response to selection for body size
presented by MacArthur. There also have been numerous reports of the effectiveness of selection in mice involving a wide variety of traits.

[16]MacArthur, J. W. 1949. *Genetics* 34:194–209.

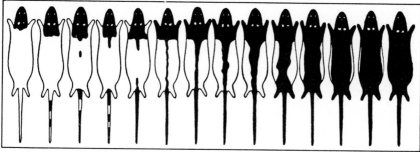

FIGURE 8-2
Series of grades for classifying the plus and minus variations of the white spotting pattern of hooded rats in Castle's selection studies. (*From Castle, W. E., "Genetics and Eugenics," Harvard University Press.*)

Selection with Farm Animals

Designed selection experiments with farm animals are less numerous. Yet experimental evidence is now growing that responses similar to those demonstrated with laboratory mammals can be achieved.

Swine   A study of response to selection for back-fat thickness by Hetzer and Harvey[17] is a classic in illustrating changes that can be realized with persistent directed selection. Three lines of Durocs were established from the same base population, with one selected for high and one for low back-fat thickness as determined by the probe of all animals at 79.4 kg live weight. The third line was an unselected control. Similar lines were established in the Yorkshire breed. Figure 8-4 gives a plot of the response for 16 generations of selection in the Durocs and for 14 generations in the Yorkshires. Realized heritabilities for live back-fat thickness were .47 and .48 for the first 10 generations of upward and downward selection in the Durocs. Similar figures for the first eight generations of selection in the Yorkshires were .38 and .43. In view of the response shown in Figure 8-4, the realized heritabilities in the later generations must have been very close to those in the earlier generations.

Representative samples of each line were slaughtered at an average live weight of 95.4 kg and carcass traits were evaluated. Line averages for live back-fat thickness and averages for a few other carcass measurements for each line after 16 and 14 generations of selection in the Durocs and Yorkshires, respectively, are shown in Table 8-2. Traits related to yield of lean meat in carcasses of animals slaughtered at a standard weight appear to be highly responsive to selection.

[17]Hetzer, H. O., and W. R. Harvey. 1967. *J. Anim. Sci.* 26:1244–1251.

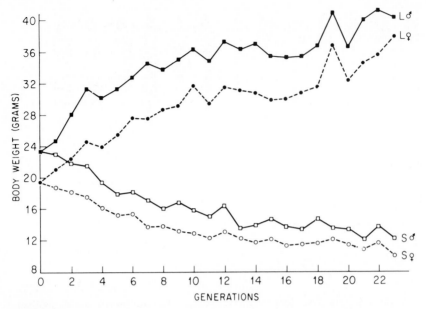

FIGURE 8-3
Response to individual selection for 60-day body weight in mice. (*MacArthur, J. W. Genetics 34:196, 1949.*)

FIGURE 8-4
Response to selection for back-fat thickness in Duroc and Yorkshire swine plotted as deviations from contemporary controls. (*From Hetzer, H. O., and L. R. Miller, J. Anim. Sci. 37:1294, 1973.*)

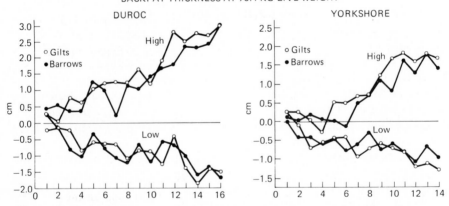

Beef Cattle   Experiments with cattle to demonstrate response to selection generally lack the definitive response noted in the above experiment with swine. Long generation intervals, limited selection intensity due to lower re-

TABLE 8-2
BACK-FAT THICKNESS AND RELATED CARCASS CHARACTERS AFTER 14 TO 16
GENERATIONS OF SELECTION*

	Lines		
	High back fat	Low back fat	Unselected control
Duroc			
Live back-fat thickness (cm)	7.2	2.5	4.0
Carcass:			
Back-fat thickness (cm)	7.8	3.0	4.6
Percent lean cuts	34.6	44.0	43.0
Percent lean in ham	41.0	64.9	58.0
Yorkshire			
Live back-fat thickness (cm)	5.0	2.3	3.4
Carcass:			
Back-fat thickness (cm)	6.4	2.9	3.9
Percent lean cuts	40.7	47.8	42.4
Percent lean in ham	53.6	67.1	57.8

*Adapted from Hetzer, H. O., and L. R. Miller, *Jour. Anim. Sci*, 37:1289–1301, 1973.

productive rates, and sex-limited traits are not conducive to such dramatic responses.

Newman et al.[18] have demonstrated positive response to selection for yearling weight in a replicated experiment with beef Shorthorns at two locations in Canada—Brandon and LaCombe. A selection and a control line were established at each location and the selection criterion was weight at 1 year of age. Their experiment covered 10 years, and the yearly values for unweighted means of the males and females combined are plotted in Figure 8-5. The authors present the data for the sexes separately. Although the realized heritabilities are slightly higher for the males, the difference between the sexes was not significant statistically. The pooled, realized heritability of yearling weight for Brandon was .50 ± .08 and for LaCombe it was .40 ± .09.

An interesting feature of the experiment at Brandon involved mating the initial selection line bulls ($S_1 \male \male$) via artificial insemination and frozen semen to the tenth-year control females ($C_{10} \female \female$). Bulls used in the tenth year for the selection line ($S_{10} \male \male$) also were mated to control cows for the tenth year ($C_{10} \female \female$). The regular matings were made providing offspring from four types of matings in the same year, ($C_{10} \times C_{10}$), ($S_1 \times C_{10}$), ($S_{10} \times C_{10}$), ($S_{10} \times S_{10}$). The regression of the performance of these offspring groups on the mean accumulated selection differentials for their parents provided another estimate of the realized heritability of yearling weight. These were .54 ± .005 for males and .62 ± .16 for females. Clearly the experiment demonstrated positive response to selection over a time interval in which many individual breeders could be involved.

[18]Newman, J. A., et al. 1973. *Can. J. Anim. Sci.* 53:1–12.

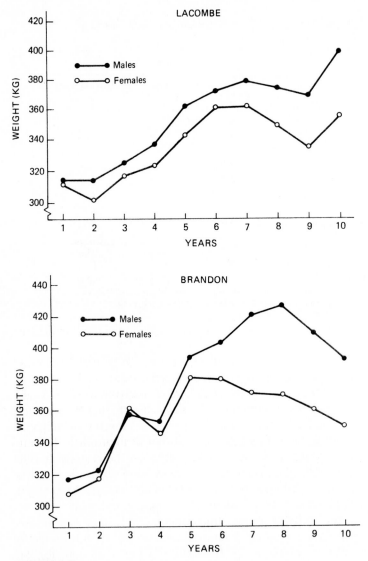

FIGURE 8-5
Response to selection for yearling weight in Shorthorn beef cattle at two
locations. (*Adapted from Newman, J. A., et al. Can. J. Anim. Sci. 53:8, 1973.*)

Several other selection experiments with randomly maintained controls
have been reported, but their goals were not so singular and their responses
not so distinctive. References to some of these reports are given at the end of
this chapter.

Dairy Cattle Selection experiments with dairy cattle, which have included
a randomly maintained control as in the experiments cited for swine and beef

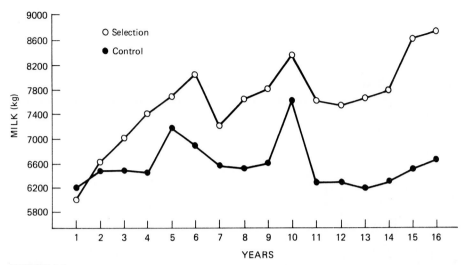

FIGURE 8-6
Response to selection with artificial insemination and progeny testing for milk yield in Holsteins. Milk yield (3.7% FCM) is expressed on a 305-day, twice daily milking basis, mature equivalent adjusted to 100 days open. (*From Legates, J. E., and R. M. Myers. J. Dairy Sci. 71:1025–1033. 1988.*)

cattle, have been reported. Overlapping generations, sequential culling on subsequent lactation performance, sex-limited expression of milk yield, plus the general use of sires proved via artificial insemination make it difficult to interpret direct responses to selection. Several statistical studies of field data from continuous milk recording in many herds have been utilized to estimate genetic trends in large breeding units. A variety of approaches have been used, some of which are described by Smith.[19] These have shown varying positive genetic trends for milk and fat yields, most of which were smaller in magnitude than would be expected for optimal selection for a single trait.

One of the two selection experiments utilizing a randomly maintained control was continued for 16 years.[20] Its objective was to assess the genetic change in milk yield attributable to selection and progeny testing via artificial insemination. The selection group was bred to sires from the North Carolina Institutional Breeding Association, which sampled young bulls for progeny testing before they were returned to service. Bulls to breed the control group were chosen randomly from within the control group. Both groups were milked and housed as one herd.

The trends in milk yield for the selection and control groups are shown in Figure 8-6. Although inbreeding in the control group was minimized by arranging for least related matings, it increased to just under 6 percent. The average yearly increase in the difference between the selection and control

[19] Smith, C. R. 1962. *Anim. Prod.* 4:239–251.
[20] Legates, J. E., and R. M. Myers. 1988. *J. Dairy Sci.* 71:1025–1033.

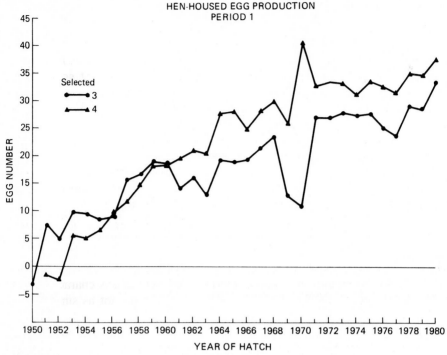

FIGURE 8-7
Hen-housed egg production to 273 days for two selected strains plotted as deviations from a contemporary control. (*From Gowe, R. S., and R. W. Fairfull. "Proceedings, Third World Congress on Genetics Applied to Livestock Production," XII: p. 155, 1987.*)

groups, adjusted for the difference in inbreeding for the two groups, was 110 ± 14 kg of milk. This represented an increase of 1.5 percent of the mean yield per year over the duration of the experiment. No important or significant correlated changes in fat percent, nonfat solids percent, size, or reproductive performance were evident.

There are a large number of well-designed experiments with poultry that support the effectiveness of selection. One long-term selection experiment for egg production reported by Gowe and Fairfull[21] is of special interest. They have reported the results of 30 years of selection where randomly maintained control flocks were maintained. The hen-housed egg production for two selected strains shown as deviations from a control strain are presented in Figure 8-7. An increase of two eggs and 0.3 g in egg size per generation was made over the 30 generations without significant changes in fertility, hatchability, or viability.

[21] Gowe, R. S., and R. W. Fairfull. 1987. "Proceedings, Third World Congress on Genetics Applied to Livestock Production," XII, pp. 152–167.

Selection Limits or Plateaus

Many generations of selection for highly heritable, quantitative traits under outcrossing systems with laboratory animals including mice, rats, the fruit fly *Drosophila melanogaster,* and the flour moth *Tribolium Castaneum,* usually have shown rapid responses for from 10 to 20 generations. At some point, change due to selection has slowed, and in some instances it appeared to have ceased. After a number of additional generations of selection, the population again has often responded to selection. The level at which the temporary cessation of response occurs is termed a *plateau.* When a population reaches a plateau, the inclination is to infer that the limit of selection has been reached.

After a careful analysis of a White Leghorn flock at Berkeley, it was first concluded that it was on a selection plateau for egg number. It was decided to induce genetic variation in the plateaued line by irradiating sperm.[22] However, the nonirradiated portion of the flock responded to selection when the experiment was resumed. Hence, the plateau was clearly not a selection limit.

Except in experimental situations it is unlikely that selection limits have been reached or are imminent in farm animals. Interbreeding populations of farm animals are rarely closed to outside breeding to the extent that is characteristic of experimental populations. Further, the selection goals are not as singular, and they are not held to for as many generations, since selection goals are apt to change due to economic and other conditions. Beyond this, the selection pressure for a given trait will likely be less intense than single-trait experimental selection. Most practical programs will emphasize more than one trait.

When plateaus actually exist or response to selection declines, some possible reasons may include the following:

1 *Exhaustion of genetic variability.* The strain may have become homozygous for genes affecting the trait under selection. This could occur if the number of loci were small. Complete homozygosity is unlikely if the number of loci affecting a given trait is in the 100's or 1,000's, as seems probable for many complex traits. Often, detailed analyses of data from plateaued populations have shown that additive genetic variance is still present, thus demonstrating that in at least some such populations, exhaustion of genetic variability is not wholly responsible for failure to respond to further selection.

2 *Negative relations between reproductive ability or viability and the trait or traits under selection.* Relations of this kind could make it impossible to select intensely for the desired trait or traits owing to impaired reproduction. In extreme cases this could result in negative selection when the most desired genotypes have much reduced reproductive ability. Selection often would need to be relaxed for a time to permit recovery of reproductive fitness. Simultaneous selection for two or more desired traits, which are negatively associated genetically, could reduce selection response for both.

---

[22] Abplanalp, H., et al. 1964. *Genetics* 50:1083–1092.

3 *Traits are under influence of genes with overdominance.* Outcrossing with selection is ineffective for fixing genes with this type of nonadditive action. This is because maximum expression of the traits depends upon heterozygosity. Although the maximum phenotypic expression does not require heterozygosity for simple dominance, selection for a dominant gene can be highly effective if its frequency is low but can be relatively ineffective at high frequencies (see Figure 6-2). With overdominance, heterozygotes will be phenotypically superior and selection would be expected to bring gene frequency of alleles to an optimum level, after which no further progress from selection within the population is realized.

SUMMARY

Realistic breeding goals involve emphasis on more than one trait. Since each additional trait reduces the selection pressure on any one trait, care must be exercised in the choice of additional traits and the weights (emphasis) to be given them. It is accepted that the goal of a breeding program is to improve net economic merit. Hence, in order to efficiently combine selection for multiple traits, the net economic value of a unit of improvement in a trait independent of change in the others for which improvement is desired must be determined. Additionally the genetic and phenotypic variances and covariances among the traits must be available to construct a selection index for efficient simultaneous selection for a multiple-trait goal. The level and nature of the environment that should be provided for breeding herds and flocks must be examined in terms of possible genotype-environmental interactions (i.e., Do certain breeds or strains respond to or are they more adapted to specific environments?). Evidence does not suggest that such interactions are of widespread importance, but a wise guide is to provide the breeding unit with an environment comparable to what their offspring would encounter in commercial situations. The maternal environment influences the expression of many traits such as weaning weight and growth soon after weaning. Such composite traits bring together the expressions of the offspring's genotype and that of its dam, as she influences a portion of the environment for the young. The single phenotypic measure is impacted by two separate and identifiable genotypes. Ample evidence has now accumulated with laboratory mammals, and also with farm animals, to confirm the reliability of fundamental selection theory and to demonstrate that desired change can be expected under most circumstances during the tenure of many breeding units. Even though progress may often appear to be slow, evidence suggests that there is sufficient genetic variation in most populations to warrant expected response to selection for traits of medium heritability for many generations.

APPENDIX: Computation of a Selection Index

The computation of a selection index is somewhat complex, but inquiring students may desire to gain an acquaintance with the techniques. A statement of the breeding goal is

the first information needed. The concept of merit based on a single trait must be replaced by merit based on a combination of traits which are economically important and
which have sufficient additively genetic variance to give a reasonable response to selection. The worth W, or value, of an individual or group provides a statement of the
breeding objective and can be defined as

$$W = (W - \overline{W}) = w = a_1x_1 + a_2x_2 + \cdots + a_nx_n = \Sigma a_ix_i$$

where the a_i are the relative increases in net worth expected from one unit of improvement in the trait x_i independent of the improvement in the other traits in w. The additive
genetic value for worth then becomes g_w in contrast to g_i for the ith trait, and it can be
defined as

$$g_w = a_1g_1 + a_2g_2 + \cdots + a_ng_n = \Sigma_ia_ig_i$$

Now the goal of our selection is to improve the linear combination of traits g_w rather
than the breeding value for any one trait g_i.

An example of defining worth can be related to individual selection in pigs with three
traits: average daily gain x_1, back-fat probe x_2, and feed efficiency x_3 designated as important.

$$w = a_1x_1 + a_2x_2 + a_3x_3$$

What are the appropriate values for the economic weights (a's)? Establishing values
that would be precisely appropriate for general use is virtually impossible. Bereskin and
Steele[23] studied a swine population and estimated that each kilogram of increase in average daily gain should be valued at \$22.50. An increase of 1 cm of back fat (measured
by probe) decreased the potential value of the carcass by \$6.30. For reducing the ratio
of feed consumed to gain by one unit, i.e., from 3.0 to 2.0, \$18.00 in feed costs would
be saved. Then,

$$W = 22.5x_1 - 6.3x_2 - 18x_3 \quad \text{and} \quad g_w = 22.5g_1 - 6.3g_2 - 18g_3$$

This provides the first item of information identified on page 187. The phenotypic
and genetic variances and covariances for the three traits are also required. Those used
by Bereskin and Steele are presented below and in Table 8-3.

Trait	σ^2x_i	σ^2g_i	h^2
Average daily gain, x_1 (kg/day)	.0040	.0010	0.25
Back-fat probe, x_2 (cm)	.0950	.0428	0.45
Feed efficiency, x_3 (units of F/G)	.0676	.0101	0.15

First, for simplicity the above statistics are used to set up equations to calculate an
index for only average daily gain and back-fat probe. In algebraic terms these are

---

[23] Bereskin, B., and N. C. Steele. 1986. U.S. Department of Agriculture Production Research
Report 184.

TABLE 8-3
GENETIC* AND PHENOTYPIC COVARIANCES AND CORRELATIONS BETWEEN AVERAGE DAILY GAIN X_1, BACK-FAT PROBE X_2, AND FEED EFFICIENCY X_3 AS USED BY BERESKIN AND STEELE†

Trait	Average daily gain, kg	Back-fat probe	Feed efficiency
		Covariances	
X_1		.002288	−.001909
X_2	.004867		.007292
X_3	−.005751	.016026	
		Correlations	
X_1		.35	−.60
X_2	.25		.35
X_3	−.35	.20	

*Genetic covariances and correlations underscored.
†Bereskin, B., and N.C. Steele. 1986. U.S. Department of Agriculture Production Research Report 184.

$$b_1\sigma^2 x_1 + b_2\sigma x_1 x_2 = a_1\sigma^2 g_1 + a_2\sigma g_1 g_2$$

$$b_1\sigma x_1 x_2 + b_2\sigma^2 x_2 = a_1\sigma g_1 g_2 + a_2\sigma^2 g_2$$

Then in terms of the values for the variances and covariances indicated above, these become

$$b_1(.0040) + b_2(.004867) = (22.5)(.0010) + (-18)(.002288)$$

$$b_1(.004867) + b_2(.0950) = (22.5)(.002288) + (-18)(.0428)$$

The partial regression coefficients b_1 and b_2 can be obtained by the solution of two simultaneous equations to give

$$b_1 = 5.14$$

$$b_2 = -2.56$$

Thus the index for selection would be

$$I_1 = 5.14X_1 - 2.56X_2$$

For practical use only the relative weights given the two traits are important. Often a constant is added to the index to avoid negative index values or adjustment is made to the index to give a specified standard deviation. After adjusting the above index to a standard deviation of 20 units it becomes

$$I_1 = 133X_1 - 66X_2$$

Even though the actual values of b_1 and b_2 differ, their relative values, i.e., b_1/b_2, are unchanged.

In the matrix algebra approach, the equations whose solution provides b values for the index can be expressed as

$$X \times b = G \times a$$

where X = Phenotypic variance-covariance matrix which includes all the variables in the index

b = Column vector of the b_i for the index

G = Genetic variance-covariance matrix for the traits included in the index

a = Column vector of the a_i indicating the relative net economic weights

The inverse of X is designated X^{-1}, and in accord with standard procedures a solution for the b_i can be obtained:

$$X^{-1}Xb = X^{-1}Ga$$

$$b = X^{-1}Ga$$

Some remarks are in order regarding the variances and covariances in matrices X and G. These should be estimated from data comparable to the population to which the index is to be applied. If selections are to be made within year and location, analyses performed to provide this information should be carried out on a within-year and location basis also.

For an index using all three traits, the equations given below in matrix form are:

$$
\begin{matrix} & X & \\ \begin{bmatrix} .0040 & .004867 & -.005751 \\ .004867 & .0950 & .016026 \\ -.005751 & .016026 & .0676 \end{bmatrix} & \begin{matrix} b \\ \begin{bmatrix} b_1 \\ b_2 \\ b_3 \end{bmatrix} \end{matrix} \end{matrix}
$$

$$
= \begin{matrix} G \\ \begin{bmatrix} .0010 & .002288 & -.001909 \\ .002288 & .0428 & .007292 \\ .001909 & .007292 & .01014 \end{bmatrix} \end{matrix} \begin{matrix} a \\ \begin{bmatrix} 22.50 \\ -6.30 \\ -18.00 \end{bmatrix} \end{matrix}
$$

$$
b = \begin{matrix} X^{-1} \\ \begin{bmatrix} 321.10416 & -21.932291 & 32.510416 \\ -21.932291 & 12.364683 & -4.791667 \\ 32.510416 & -4.791667 & 18.5625 \end{bmatrix} \end{matrix} \begin{matrix} Ga \\ \begin{bmatrix} .0424476 \\ -.349416 \\ -.2714121 \end{bmatrix} \end{matrix}
$$

$b_1 = 12.47$

$b_2 = -3.95$

$b_3 = -1.98$

The index would thus be

$$I = 12.47X_1 - 3.95X_2 - 1.98X_3$$

For routine use it would probably be best to round the numbers and perhaps make other adjustments. Bereskin and Steele rounded and adjusted to obtain an index with a standard deviation of 20. The index then became

$$I = 160X_1 - 51X_2 - 25X_3$$

It should be apparent that inadequate relative net economic values or variances and covariances required to compute an index could introduce errors into the index. Large volumes of accurate data representative of the population to which the index is applied are necessary to provide sound information to compute the selection index. Generally the genetic covariances (cov g_1g_2, etc.) are the least adequately estimated items of information needed in the computations. Studies have shown that indexes computed from inadequate information can be ineffective in selection. When the phenotypic and genetic parameters are poorly estimated, weighting the traits on the basis of the values for their relative net economic worth may be more desirable.

Considerable theory relating to selection indexes has been developed. The reader is directed to the "Suggestions for Further Reading" at the end of the chapter for a coverage of some of the more advanced topics.

SUGGESTIONS FOR FURTHER READING

Books

Dudley, J. W. (ed.). 1974. "Seventy Generations of Selection for Oil and Protein in Maize," Crop Science Society of America, Madison, Wisconsin.

Hammond, J. 1952. "Farm Animals. Their Breeding, Growth and Inheritance," Edward Arnold and Company, London.

Johansson, I., and J. Rendel. 1968. "Genetics and Animal Breeding," W. H. Freeman and Company, San Francisco.

Lerner, I. M. 1951. "Population Genetics and Animal Improvement," Cambridge University Press, Cambridge.

Lerner, I. M. 1958. "The Genetic Basis of Selection," John Wiley & Sons, Inc., New York.

Lush, J. L. 1945. "Animal Breeding Plans," 3d ed., Collegiate Press, Inc., of Iowa State University, Ames, Iowa.

———. 1951. Genetics and Animal Breeding, in Dunn, L. C. (ed.). "Genetics in the 20th Century," The Macmillan Company, New York.

Pirchner, F. 1983. "Population Genetics in Animal Breeding," Plenum Press, New York.

Van Vleck, L. D., E. J. Pollak, and E. A. B. Oltenacu. 1987. "Genetics for the Animal Sciences," W. H. Freeman and Company, San Francisco.

Articles

Aaron, D. K., R. R. Frahm, and D. S. Buchanan. 1986. Direct and Correlated Responses to Selection for Increased Weaning or Yearling Weight in Angus Cattle. I. Measurement of Selection Applied. *J. Anim. Sci.* 62:54–65; II. Evaluation of Response. *J. Anim. Sci.* 62:66–76.

Burns, W. C., M. Koger, W. T. Butts, O. F. Pahnish, and R. L. Blackwell. 1979. Genotype by Environmental Interaction in Beef Cattle. II. Birth and Weaning Traits. *J. Anim. Sci.* 49:403–409.

Dickerson, G. E. 1970. Efficiency of Animal Production—Molding the Biological Components. *J. Anim. Sci.* 30:849–859.

Frahm, R. R., C. G. Nichols, and D. S. Buchanan. 1985. Selection for Increased Weaning or Yearling Weight in Hereford Cattle. I. Measurement of Selection Applied. *J. Anim. Sci.* 60:1373–1384; II. Direct and Correlated Response. *J. Anim. Sci.* 60:1385–1395.

Gowe, R. E., and R. W. Fairfull. 1987. Long Term Selection for Egg Production in Chickens, In "Proceedings, Third World Congress on Genetics Applied to Livestock Production." XII, pp. 152–167.

Hazel, L. N. 1943. The Genetic Basis for Constructing Selection Indexes, *Genetics* 28:476–490.

Hazel, L. N. and J. L. Lush. 1942. The Efficiency of Three Methods of Selection. *J. Hered.* 33:393–399.

Henderson, C. R. 1963. Selection Index and Expected Genetic Advance, in Hanson, W. D., and H. F. Robinson (eds.). "Symposium on Statistical Genetics and Plant Breeding," pp. 141–163, National Academy of Sciences, National Research Council, Washington.

Hetzer, H. O., and L. R. Miller. 1973. Selection for High and Low Fatness in Swine: Correlated Responses of Various Carcass Traits. *J. Anim. Sci.* 37:1289–1301.

Koger, M., W. C. Burns, O. F. Pahnish, and W. T. Butts. 1979. Genotype by Environmental Interaction in Hereford Cattle. I. Reproductive Traits. *J. Anim. Sci.* 49:396–402.

Legates, J. E., and R. M. Myers. 1988. Measuring Genetic Change in a Dairy Herd Using a Control Population. *J. Dairy Sci.* 71:1025–1033.

Meland, O. M., R. E. Pearson, J. M. White, and W. E. Vinson. 1982. Response to Selection for Milk Yield in Holsteins. *J. Dairy Sci.* 65:2131–2140.

Mrode, R. A. 1988. Selection Experiments in Beef Cattle. Part 1. A Review of Design and Analysis. *Anim. Breed. Abs.* 56:67–73; Part 2. A Review of Responses and Correlated Responses. *Anim. Breed. Abs.* 56:155–167.

Newman, J. A., G. W. Rahnefeld, and A. T. Fredeen. 1973. Selection Intensity and Response to Selection for Yearling Weight in Beef Cattle. *Can. J. Anim. Sci.* 53:1–12.

Pahnish, O. F., J. J. Urick, W. C. Burns, W. T. Butts, and G. V. Richardson. 1983. Genotype by Environmental Interactions in Hereford Cattle. III. Postweaning Traits of Heifers. *J. Anim. Sci.* 56:1039–1046.

Pahnish, O. F., J. J. Urik, W. C. Burns, W. T. Butts, M. Koger, and R. L. Blackwell. 1985. Genotype of Environmental Interaction in Hereford Cattle. IV. Postweaning Traits of Bulls. *J. Anim. Sci.* 61:1146–1153.

Roberts, R. C. 1966. The Limits to Artificial Selection for Body Weight in the Mouse. I. The Limits Attained in Earlier Experiments. *Genet. Res.* 8:347–360; II. The Genetic Nature of the Limits. *Genet. Res.* 8:361–375.

CHAPTER 9

INBREEDING AND RELATIONSHIP

The forces which can produce changes in the frequency of the genes in a herd or population, such as selection, mutation, migration, and change, were enumerated in our consideration of population genetics. Selection is the most important of these forces, as a breeder seeks to develop animals to meet established goals. Some control over the distribution of the zygotes in a population can be exercised by the breeder's choice of a mating system. By mating individuals which are alike in pedigree, or even in phenotypic similarity, homozygosity tends to increase. Conversely, by mating individuals which are not alike, from either a pedigree or a phenotypic standpoint, heterozygosity increases. In both cases, choosing mates on a pedigree basis has a stronger influence on the zygotic makeup.

When the breeding practices of most of the outstanding early breeders have been examined, it will be seen that some close matings or some inbreeding has been practiced, although generally the inbreeding was mild. The genetic consequences of inbreeding were not fully understood when most of our current breeds were formed, but practical people gained an appreciation for it through their experiences.

INBREEDING

Inbreeding is a system of mating whereby the mates are more closely related than are average members of the breed or population being intermated. Previously we considered random mating in which, after selections are completed, no special effort is made to assign the particular mates. With inbreeding, the mates are chosen because of a common ancestral (pedigree) background or relationship. For very mild inbreeding the parents may be related only as second cousins, but intense inbreeding may involve the mating of full brothers with full sisters or of parent with offspring.

When related individuals are mated, the offspring tend to become more homozygous. This increase in homozygosity, and the accompanying decrease in heterozygosity, is the underlying reason for the genotypic and phenotypic changes which are associated with inbreeding.

The increase in homozygosity can be most vividly illustrated by what happens in self-fertilization, the mating of an individual with itself. Many species of plants normally reproduce by self-fertilization; and more normally cross-fertilized species, such as corn, produce both male and female gametes on the same plant. These can be self-fertilized (selfed) by artificially putting pollen grains (bearers of male germ cells) on the female portion of the flower.

The expected increase in homozygosity for a single pair of genes when a population which is heterozygous for this pair of genes is selfed is shown in Table 9-1. Selfing of a *Bb* individual is equivalent to a *Bb* × *Bb* mating and produces on the average ¼ *BB*, ½ *Bb*, and ¼ *bb* offspring. In subsequent generations selfing of *BB* individuals (*BB* × *BB*) will produce only *BB* offspring. Selfing *bb* individuals will likewise produce only *bb* individuals. Continued self-fertilization removes genes from the *Bb* group and does not return any. In a few generations the population is virtually all of the *BB* or *bb* types.

This same process occurs with inbreeding in animals. However, the increase in homozygosity takes place much more slowly, since inbreeding cannot be as intense as selfing. Even with full brother–sister matings, which is the most intense type of inbreeding possible with animals, the increase in homozygosity is less than half as rapid as with self-fertilization in plants.

DEGREE OF INBREEDING

Although certain of the consequences of inbreeding were realized during the early period of breed formation, a serious handicap existed in that there was

TABLE 9-1
INCREASE IN HOMOZYGOSITY IN THE DESCENDANTS PRODUCED BY CONTINUED
SELF-FERTILIZATION FROM A SINGLE HETEROZYGOUS INDIVIDUAL

| Generation | Proportion of each genotype | | | Percent homozygous | Inbreeding coefficient |
	BB	*Bb*	*bb*		
1	1	2	1	50.0	0
2	3	2	3	75.0	.500
3	7	2	7	87.5	.750
4	15	2	15	93.8	.875
5	31	2	31	96.9	.938
6	63	2	63	98.4	.969
n	$2^n - 1$	2	$2^n - 1$	$1 - (\frac{1}{2})^n$	$1 - (\frac{1}{2})^{n-1}$

no realistic method for expressing the degree of intensity of inbreeding. In laboratory animals such as mice, the amount of inbreeding was expressed in terms of the number of generations of full brother–sister mating. In plant species, inbreeding could be reckoned on the basis of the number of generations of selfing that had occurred.

Although several attempts were made to provide a measure for the intensity of inbreeding, the coefficient of inbreeding (F) proposed by Wright[1] in 1921 continues in general usage. Even before considering the definition of the inbreeding coefficient, we should reemphasize that the underlying genetic consequence of inbreeding is an increase in homozygosity. As related individuals are mated, genes which trace to ancestors common to each parent are recombined in the resulting offspring. Bringing together the two samples of genes (gametes) which arise at least in part from an earlier common gene pool increases the opportunity for loci in the offspring to become homozygous. Thus it appears logical that the measure of the intensity of inbreeding should reflect this probable increase in homozygosity.

Coefficient of Inbreeding

The coefficient of inbreeding represents the probable increase in homozygosity resulting from the mating of individuals more closely related than the average for the population. It may range in value from 0 to 1.0. It also is often referred to as a percentage ranging from 0 to 100. The intensity of inbreeding as measured by the inbreeding coefficient is relative to a particular breed or population at a specified time. If pedigrees are traced back six generations, the inbreeding coefficient then represents the probable increase in homozygosity which has occurred as a result of the mating of related individuals in the population since the reference date six generations previously. Conversely, as the value of F increases, the relative proportion of heterozygotes, again measured from the specific base date, declines a proportionate amount equal to $(1 - F)$.

This increase in homozygosity expressed by the inbreeding coefficient is a most probable result, or it is that which is expected on the average. Many loci would be involved, and by chance either a larger or smaller proportion of the loci might have become homozygous than the computed value of F indicates. Even though two individuals from a line may have the same inbreeding coefficients, they would not necessarily have become homozygous for the same loci. What is expected is that the same proportion of heterozygous loci would have become homozygous for the two individuals. There are two alternatives at each locus, and F represents the proportion of loci originally heterozygous that have become homozygous. The second alternative is that the loci are still heterozygous, and $1 - F$ represents the probable proportion of such loci. Thus the standard deviation for a computed value of F would be $[F(1 - F)/n]^{1/2}$

---

[1] Wright, S. 1921. *Am. Naturalist* 56:330–338.

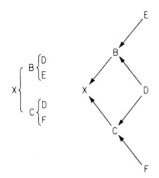

FIGURE 9-1
Bracket and arrow-style pedigrees of an animal (X) produced
by mating a half brother and half sister.

where n would be the effective number of independently segregating loci in the population at the base time.

When viewed from the standpoint of an individual locus, the inbreeding coefficient represents the probability that a random locus originally heterozygous has become homozygous following inbreeding. Reasoning in terms of probability leads to a very useful concept, and it is the basis of the definition of inbreeding that was introduced by Malecot.[2] In an inbred population an individual may possess two genes at a locus that are the same (homozygous). The homozygosity which is exhibited may basically arise from two sources. In one situation homozygosity may be present because the two genes are "alike in state," that is, they may be randomly drawn from the same population, and both happen to be either *B* or *b*. Such would be the case in random mating where $p = .5$. In accord with the Hardy-Weinberg law, the probability that these genes are alike in state would be .50 (.25*AA* + 25*aa*). With inbreeding there is an additional probability of homozygosity because the two genes arise from the direct replication of the same gene from a common ancestor in an earlier generation. Two such genes are said to be "identical by descent." The additional homozygosity contributed by gene pairs which are "identical by descent," represents the impact of inbreeding. Thus the inbreeding coefficient also can be defined as the probability that the two alleles at a locus in an individual are "identical by descent."

Calculation of Inbreeding Coefficient

To illustrate the calculation of the inbreeding coefficient, consider the pedigree in Figure 9-1. We are interested in determining the inbreeding coefficient of individual X. We desire to determine the probability that X received genes that were identical by descent from D transmitted through B and C. In other words, what is the probable proportion of loci in X that is

---

[2] Malecot, G. 1948. "Les Mathématiques de l'Hérédité," Masson et Cie., Paris.

homozygous because X received two replicates of a gene at these loci from D? Consider B and C; the probability that they received replicates of the same gene at a given locus from D is $\frac{1}{2}$, and the probability that they received different genes is also $\frac{1}{2}$. If D is inbred, the probability that both B and C will receive replicates of identical genes from D becomes $\frac{1}{2}(1 + F_D)$ rather than merely $\frac{1}{2}$. Considering the remainder of the paths, the probability that B passes the gene it got from D to X is $\frac{1}{2}$, and the probability that C passes the gene it got from D on to X is also $\frac{1}{2}$. Obtaining the product of these probabilities (paths), we find that the probability that X received replicates of a specific gene from D is

$$(\tfrac{1}{2})(\tfrac{1}{2})(\tfrac{1}{2})(1 + F_D) = (\tfrac{1}{2})^{n + 1}(1 + F_D) = F_X$$

where n represents the number of segregations included in the chain of inheritance from the two parents of X traced through the common ancestor.[3,4]

In the above example $n = 2$ (1 for D to B and 1 for D to C). If the common ancestor is noninbred ($F = 0$), then $F_X = (\tfrac{1}{2})^3 = .125$. This can be interpreted as meaning that one-eighth of the loci in individual X are probably homozygous because X received replicates of genes from D. The pedigree in Figure 9-1 represents the mating of paternal half sibs; hence, only one common ancestor and only one path connect B and C.

Another approach following a single heterozygous locus (Rr) in D through segregations to X may be instructive (Figure 9-1). We are not concerned with E and F, since they are presumably unrelated. Genes of individual B at the R locus are designated as B_1B_2 and those at the R locus for individual C as C_1C_2. If B is heterozygous at the R locus (Rr), the possible genotypes and gametes produced in equal proportions by individual B[5] are:

Genotypes	Gametes
B_1R	B_1
B_1r	B_2
B_2R	R
B_2r	r

[3] One also can consider n to represent the sum of the number of generations from the sire to the common ancestor plus the number of generations from the dam of the individual to the common ancestor.

[4] The word *common* as used here merely means "shared" and in no sense implies that the ancestor was ordinary or inferior.

[5] The B_1, B_2, C_1, C_2 represent genes other than R or r that could be present at this locus in individuals B and C.

Those for individual C are:

Genotypes	Gametes
C_1R	C_1
C_1r	C_2
C_2R	R
C_2r	r

The 16 possible combinations of genes that would occupy the R locus in individual X are:

Genotypes	Gametes
B_1C_1	Rc_1
B_1C_2	RC_2
B_1R	RR
B_1r	Rr
B_2C_1	rc_1
B_2C_2	rc_2
B_2R	rR
B_2r	rr

Note that two of the 16 combinations (RR and rr) are homozygous due to X having received these genes from D. This is the most probable expectation; hence the inbreeding coefficient of X is .125.

The complete expression for the inbreeding coefficient must take into account the possibility that there may be more than one common ancestor and that more than a single path may connect the sire and the dam of the individual (X) whose inbreeding coefficient is being determined. This complete formula can be expressed as

$$F_X = \Sigma(\tfrac{1}{2})^{n+1}(1 + F_A)$$

where F_X = Inbreeding coefficient of individual X

Σ = Summation of all independent paths of inheritance which connect sire and dam of X

n = Number of segregations in a specific path between sire and dam of X

F_A = Inbreeding coefficient of common ancestor for each path

In the more complicated pedigrees it is often convenient to set the pedigree up in the "arrow" style rather than in the conventional bracket form when relationships are being studied. In the arrow style each common ancestor is included only once, with lines drawn to each of his or her offspring in the ped-

TABLE 9-2
EXPONENTIAL VALUES OF $(\tfrac{1}{2})^n$

$(\tfrac{1}{2})^2 = .250000$	$(\tfrac{1}{2})^6 = .015625$
$(\tfrac{1}{2})^3 = .125000$	$(\tfrac{1}{2})^7 = .007812$
$(\tfrac{1}{2})^4 = .062500$	$(\tfrac{1}{2})^8 = .003906$
$(\tfrac{1}{2})^5 = .031250$	$(\tfrac{1}{2})^9 = .001953$

igree. These lines represent the paths of inheritance by which genes are transmitted.

Full Brothers and Full Sisters

Matings of full brothers and full sisters represent a more intense system of inbreeding than does the half-sib mating shown in Figure 9-1. The pedigree in both the bracket and arrow styles of an individual resulting from a full brother–sister mating is given in Figure 9-2. Here we have two common ancestors and two independent paths of inheritance to evaluate. The calculations required to compute F, assuming C and D are noninbred, are summarized in Table 9-3. This is as intense a form of inbreeding as is possible in mammals. The inbreeding coefficient of X is .25. Parent-offspring matings are equal to full brother–sister matings in intensity of inbreeding. This is illustrated in Figure 9-3 and Table 9-4. The only common ancestor is T, and $F_R = .25$.

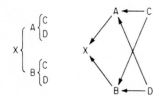

FIGURE 9-2
Bracket and arrow-style pedigrees of an animal resulting from a mating of full brother and full sister.

TABLE 9-3
CALCULATION OF INBREEDING OF INDIVIDUAL X IN FIGURE 9-2

Common ancestor	n	Contributions
C	2	$(\tfrac{1}{2})^{2+1} = (\tfrac{1}{2})^3 = .125$
D	2	$(\tfrac{1}{2})^{2+1} = (\tfrac{1}{2})^3 = \underline{.125}$
		Sum $= .25^*$

$^*F_X = 25$ percent

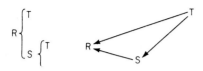

FIGURE 9-3
Bracket and arrow-style pedigrees of an animal resulting from the mating of parent offspring (sire to daughter).

TABLE 9-4
INBREEDING OF INDIVIDUAL R IN FIGURE 9-3

Common ancestor	n	Contributions
T	1	$(\tfrac{1}{2})^{1+1} = .25^*$

$^*F_R$ = 25 percent

Further Example of Inbreeding

The above examples were kept simple to illustrate the principles involved in computing inbreeding coefficients. The principles apply equally well to more complicated pedigrees. As we scan Figure 9-4 we see that animals B, E, and F occur on both sides (top and bottom) of the pedigree. We also note that B is himself 12.5 percent inbred, since he is the result of a half brother–sister mating. Table 9-5 shows the calculations necessary to compute the inbreeding of A as 35.9 percent.

In Table 9-5, individual B is listed twice because the same gene or genes from B might have been transmitted to individual A directly and also indirectly by either of two routes—through J or through I. Segregation of allelic genes occurs in the formation of each germ cell, and so these two routes represent independent chances for A to receive B's genes in homozygous form.

FIGURE 9-4
Bracket and arrow-style pedigrees of an animal also having an inbred common ancestor.

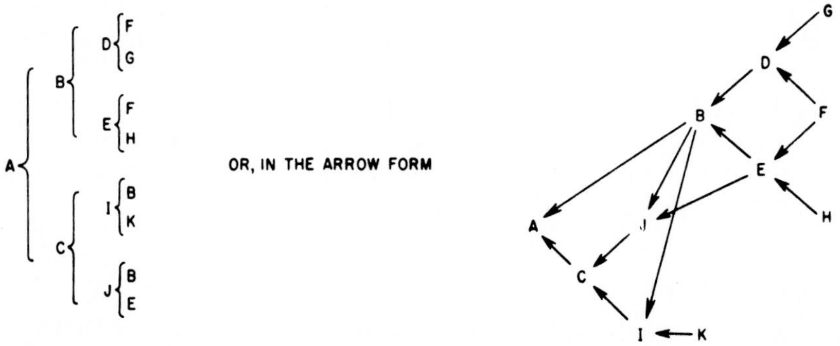

OR, IN THE ARROW FORM

TABLE 9-5
CALCULATION OF INBREEDING OF INDIVIDUAL A IN FIGURE 9-4

Common ancestor	n	$1 + F_B$	Contributions
B (as sire of A and of I)	2	1.125	$(\frac{1}{2})^3(1.125) = .1406$
B (as sire of A and of J)	2	1.125	$(\frac{1}{2})^3(1.125) = .1406$
E (as dam of B and of J)	3		$(\frac{1}{2})^4 = .0625$
F (as sire of D and of E)	5		$(\frac{1}{2})^6 = \underline{.0156}$
			Sum = .3593*

*F_A = 35.9 percent.

RELATIONSHIP

Being *related* means that two individuals have one or more ancestors in common. Actually any two animals in a breed are usually related in this sense. If the pedigrees of a pair of Holsteins, Herefords, Hampshires, or Hackneys were traced far-enough back, each pair would have some common ancestors.

With regard to the human family, we might be surprised at the duplications in each of our pedigrees, if we could trace them back to the time of Christ. Allowing an average of 28 years per generation, we find that we are more than 70 generations removed from that historic period. Since the number of ancestors doubles at each generation, there would be a total of 2^{70}, or about 1,200,000,000,000,000,000, individuals in each of our pedigrees in the single generation around 1 A.D. It is obvious that that many people were not alive at that time (there are more people in the world now than ever before, over 5 billion). Thus, the same individuals must have occurred many times in the initial and other generations of each of our pedigrees.

Obviously, in a relatively few generations the number of ancestors in the pedigree of any farm animal is larger than the total number of animals which were alive in the breed at that time. For example, the pedigree of an individual traced back 20 generations would contain over 2 million individuals. Thus, in a broad sense, all the animals in a breed are related. In farm animals, however, we use the term *related* in a more restricted sense to mean that the animals which are mated are more closely related than average animals of their breed. This usually means that there are common ancestors at least in the first four to six generations of their pedigrees. If two animals had an ancestor in common in the tenth generation, that common ancestor's inheritance would have been halved 10 times in getting down through the ten generations to each of the animals in question. Obviously after ten halvings of that remote ancestor's heredity, the two animals would have little genetic relationship because of the common remote ancestor. But if, for example, the shared ancestor is only two generations removed, its inheritance has been halved only twice in getting to each of them.

Measurement of Relationship

The coefficient for expressing the degree of relationship between two individuals also was developed by Wright. It measures the probable proportion of genes that are the same for two individuals due to their common ancestry, over and above that in the base population.

Parent-offspring relationships are the simplest. They are fundamental to all other degrees of relationships as these represent combinations of several parent-offspring relationships. Since half the genes of any animal come from its sire and half from its dam, any offspring is 50 percent related to each parent. Since each parent in turn received half its genes from its sire and half from its dam and since a sample half is transmitted to each offspring, on the average 25 percent of the genes of any animal originally came from each grandparent. Thus, on the average, an animal is 25 percent related to each of its grandparents. Again it should be kept in mind that the relationship between two individuals is the extra similarity in the genes they possess due to their common ancestry. Many of their genes will already be alike because of the high frequency of these genes in the populations (breed).

The key to measuring relationship is the determination of the number of generations between the two animals being studied and their common ancestors. The first step in computing a relationship coefficient consists of counting the number of generations intervening between some common ancestor and the two descendants in question.

In Figure 9-5, animal C is a grandsire of both A and Z. In other words, A and Z are more closely related than average animals of their breed because they have an ancestor in common close up in their pedigrees. To find their degree of relationship (because of common grandsire C), we count the generations from C to A, which is 2, and from C to Z, which is also 2. Heredity is a halving process, and so we see that C's inheritance has been halved twice in getting to A and twice in getting to Z. In short, it has been halved a total of 4 times. So the relationship of A and Z is C's inheritance halved 4 times, or $(½)^4$, which is $½ \times ½ \times ½ \times ½$, or $\frac{1}{16}$, or 6.25 percent. The relationship between A and Z is 6.25 percent. This simply means that about 6 percent more of A's and Z's genes are the same than would be the case with average animals of their breed. This is true since they both receive genes from the same grandsire C. This is expressed more concisely as follows:

$$R_{AZ} = (½)^4$$

The foregoing example involved the relationship between animals that are related because they are descendants of some of the same animals. A and Z were *single first cousins*, since they had one grandparent in common. The relationship of animals (or people) with two grandparents in common is 12.5 percent. These animals are termed *double first cousins*. Such relationships are called *collateral*. The other possible type of relationship occurs between indi-

$$A \Big\{ {}^{B \big\{ {}^{C}}} \qquad Z \Big\{ {}^{Y \big\{ {}^{C}}}$$

FIGURE 9-5
Pedigrees of two animals with a common grandsire.

viduals when one is a descendant of the other. This is called *direct relationship* as in parent and offspring.

Calculation of the relationship of individuals X and Y (Figure 9-6) is somewhat more complicated than in the previous examples because X and Y have two common ancestors, S and D. When there are two or more common ancestors, the contributions of each are added to arrive at the complete coefficient of relationship. In such cases it is convenient to set up the calculations as in Table 9-6.

Thus X and Y have .5 of their genes in common, and they are related 50 percent. You can readily calculate the relationship of half brothers as 25 percent. In the above example you may wonder why individuals A, B, C, and E were not considered as common ancestors of X and Y. The answer is that they make their contributions only through S or D, not through both. For example, the only way A is an ancestor of both X and Y is through S, and by including S as a common ancestor we automatically take care of A's contributions. If A had been an ancestor of both S and D, we would have included this animal as a common ancestor since it could have contributed the same genes to X and Y through both S and D.

Coefficient of Relationship

Because inbreeding increases homozygosity, an inbred animal will transmit similar genes to each of its offspring more frequently than will a noninbred individual. If an inbred animal is the common ancestor of two related individuals, they will therefore have more genes in common and thus be more highly related than if the common ancestor had not been inbred. To take account of this, the contribution of each inbred common ancestor must be multiplied by $(1 + F_A)$.

Inbreeding also makes a population more variable by producing separate inbred strains. Inbred descendants of any animal will be homozygous in a larger percentage of their gene pairs than if they were not inbred, but they may be homozygous for different alleles of the same gene pair and thus less related

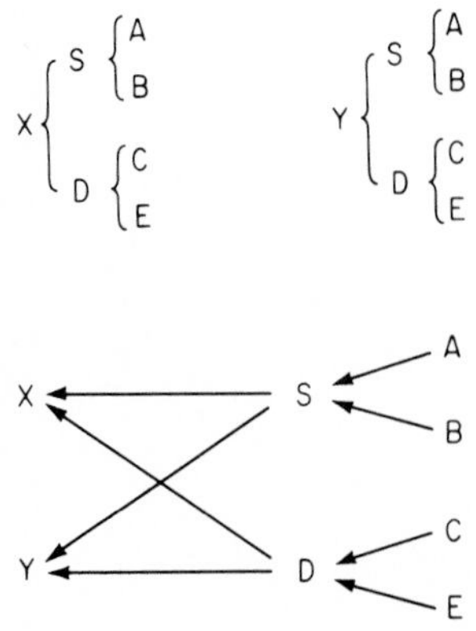

FIGURE 9-6
Bracket and arrow-style pedigrees for two full sibs (X and Y).

TABLE 9-6
RELATIONSHIP OF INDIVIDUALS X AND Y FROM FIGURE 9-6

Common ancestor	n	Contributions
S	2	$(\frac{1}{2})^2 = (\frac{1}{2})^2 = .25$
D	2	$(\frac{1}{2})^2 = (\frac{1}{2})^2 = \underline{.25}$
		Sum $= .50$

than if they were not inbred. The denominator for the relationship formula takes this into account making the complete formula:

$$R_{XY} = \frac{\Sigma[(1/2)^n(1 + F_A)]}{[(1 + F_X)(1 + F_Y)]^{\frac{1}{2}}}$$

This correction makes the relationship coefficient a measure of the degree to which the genotypes of related animals are similar rather than leave it in terms of the proportions of genes from a common source.

As an example of the use of this formula, consider the pedigree shown in Figure 9-7 of an animal produced by mating an inbred sire to his daughter. S is 12.5 percent inbred since he is the product of a half brother–sister mating.

From Table 9-7 the value of the numerator of the relationship coefficient is .84375. The complete relationship is

$$R_{XS} = \frac{.84375}{[(1 + F_X)(1 + F_S)]^{\frac{1}{2}}}$$

$$R_{XS} = \frac{.84375}{[(1.28125)(1.12500)]^{\frac{1}{2}}}$$

$$R_{XS} = .7028$$

FIGURE 9-7
Pedigree where inbreeding of two related individuals must be taken into account.

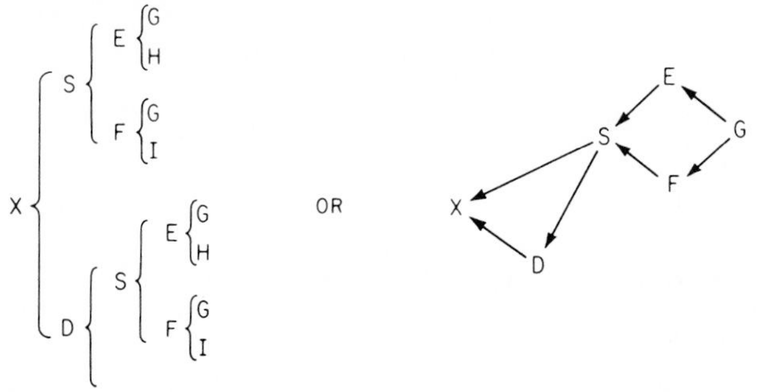

TABLE 9-7
CALCULATION OF RELATIONSHIP OF X TO S IN FIGURE 9-7

Common ancestor	n	n'	$1 + F_A$	Contribution
S	0	1	1.125	$(\frac{1}{2})(1.125) = .5625$
S	0	2	1.125	$(\frac{1}{2})^2(1.125) = .28125$
				Sum = .84375

In the above parent-offspring mating, it is most likely that 75 percent of the genes of individual X came from S. Since X is more highly inbred than S, however, X will probably have some homozygous gene pairs that were heterozygous in S, thus reducing the similarity of their genotypes.

There has been so little inbreeding in most of our farm animals that the denominator of the relationship coefficient seldom is very much larger than 1, and can therefore usually be omitted.

In the original derivation of the relationship coefficient by Wright, the path coefficient approach was used. Accordingly the relationship coefficient was derived as the correlation between the breeding values of two individuals. This concept is still expressed in the literature. A higher relationship (correlation) between two individuals reflects a higher likelihood that the two individuals have the same genes.

SIMILARITY BETWEEN INBREEDING AND RELATIONSHIP

The inbreeding and relationship coefficients can both be considered as probability statements. Inbreeding is the probability that two genes at a given locus in an individual are alike by descent; while relationship is the probability that two individuals related by descent possess more of the same genes than unrelated individuals from the population.

Inspection of the formula for the inbreeding coefficient and the complete expression for the coefficient of relationship reveals a similarity. The inbreeding coefficient is one-half the numerator of the relationship coefficient for the sire and dam of an individual, or one-half the relationship coefficient of the sire and dam when the two related individuals are not inbred.

$$F = \Sigma(\tfrac{1}{2})^n (1 + F_A) \cdot (\tfrac{1}{2})$$

When the related individuals are the sire and dam, the inbreeding of the offspring can be expressed as one-half the numerator of the relationship coefficient between the sire and dam.

F (offspring) = $\frac{1}{2}$ (numerator relationship for sire and dam)

In the original derivation of the relationship coefficient by Wright, it was viewed as a genetic correlation, expressing the correlation between two related genotypes. As a corollary to the simple correlation coefficient, the numerator relationship has been termed the *covariance*. The above relations have been drawn upon to reduce the labor of computing inbreeding and relationship coefficients in closed herds or lines. Convenient reference tables of covariance values (numerator relationships) can be prepared for all potential pairs of breeding individuals in the closed herd or inbred line. Such calculation of covariance tables has been a routine procedure with Lush and coworkers since the 1930s. The table of covariance permits choosing least related or most closely related matings, whichever may be desired. It also allows for easy computation of all necessary inbreeding coefficients.

In essence, the procedure of covariance is practically workable because covariance values among the parents can be used to compute the values for the offspring. The tedium of counting independent paths, with its concomitant errors, is reduced to a simple addition of covariances. Once one becomes familiar with the process, it generally is preferable to compute inbreeding coefficients for complex extended pedigrees by using this procedure rather than by tracing out the independent paths.

The above pedigree of individual H (Figure 9-8) can be used to illustrate the calculation of covariance tables. The principle of averaging covariances is most helpful and necessary. From the above pedigree the covariance between F and G is needed to compute the inbreeding of H ($F = $ cov FG $\cdot$ ½). Several relationships can be drawn upon to compute the covariance between F and G. Two forms are given below.

$$\text{Cov FG} = \text{¼ (cov ED + cov EC + cov CD + cov CC)}$$

$$= \text{½ (cov FE + cov FC)} = \text{½ (cov GD + cov GC)}$$

When the covariances are compiled by hand, generally the second of the above two relationships is used. In situations where both parents may not be in the same generation, it is important to take the average of the two covariances between the older animal with the parents of the younger. Consider the pedigree below to illustrate the need to apply this rule to avoid duplicating contributions.

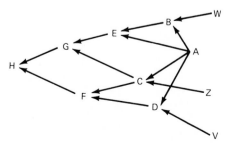

FIGURE 9-8
Arrow-style pedigree of a closed line showing the inbreeding buildup for individual H.

Cov CG = ½ (cov CC + cov CE), but cov CG ≠ ½ (cov AG + cov GZ). The necessary covariances can be listed in tabular form as in Table 9-8.

The following sample calculations will be helpful to illustrate the technique.

Cov AB = 1/2 (cov AA + cov AW) = 1/2 (1.000 + 0) = .500
Cov AC = 1/2 (cov AA + cov AZ) = 1/2 (1.000 + 0) = .500
Cov AD = 1/2 (cov AA + cov AV) = 1/2 (1.000 + 0) = .500
Cov AE = 1/2 (cov AA + cov AB) = 1/2 (1.000 + .500) = .750
Cov BC = 1/2 (cov BA + cov BZ) = 1/2 (.500 + 0) = .250
Cov BD = 1/2 (cov BA + cov BV) = 1/2 (.500 + 0) = .250
Cov BE = 1/2 (cov BB + cov BA) = 1/2 (1.000 + .500) = .750
Cov CD = 1/2 (cov DA + cov DZ) = 1/2 (.500 + 0) = .250
Cov CE = 1/2 (cov BC + cov AC) = 1/2 (.250 + .500) = .375
Cov DE = 1/2 (cov BD + cov AD) = 1/2 (.250 + .500) = .375

Since the relationship of an individual with itself is by definition 1.00, the numerator relationship, or covariance, of an individual with itself (X) is $1 + F_X$

$$R_{XX} = \frac{1 + F_X}{(1 + F_X)^{1/2}(1 + F_X)^{1/2}}$$

Thus the diagonal values in Table 9-8 are 1.000 plus the inbreeding coefficient of the individual. The inbreeding coefficient of the individual is one-half the covariance between its parents. For individual E, its inbreeding coefficient is

$$F_E = \text{½ (cov AB)} = \text{½ (.500)} = .250$$

The remaining values in Table 9-8 can be verified by the reader in order to gain additional experience with this approach.

TABLE 9-8
COVARIANCE TABLE FOR INDIVIDUALS IN THE PEDIGREE OF H

	A	B	C	D	E	F	G	H
A	1.000							
B	.500	1.000						
C	.500	.250	1.000					
D	.500	.250	.250	1.000				
E	.750	.750	.375	.375	1.250			
F	.500	.250	.625	.625	.375	1.125		
G	.625	.500	.688	.312	.813	.500	1.187	
H	.562	.375	.656	.469	.594	.812	.844	1.250

Relationship coefficients between any two individuals included in the table also can be computed:

$$R_{DE} = \frac{.375}{(1.000)^{\frac{1}{2}}(1.250)^{\frac{1}{2}}} = \frac{.375}{(1.250)^{\frac{1}{2}}} = .335$$

When the two individuals are not inbred, the relationship can be read directly from the covariance table.

IRREGULAR INBREEDING SYSTEMS

The rate of increase in homozygosity with inbreeding is dependent on the closeness of the relationship of the individuals which are mated. Previously it was stated that full brother–sister matings are the closest that can be made in mammals. The mating of sire to daughter or of son to dam gives the same rate of increase in homozygosis. However it is much more difficult to carry out these latter schemes than the full brother–sister mating for an extended time. Expressions have been developed to compute the inbreeding coefficient for such regular inbreeding systems. The inbreeding coefficients for successive generations of full brother–sister matings are given in Table 9-9. Note that the increase in the inbreeding coefficient is much less rapid than for self-fertilization. Actually about six generations of brother-sister mating is required to attain the degree of inbreeding that would be attained in two generations of selfing.

A plan of inbreeding that has been useful in the formation of families with a high relationship has been to mate a sire to his daughters, as shown in Figure 9-9. Such a mating plan would start first with the choice of a desirable male and a number of his female progeny. With frozen semen, this plan for inbreeding could be continued for several generations before a new sire would need to be chosen. The main advantage of such a plan is that the inbreeding can be increased rapidly to above .40 in three generations. Following these early generations, the system becomes more like a one-sire herd or line. This is especially so when a new foundation sire (son) must be chosen to head up the line.

TABLE 9-9
INBREEDING COEFFICIENTS IN VARIOUS GENERATIONS OF DIFFERENT INTENSITIES OF INBREEDING

Generation	Selfing	Full-sibling	Sire back to offspring*	One-sire herd	Two-sire herd	Three-sire herd
1	.500	.250	.250	.125	.066	.042
2	.750	.375	.375	.218	.128	.082
3	.875	.500	.438	.304	.185	.119
4	.938	.594	.469	.380	.239	.155
5	.969	.672	.484	.448	.289	.189

*See Figure 9-9.

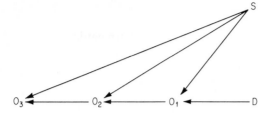

FIGURE 9-9
Diagram of inbreeding resulting from the mating of a sire back to his offspring in successive generations. This mating plan allows for a comparatively rapid increase in inbreeding in cattle, where frozen semen would extend the serviceable life of the sire (S).

Such extreme inbreeding is largely relegated to experimental studies. In practical breeding with livestock, the breeding plan cannot usually be specified regularly over an extended period. More commonly a herd may be closed to outside introductions and a limited number of sires placed in service. Often the breeding population is described in terms of a one-, two-, or three-sire herd. On occasion a breeder may desire to close the herd, and yet would like to keep the loss of heterozygosity to what might be termed a "reasonable level."

Wright[6] has shown that approximately $1/8N_m + 1/8N_f$ of the heterozygosity present in a closed population would be lost each generation. The number of males used each generation is represented by N_m, and N_f represents the number of females in the breeding herd. In a herd of 3 sires and 100 females, $1/24 + 1/800$ of the loci that are heterozygous would be expected to have become homozygous each generation. Since the value for N_f will almost invariably be many times larger than N_m, the $1/8N_f$ term can be neglected with little error. Using the expression $1/8N_m$ the loss of heterozygosis (gain in homozygosis) for one-sire, two-sire, and three-sire herds is approximately 12, 6, and 4 percent, respectively, per generation. The expected increase in the inbreeding for successive generations in such populations is given in Table 9-9.

The above values are based on random mating within the herd. With a one-sire herd, rather intense inbreeding cannot be avoided; however, when several sires are used, the inbreeding can be reduced by avoiding the mating of individuals as closely related as half sibs. Breeders often accomplish this by mating females from one sire line to males from another sire line. When the mating plan is devised to give the maximum avoidance of inbreeding, the rate of increase in homozygosity can be reduced to approximately one-half that given by the formula. In a control herd of dairy cattle of 6 males and approximately 20 females per generation with maximum avoidance of inbreeding consistent with proportionate representation of each sire, inbreeding increased to 5.6 percent over six generations or approximately 1 percent per generation.[7] By the approximate expression of Wright (op. cit.) with random mating within the herd, it would be expected that over 2 percent of the heterozygosis would be lost per generation. Consequently, the closing of a herd does not automatically mean that a high level of inbreeding will develop. With several sires a herd can

[6] Wright, S. 1931. *Genetics* 16:107–111.
[7] Legates, J. E., and R. M. Myers. 1988. *J. Dairy Sci.* 71:1025–1033.

be closed for several generations before the average inbreeding reaches a high level.

Robert Bakewell, the pioneer English breeder of Longhorn cattle, Leicester sheep, and Shire horses is reported to have inbred intensively with apparent good results. Culley, writing in 1794, said:

> Mr. Bakewell has not had a cross for upwards of 20 years, his best stock has been inbred by the nearest affinities, yet they have not decreased in size, neither are they less hardy, or more liable to disorder, but, on the contrary, they have kept in a progressive state of improvement.

Although many highly inbred animals undoubtedly were developed by Bakewell, he is also reported to have loaned out many sires for possible later use in his herd. Thus even though he had not "had a cross for upwards of 20 years," the average inbreeding coefficients apparently did not become excessively high.

GENETIC EFFECTS OF INBREEDING

Inbreeding increases the proportion of gene pairs that are homozygous and decreases the proportion that are heterozygous. Inbreeding itself does not change gene frequency, but in small populations some alleles may be lost by chance (Chapter 6) as inbreeding progresses. In a small population the gene frequency can fluctuate rather extremely, and by chance certain genes would be lost as others become fixed (homozygosis). In large populations the fluctuations of gene frequency would be less extreme, and few of the alleles would be lost. As inbreeding

FIGURE 9-10
Comet (155). A Shorthorn bull bred by Thomas Bates and reported to have sold for as much as $5,000 in about 1804. His pedigree (Figure 9-11) indicates that he was highly inbred. (*From Sanders, Shorthorn Cattle, The Breeders Gazette.*)

progressed without selection, the large population would become a series of subpopulations, each likely to be homozygous for different alleles.

Assumptions

Certain conditions were accepted in the development of the coefficients of inbreeding and relationship that we should clarify. The expressions were derived on the assumptions that mutation and selection were not important in the population. Such an assumption for mutation may be practically satisfactory over the few generations normally involved in the compilation of the inbreeding coefficients of farm animals. Available evidence suggests that mutation rates are low; hence, few loci expected to be homozygous from inbreeding would be rendered heterozygous by mutation.

Acceptance of the assumption that selection does not measurably influence the degree of homozygosity expressed by the inbreeding coefficient may be

FIGURE 9-11
Pedigree of the Shorthorn bull Comet. His inbreeding coefficient is .4687 and that of his sire (Favorite, 252) is .1875.

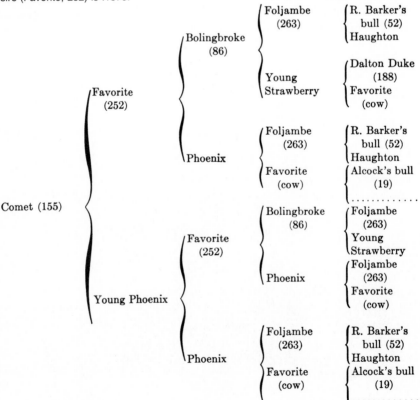

more difficult. Natural selection is at work at all times in all populations. Certain of the weaker, less vigorous animals die in every herd or flock.

Just what effect selection has on the degree of homozygosity in a specific situation is problematical. Situations have been reported in which heterozygosity in inbred lines is apparently greater than would be expected from the computed inbreeding coefficients. These cases point up the possibility that selection during inbreeding favored the heterozygotes over either homozygote.[8] How frequently heterozygote superiority is manifested is not known. It would likely be more frequent for traits influencing reproductive capacity.

Since inbreeding on the average does tend to depress performance and since there may be heterozygote superiority, it is probable that selection for performance traits during inbreeding favors animals that are more heterozygous than average. This selection process, continued over a period of several generations, might result in substantially more heterozygosity in an inbred line than inbreeding coefficients would indicate if (1) gene pairs with major effects exhibited heterozygote superiority and (2) the inbreeding system was a mild one permitting selection on traits other than reproductive fitness.

Basis for Inbreeding Effects

What accounts for the depression of performance which accompanies inbreeding? If all the gene effects were additive, inbreeding would not be expected to produce a reduction in mean performance. Alleles become more homozygous without regard to their effects, and the more desirable and the less desirable gene effects would tend to balance. On the contrary, if some degree of dominance is expressed, an increase in homozygosity would produce a decline in average merit, since the proportion of homozygous recessive loci would increase. Thus, the hypothesis of dominance of favorable genes was advanced early in explaining the consequences of inbreeding.

Another possible genetic reason for the harmful effects of inbreeding is that, in at least some cases, the heterozygote may be better than either homozygote. For example, a pair of genes B and b might have effects such that BB is better than bb but that Bb is better than either. Lerner[9] reviewed the experimental evidence and made a strong case for the existence of genetic mechanisms such that animals heterozygous for either single gene pairs or gene complexes have superior fitness and performance characteristics. The term *overdominance* has been applied to describe this phenomenon.

If an appreciable fraction of the gene pairs has this effect, the inbreds lack vigor and productiveness because of homozygosity itself. Development of highly productive inbreds would therefore be impossible, and the ultimate in productive ability could occur only in crosses or hybrids produced by breeding systems giving maximum heterozygosity.

[8] Shultz, F. T., and W. E. Briles. 1953. *Genetics* 38:34–50.
[9] Lerner, I. M. 1954. "Genetic Homeostasis," John Wiley & Sons, Inc., New York.

Either of the above theories would fit observed results to date reasonably well. The most probable situation would appear to be that both hypotheses are partially correct. Certainly inbreeding brings many recessives to light. For many of these there has not been any suggestion of the heterozygote having a selective advantage. This indicates that the dominance hypothesis is important. Neither can the apparent evidences of overdominance in some cases be overlooked. Further work will doubtless clarify the situation. Obviously the genetic basis for inbreeding depression and heterosis are clearly related, since the same loci are involved. Those traits which show a marked inbreeding depression also exhibit heterosis when genetically diverse stocks are crossed.

PHENOTYPIC EFFECTS OF INBREEDING

The outward effects of inbreeding were noted and recorded long before the experimental approach was undertaken. Darwin's statement that "the consequences of close interbreeding carried on for too long a time are, as is generally believed, loss of size, constitutional vigor and fertility, sometimes accompanied by a tendency to malformation"[10] is pertinent today. For many years the belief persisted that the detrimental effects of inbreeding resulted from inbreeding per se. This viewpoint was corrected when the Mendelian nature of inheritance was elaborated. Detrimental effects were a consequence of inbreeding, but inbreeding itself was not the culprit. It merely uncovered the latent deleterious tendencies and permitted them to express themselves.

Before the turn of the nineteenth century, Crampe[11] and Ritzema-Bos[12] demonstrated the adverse effect of inbreeding on fertility and litter size in rats. Numerous abnormalities also appeared as inbreeding progressed. Crampe's rats were inbred for about 17 generations and those of Ritzema-Bos about 30 generations; however, since the inbreeding was irregular, a reliable estimate of the inbreeding coefficients could not be made.

At the Wistar Institute, rats have been inbred by brother-sister mating for more than 125 generations. The results of the research of King[13] were different from those of the earlier workers. By rigorous selection along with inbreeding neither fertility nor constitutional vigor were reduced by inbreeding. Her foundation rats came from a previously interbred colony, and special care was given to the nutritional requirements of the inbreds. Wistar rats continue to be one of the most widely used strains of laboratory animals in the United States.

Extensive inbreeding research with guinea pigs, initiated by G. M. Rommell of the U.S. Department of Agriculture in 1906, has resulted in major contributions to our understanding of the consequences of inbreeding in livestock. Later, upon joining the department, Sewall Wright became responsible for this

---

[10] Darwin, C. 1868. "The Variation of Plants and Animals under Domestication," London.
[11] Crampe, H. 1883. *Landw. Jahrb.* 12:389–458.
[12] Ritzema-Bos, 1894. *J. Biol. Cent.* 16:75–81.
[13] King, H. D. 1918. *J. Exp. Zool.* 26:1–98.

research. A brief quotation from Wright's summary of the inbreeding studies with guinea pigs shows the remarkable general agreement between these results and those since reported in farm animals.

> There has been an average decline in vigor in all characteristics during the course of 13 years of inbreeding of guinea pigs, brother with sister. The decline is most marked in the frequency and size of litter, in which it is so great that it would have to be accounted for even though the decline in other respects was assumed to be due wholly to a deterioration in the environmental conditions. The decline is greater in the gains after birth than in the birth weight, and greater in the percentage raised of the young born alive than in the percentage born alive. The ability to raise larger litters has fallen off much more than ability to raise small litters.
>
> A comparison of the inbred guinea pigs with a control stock, raised under identical conditions without inbreeding, and derived in the main from the same linebred stock as the inbred families, indicates that the inbreds have suffered a genetic decline in vigor in all characteristics. The decline in fertility is again shown to be marked. Experimental inoculation with tuberculosis has shown that the inbreds were inferior, on the average, to the controls in disease resistance. A study of sex ratio yields results in marked contrast to those obtained in connection with the other characters. There are no significant fluctuations from year to year, no contrast between inbreds and controls, and no indications of change due to inbreeding.[14]

Impact of Inbreeding on Farm Animals

Inbreeding experiments with cattle and even with the smaller farm animals are expensive and time-consuming. Nevertheless, results of experiments have been most valuable in corroborating results from laboratory mammals. Available results have provided guides to the expected decline in performance that might be expected when selection is accompanied by mild inbreeding.

Beef Cattle Inbreeding effects have been variable in different beef cattle studies,[15] but on average, weaning weight is depressed about .3 kg for each 1 percent inbreeding in the calf (with little difference between sexes) and an additional .25 to .50 kg for each 1 percent inbreeding of the dam. Bull calves are more affected than heifers by inbreeding of the dam. The postulated reason is that males have more growth potential and are therefore handicapped to a greater extent by reduced milk production of inbred dams. Inbreeding has a detrimental effect on gains after weaning amounting on average to about $-.0015$ kg per day in both sexes per 1 percent inbreeding. Effects of inbreeding on growth before and after weaning vary markedly between lines with some of the lines showing no detrimental effects.

Fertility and calf viability are reduced more drastically than other traits. On average, a 10 percent increase in inbreeding of the dam results in an increase

[14] Wright, S. 1922. U.S. Department of Agriculture Bulletin 1080.
[15] See Brinks, J. S., and B. W. Knapp. 1975. Colorado Agr. Exp. Sta. Tech. Bul. 123.

of over 2 percent in nonpregnant cows and a reduction of over 1 percent in calf weaning rate of pregnant cows. Similar figures for inbreeding of the calf (or potential calf) of a mating are 1.3 and 1.6 percent. In total this means that calf crop weaned would be reduced by the total of these figures or about 6 percent for matings of 10 percent inbred cows to produce calves also 10 percent inbred. There may be proportionally more severe effects at higher levels of inbreeding. Even if effects are linear, at 50 percent inbreeding of both dam and calf, a reduction in calf crop 5 times as severe as at 10 percent inbreeding (30 percent reduction) could endanger the survival of a line. As with growth, some lines are more severely affected than others. Inbred bulls produce lower-quality semen than outbreds at young ages. Their slower attainment of puberty and the possibility of lower later fertility is a problem in producing inbred bulls for sale to commercial producers.

Dairy Cattle   Inbreeding experiments with Guernseys, Holsteins, and Jerseys have been conducted at a number of locations in the United States since the 1930s. Early U.S. Department of Agriculture studies with Guernseys and Holsteins showed that the birth weight of the calves decreased markedly and that mortality for the first year after birth was 15 percent for the inbreds and 8 percent for the outbreeds. With moderate inbreeding there was little apparent decrease in production, but as inbreeding became more intensive, production declined rather drastically. California workers found decreases of 95 kg of milk and 2.2 kg of fat per 1 percent inbreeding in Holsteins. Fat percentage also declined with inbreeding. In the Iowa State Holstein herd, for each 1 percent increase in inbreeding, the milk yield for the first lactation decreased 25 kg. Early growth was depressed, and the attainment of mature size was delayed by inbreeding in both the Iowa and California herds.

Young et al. have compiled an excellent summary of inbreeding investigations with dairy cattle. Regressions of milk yield, fat yield, and fat percentage for several studies are given in Table 9-10. These results are all from experimental in-

TABLE 9-10
REGRESSIONS OF MILK YIELD, FAT YIELD, AND FAT PERCENTAGE ON PERCENT OF INBREEDING (*F*)*

				Regression on percent of inbreeding		
Location	Breed	Lactation	Average *F*	Milk, kg	Fat, kg	Fat, %
Iowa	Holstein	First	9.9	−24.0	−.8	.000
Iowa	Holstein	All	7.4	−24.5	−.8	.003
Iowa	Guernsey	All	6.4	−16.3	−.4	.001
Michigan	Jersey	All	18.0	−9.5	−.8	.005
Ohio	Holstein	First	3.3	−18.6	−.5	
Ohio	Holstein	Second	2.8	−19.5	−.5	

*Young, C. W., et al. 1969. University of Minnesota Technical Bulletin 266, p. 6.

breeding at relatively low levels. As expected, the regressions for milk and fat are negative. Those for fat percentage are zero or insignificantly positive.

Studies in Wisconsin and Nebraska and in Great Britain have also shown that milk yields are depressed by inbreeding. The results show expected decreases in production ranging from $-.25$ to $-.60$ percent of the mean yield per 1 percent inbreeding. While formation of inbred lines for eventual crossing does not appear to be practically feasible, knowledge of the magnitude of inbreeding depression is important in projecting progress from selection.

Swine As with other livestock, on the average, inbreeding in swine is detrimental. From extensive data of the Regional Swine Breeding Laboratory,[16] it has been accompanied by reductions of .3 to .6 pigs per litter at birth, .5 to .7 pigs per litter at weaning, and 1.4 to 2.7 kg per pig at 154 days of age. The reduction in litter size is much more important economically than the reduction in growth rate.

Efforts to find the physiological bases for reduced reproductive efficiency in inbred swine have shown that in some lines inbreeding delays testicular development, delays puberty in both sexes, reduces the number of ova shed by females, and increases early embryonic death rates. Observations indicate that the percentage of boars which refuse to breed and those which are slow breeders increases with inbreeding. In spite of the rather drastic average effects of inbreeding on reproductive performance in swine, some lines have performed reasonably well for several generations of inbreeding.

Carcass characters have been affected to only minor degrees by inbreeding, with different studies giving somewhat different trends.

The experimental results are clear in indicating that the effects of inbreeding in swine are detrimental and that the detrimental effects cannot *on the average* be overcome by selection in herds in which inbreeding is increasing very much—probably as little as 3 to 5 percent per generation. Thus, inbreeding in even mild forms should be avoided in commercial herds.

However, the effects of inbreeding are not so drastic but that some lines can be maintained to very high levels of inbreeding and a majority of lines can be maintained to inbreeding levels of perhaps at least 30 to 50 percent. The formation and use of inbred lines is thus a possible breeding system for developing stocks to be used commercially in crosses and in combinations for future breed improvement as proposed by Wright many years ago. It is doubtful, however, that the results will justify the expense of inbred line development for direct use in crosses for commercial production.

Sheep In extensive studies at the Western Sheep Breeding Laboratory,[17] in Dubois, Idaho, general decreases were observed in both weaning and yearling

---

[16] Craft, W. A. 1958. *J. Anim. Sci.* 17:960–980; and Dickerson, G. E., et al. 1954. Mo. Agr. Expt. Sta. Res. Bul. 551.

[17] See Terrill, C. E., et al. 1948. *J. Anim. Sci.* 7:181–190; Ercanbrack, S. K., and A. D. Knight. 1981. *J. Anim. Sci.* 52: 977–988, and 1983. *J. Anim. Sci.* 56: 316–329.

weights of inbred Rambouillet, Columbia, and Targhee sheep. The decreases varied from group to group but averaged .15 kg for weanling lambs, .12 kg for yearling ewes, and .22 kg for yearling rams per 1 percent inbreeding. Decreases in size of Australian Merinos have occurred with inbreeding. For fleece characters, the Dubois studies, as well as studies in Australia on Merinos, have shown a trend toward decreased staple length and fleece weight with increases in inbreeding. The decreases were large enough in some cases to be statistically significant but of less practical importance than several other characters. The maximum figure reported indicated a decrease of .6 lb in grease-fleece weight per 10 percent increase in inbreeding. Reduced fleece weight with inbreeding may be a consequence of reduced body size and not a specific effect of inbreeding on wool growth.

Inbreeding has rather severe effects on the fertility and survival of lambs. The trends are thus consistent with those of other species in showing greater reductions in less heritable traits.

Poultry The higher reproductive rate in poultry has permitted much more intense inbreeding than in other farm animals. Large families also allow regular full-sib inbreeding, and some lines with inbreeding coefficients in the high nineties have survived. Declines in egg production, hatchability, and viability accompany increases in inbreeding as with other farm animals. Egg production has decreased by as much as one egg for each 1 percent inbreeding. Mortality increased by .55 percent for each 1 percent increase in inbreeding in one study.[18] These authors concluded that when all losses accompanying inbreeding are considered, inbreeding at a level of $F = .5$ raised the cost of egg production one and one-half times compared to a level of zero inbreeding.

Most commercial poultry are crosses among lines, breeds, or strains that have been inbred to varying degrees. The above analyses suggest the burden required to develop and maintain such foundation stocks. Abplanalp (1974) has provided a summary (Table 9-11) of impact of inbreeding on performance of both chickens and turkeys.

USEFULNESS OF INBREEDING

In spite of the occasional good results obtained from inbreeding, it is becoming increasingly apparent that the development of highly productive inbreds in our farm animals is not likely within the foreseeable future. Rather, any positive contributions from inbreeding must come as an aid in producing seed stock which can be used with predictable results as parents for outbred or crossbred commercial animals. The use of inbreds for commercial production is being exploited more widely in poultry than with other farm animals. Despite much

[18] MacLaury, D. W., and A. W. Nordskoy. 1956. *Poultry Sci.* 33:704.

TABLE 9-11

EFFECT OF 25 PERCENT INBREEDING ON THE RELATIVE PERFORMANCE OF CHICKENS AND TURKEYS*

Trait	Performance as percentage of noninbreds†	
	Chicken	Turkey
Egg production	90.4	89.5
Fertility	99.1‡	98.8‡
Hatchability (embryo inbred)	90.9	83.4
Viability of females	94.3	90.7
Total reproduction	74.4	61.6
Body weight	95.0	89.9
Egg weight	100.0	95.9
Age at first egg	100.0	100.0

*Adapted from Abplanalp, H. A. 1974. "Proceedings, First World Congress on Genetics Applied to Livestock Production." 1:900.
†Performance of noninbreds equals 100.
‡With artificial insemination.

hope, little solid evidence is available to support the use of deliberate inbreeding systems in our larger farm animals.

Linebreeding

Most breeders, as well as commercial producers, are wary of intense inbreeding, but they are often willing to undertake mild inbreeding to maintain a high relationship to a supposedly outstanding ancestor. Although linebreeding is inbreeding in the fundamental sense, its primary purpose is not to increase homozygosity but to retain a goodly proportion of the genes of the designated individual. In addition to trying to retain desired genes and gene combinations, linebreeding is often used when there is a high likelihood of reducing the merit of the herd when outside sires are introduced. Thus, only breeders with superior herds should consider linebreeding.

The structure of a linebreeding program can take a variety of forms. A plan based on maintaining a high relationship to an outstanding sire is practically most workable. Males can leave many more offspring than females, and the use of artificial insemination and frozen semen can extend their useful life. Occasional examples of successful linebreeding to females have been recorded, but generally their influence has been carried forward through their sons and grandsons.

Disagreement over terminology arises because some choose to define linebreeding merely as the mating of distantly related animals, regardless of whether it is directed toward a single favored ancestor or not. Actually closebreeding and linebreeding are varieties of inbreeding and will be so considered here. Mating of sire to daughter (or conversely, son to dam) is one of the closest forms of inbreeding and is therefore classed as *closebreeding*. It is also the most effective type of mating for concentrating the hereditary material of the sire and might logically be called "linebreeding," or perhaps "close linebreeding." A pedigree which illustrates this intense type of linebreeding is shown in Figure 9-12.

Development of Families

Inbreeding tends to develop distinct lines or families as the breeding plan continues. If family relationships in excess of .50 are to be developed, some inbreeding must be undertaken. Such family formation makes possible family selection for traits such as mortality and carcass merit. The mating of a sire to his daughter, as discussed earlier in this chapter, is the most extreme approach to forming distinct families by inbreeding.

Inbreeding of itself tends to increase the additively genetic variance in the total population by $(1 + F)$, so that as F approaches 1.0, the additive genetic variance would become twice that of the original random mating population. Additionally as inbreeding progresses, an increasing proportion of the additively genetic variance would be observed among families, thus adding to the accuracy of family selection (see Chapter 7). At the same time the additively genetic variance within families is reduced, approaching zero as F approaches 1.0. Nonetheless, deliberately planned inbreeding to aid family selection does not appear promising. The restriction to selection among inbred families lowers the potential selection intensity, and only a limited increase in average gene effects is expected from progeny of the inbred matings of highly selected parents. Hence, inbreeding of itself appears to have little to offer, be it termed "linebreeding" or "inbreeding," unless the heritability of the trait(s) concerned is low and the environmental correlation among family members is small.

Elimination of Undesirable Recessives

In many inbreeding experiments, hereditary abnormalities or lethal genes have appeared. Such traits are almost always recessive in inheritance. Genes for them may be present in low frequencies in outbred herds but remain hidden and unsuspected, usually or always being covered by their dominant alleles. When inbreeding occurs, the probability of these genes appearing in homozygous form is increased, just as the probability of all genes being homozygous is increased. Inbreeding does not create such factors; it merely permits them to be expressed and identified. Several examples were presented in Chapter 4.

At one stage in corn improvement it was thought that inbreeding should be

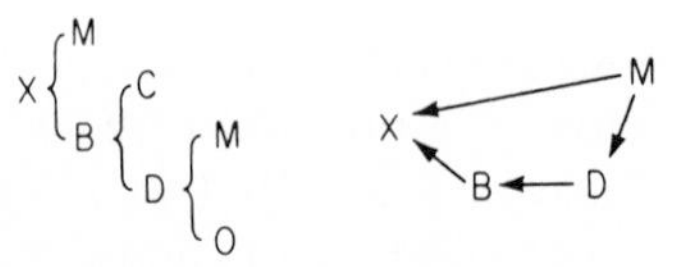

FIGURE 9-12
Bracket and arrow-style pedigrees of a common linebred mating, sire to granddaughter.

used to remove deleterious genes from the breeding stocks. Then the stocks could be recombined, and selection could progress without the hindrance of the deleterious genes. Such an approach did not gain acceptance, but it is true that stocks which have undergone inbreeding should have less likelihood of transmitting these undesirable alleles. The practice of mating a sire to 15 to 20 of his daughters is used in some situations to test for the presence of deleterious genes which are expressed early in life. Suspected carriers of lethal or semilethal genes in many cases should be tested in some manner before being used extensively. When the sire is mated to his daughters rather than to specific tester animals, all the deleterious genes rather than only those which the special tester carries have a chance to be expressed. This is a rather drastic testing procedure since the inbred offspring ($F = .25$) could be a distinct burden even though no lethals or sublethals were detected.

Homozygosity and Prepotency

Those animals which perform even reasonably well in an inbreeding program should be homozygous for a larger-than-average number of desirable genes. Their increased homozygosity means that the composition of their gametes will be more uniform than those of outbred animals. Their offspring should be more uniform.

We would also expect increased uniformity of productive characters, such as growth rates and litter size in inbreds as compared with outbreds. Actual data on this point are limited, but as far as they go, they seem to indicate no striking increase in uniformity. Perhaps the increased genetic uniformity is offset by an increased susceptibility to adverse environmental factors which affect some individuals more than others and which thus increase environmental variation. However, even if there were no increase in variation due to environmental factors, the decrease in variation in quantitative characters with inbreeding would be relatively small and might well remain undetected except at high levels of inbreeding. For characters only 20 to 30 percent heritable, a reduction of 20 or even 40 percent in the heritable portion of the variation would have little measurable effect on the total variance observed.

Since desirable genes are often dominant, good inbred animals are expected to be prepotent, that is, to stamp their own characteristics on the offspring to the exclusion of those of the other parent. Inbreeding is the only known method of increasing prepotency. Prepotency depends upon the homozygosity of dominant genes.

SUMMARY

Inbreeding is the mating of individuals which are more closely related than are average members of a breed or population. Inbreeding increases homozygosity, and the genetic consequence of inbreeding results directly from the increased homozygosity. The inbreeding coefficient measures the proportion of the loci which was heterozygous in the base population, which probably has become homozygous due to inbreeding. Uncovering deleterious recessives and reducing the proportion of heterozygotes are primarily responsible for the reduced fertility and the loss in vigor and productivity associated with inbreeding. Lowered performance with inbreeding makes it unwise to follow an inbreeding program in a commercial herd. Nevertheless, it may be used in breeding herds to develop distinct families for family selection. The increased homozygosity for dominant genes of breeding animals would increase prepotency, but any increase in uniformity due to inbreeding for most quantitative traits would be difficult to detect.

Relationship is based on the probability that two individuals are alike in more of their genes than are random individuals from the breed of population. The coefficient of relationship measures the probable proportion of genes that are alike for two individuals due to their common ancestry. Since higher relationships do reflect a higher likelihood that two individuals have the same genes, this information is useful in weighting evidence from relatives when appraising an individual's merit.

SUGGESTIONS FOR FURTHER READING

Books

East, E. M., and D. F. Jones. 1919. "Inbreeding and Outbreeding," J. B. Lippincott Co., Philadelphia.
Falconer, D. S. 1981. "Introduction to Quantitative Genetics," 2d ed., Longmans Inc., New York.
Gowen, J. W. (ed.). 1952. "Heterosis," The Iowa State University Press, Ames, Iowa.
Lerner, I. M. 1954. "Genetic Homeostasis," John Wiley & Sons, Inc., New York.
Lush, J. L. 1945. "Animal Breeding Plans," 3d ed., The Iowa State University Press, Ames, Iowa.
Winters, L. M. 1954. "Animal Breeding," 5th ed., John Wiley & Sons, Inc., New York.

Articles

Dickerson, G. E. 1973. Inbreeding and Heterosis in Animals. Pp. 54–57, in "Proceedings, Animal Breeding and Genetics Symposium in Honor of Jay L. Lush," American Society of Animal Science.
Lush, J. L. 1946. Chance as a Cause of Changes in Gene Frequency within Pure Breeds of Livestock. *Am. Naturalist* 80:318–342.
McPhee, H. S., and S. Wright. 1925. Mendelian Analysis of the Pure Breeds of Livestock. III. The Shorthorns. *J. Hered.* 16:205–215.
Wright, S. 1921. Systems of Mating. *Genetics* 6:111–178.
Wright, S. 1922. Coefficients of Inbreeding and Relationship. *Am. Naturalist* 56:330–338.
Wright, S. 1922a. The Effects of Inbreeding and Crossbreeding on Guinea Pigs. I. Decline in Vigor. II. Differentiation among Inbred Families. U.S. Department of Agriculture Bulletin 1090.
Wright, S. 1922b. The Effects of Inbreeding and Crossbreeding on Guinea Pigs. III. Crosses between Highly Inbred Families. U.S. Department of Agriculture Bulletin 1121.

CHAPTER 10

HETEROSIS AND OUTBREEDING

Whereas inbreeding is a system in which the mates are more closely related than average members of the breed or population, *outbreeding* is a breeding system in which mates are less closely related than the average of the breed or population. The broad heading of outbreeding includes the mating of unrelated animals within breeds, grading, crossing of inbred lines, crossbreeding, and the more extreme crosses between animals of different species, such as the mating of the ass with the mare to produce the mule.

The effects of outbreeding are generally opposite those of inbreeding, since with outbreeding heterozygosity is increased. For the most part, the practical usefulness of outbreeding results from the fact that genes with favorable effects generally express some dominance over their alleles. In crossing two diverse lines, strains, or breeds, an increase in heterozygosity is realized. With the increased heterozygosity, "hybrid vigor" is expressed when the average of the offspring exceeds the average of their parents. *Heterosis* is a more general expression for the difference between the average of the offspring and the parental average, with the potential for positive, negative, or no heterosis. The percent heterosis for crossbreeding is determined by the following expression:

$$\text{Percent heterosis} = \frac{(\text{average of crossbred progeny} - \text{average of parents})}{\text{average of parents}} \times 100$$

The increase in heterozygosity may promote individual superiority, but it reduces average breeding values. Thus, the difference in the goals for seedstock and commercial herds must be recognized in appraising the usefulness of specific outbreeding systems.

HETEROSIS AND ITS GENETIC BASIS

Various outbreeding systems are useful in obtaining desired zygotic combinations in the offspring. The decline in performance accompanied by inbreeding, and its resultant homozygosity, can be overcome by a judicious outcross. Outbreeding systems accomplish two major objectives. First, they bring together a desired combination of genes from two parents more rapidly than can be accomplished with selection. Thus, advantage can be taken of the complementarity among breeds, strains, or lines. Second, increased heterozygosity provides the genetic base for the expression of heterosis.

The genetic or physiological explanation for heterosis is not known with certainty. As a species, breed, strain, or line develops, it becomes by chance homozygous for some favorable dominant genes and probably for some unfavorable recessive ones. Another breed or species probably would not be expected to become homozygous for the same undesirable recessive genes. Thus, when two breeds are crossed, $AAbb \times aaBB$ (two pairs are assumed here purely for illustrative purposes), the offspring would be $AaBb$. Since the offspring have one dominant gene of each pair, they would be expected to be more vigorous or productive than either parental type. This explanation of heterosis assumes that it is largely a result of covering up the effects of recessive genes by alleles with dominant effects.

A second possible explanation of heterosis is that for some gene pairs the heterozygote may be more vigorous than either homozygote. That is, for a given gene pair such as A and a, Aa might be superior to either AA or aa. This situation is called *overdominance*. Much indirect evidence in both plants and animals suggests that overdominance may be an important factor in heterosis in at least some species and for some traits. If this is the case, it would be impossible to produce homozygous populations which could produce at maximum levels. Maximum productivity could be attained only in crosses.

When subpopulations develop, isolation would be expected to permit gene frequencies to drift, and the populations would differentiate genetically. Selection under different habitats and to meet different human needs would also allow gene frequencies to diverge. A few loci may become either fixed or lost, but a difference in gene frequencies may develop for several influential loci. The difference in gene frequency at each locus and the degree of dominance expressed determine the heterosis expressed when only dominance is involved. Lush and De Alba[1] state that the heterotic effect for each pair of genes is expressed by the relationship $(k - 1)(p_1 - p_2)^2 a$, where k is a measure of degree of dominance and is 1 for no dominance, 2 for complete dominance and greater than 2 for overdominance; p_1 and p_2 are the gene frequencies for the dominant genes in the two parent stocks; and a is one-half the average phenotypic difference between the homozygotes of the two contrasting alleles in the

---

[1] Lush, J. L., and J. De Alba. 1964. "UN Conference on Application of Science and Technology," pp. 215–277.

population (*BB* versus *bb*) as in Table 6-10. The above expression points up the fact that the degree of dominance and the difference in gene frequency for the two parental populations are keys to the amount of heterosis expressed for individual loci where dominance is expressed.

Another explanation sometimes thought to account for heterosis, at least to a degree, is that of interallelic interactions, or epistasis. Going back to our previous simplified illustration, this possible basis for heterosis would be provided if the combination *AaBb* results in an interaction such that the presence of *both A* and *B* gives a more desirable phenotype than would be expected from average phenotypes of *AAbb* and *aaBB*. A classic illustration of this theory comes from plants in which it was observed that one dominant gene resulted in *long* internodes and another resulted in *many* internodes. The net effect of the presence of both dominant genes tends to be multiplicative and results in offspring taller than would be expected from averaging the sizes of parental types.

Many other types of interallelic interactions which might result in heterosis can be visualized. Some might at least theoretically result in *negative heterosis*. As a hypothetical example, suppose one line of animals carried a dominant gene resulting in excessive production of a specific body chemical, but with the excessive production not exerting a serious detrimental effect in animals that had a reserve capacity for disposing of it. Suppose a second line carried a dominant gene resulting in inefficient metabolism of the same chemical. Even though the second line was also able to perform more or less normally as long as production of the chemical was not excessive, it can readily be seen that a cross of the two lines could be disastrous. Fortunately, most genes with undesirable effects are recessive.

Negative heterosis seems to be rare in animals. There are, however, some apparent examples in laboratory organisms.[2] It should be noted that what is desired in a trait may determine whether heterosis is positive or negative in an economic sense. For example, if smallness were desired in a species for some specific purpose and if crosses resulted in size increases, the effect would be economically negative, even though we might still think of it as biologically positive.

As a general rule, in living organisms genes with some degree of dominance are more often favorable than are recessives. Thus, crosses would usually be expected to exhibit increased rather than decreased vigor.

The difficulties in discriminating between the theories that heterosis is dependent upon covering the effects of recessive genes or interallelic interactions between dominant genes are readily apparent. In both cases maximum heterosis would depend upon the presence of at least one dominant for each contributing allelic pair. If either explanation or a combination of both were wholly responsible for heterosis, it should theoretically be possible to select pure lines or breeds which would carry all the desirable dominants in homozygous form and thus have maximum performance. If we concede that

---

[2] See Mason, R. W., et al. 1960. In: Kempthorne, O. (ed.), "Biometrical Genetics," Pergamon Press, New York; and Stern, C. 1948. *Genetics* 33:215–219.

this is theoretically possible, we must also recognize that the development of such lines or breeds would be an extremely slow and difficult process. The principle reason for this is the large number of gene pairs which affect most productive characters. Two animals heterozygous for n pairs of genes can theoretically produce 3^n types of offspring. If only seven pairs of genes were heterozygous, 3^7, or 2,187 different combinations could be expected in the offspring. If ten pairs were heterozygous, 59,049 different combinations would be possible. Actually, the number of effective gene pairs is probably much larger than this. Animals usually have only a relatively small number of offspring. Thus, the probability of the perfect combination being produced is small even with many generations of selection. The effects of the environment in reducing heritability would make it impossible to identify the perfect combinations with certainty even if they occurred. Further, there is an excellent chance that some undesirable recessives would be fixed in each line or breed purely as a result of genetic drift. Another fact which could slow progress in fixing desired combinations is that desirable and undesirable genes could be linked in unfavorable combinations in the foundation stock with which selection began. Many generations might be required to reach a linkage equilibrium.

So far as is known, no one has as yet succeeded in producing lines, breeds, or varieties of plants or animals which in themselves have all desirable traits and which show no heterosis in crosses. For the reasons given above, this is not surprising, even if the above kinds of gene actions should be largely or wholly responsible for heterosis.

The most plausible hypothesis at present is that all the foregoing possibilities may be factors in heterosis. Different genetic mechanisms may be of predominant importance in heterosis in different species of plants and animals and for different traits in the same species. A full understanding of the genetic basis of heterosis will probably have to await extensions of knowledge of the physiology of gene action and the physiological reactions resulting in heterosis. To date, too little is known of the physiology of heterosis—even for simple traits of laboratory organisms—to permit the exposition of theories or hypotheses that would likely explain heterosis in domestic animals.

The fact that the genetic and physiological bases for heterosis are imperfectly understood stands as a challenge to scientists and probably to the optimum use of heterosis in practical situations. It does not, however, hinder plant and animal breeders from taking advantage of heterosis through the use of crosses of breeds, varieties, lines, or even species, which have proved through either experimentation or experience to produce at a higher level than the average of their parental types.

CROSSBREEDING

Crossbreeding is the mating of animals from different established breeds. Technically it applies only to the first crosses of purebreds, but it is generally applied also to the more widely used systems involving crisscrossing of two

breeding levels has been appreciable. Although much genetic variability remains, the increase in inbreeding, together with chance or random fluctuation in gene frequency, has undoubtedly resulted in the fixation in homozygous form of some genes in most individuals of each breed.

In a very mild way, breeds may exhibit some of the same characteristics of inbred lines. Since the outstanding outward effect of inbreeding is to lower performance, especially for reproductive and fitness traits, we could well expect animals of the various pure breeds to have somewhat less than maximum performance levels.

Because of the random nature of gene fixation with inbreeding and the low intensity of selection, it would not be unlikely that the different breeds would have the same desirable or undesirable alleles fixed in homozygous form. This would be most true of long-separated breeds with diverse origins and less true of breeds developed from similar foundations within comparatively recent times. Crossing breeds will tend to cover the undesirable recessive genes and also increase heterozygosity. Since both of these are probably important for heterosis, the performance of crossbreds would be expected to exceed that of the purebred parents to some degree. The increased performance should be marked in those traits most depressed by inbreeding, e.g., (1) reproductive qualities including fertility and fecundity of both sexes and maternal qualities of females and (2) vigor as measured by mortality rates. The largest increases in performance would be expected from crosses of breeds having the broadest genetic diversity.

Most popular breeds have reasonably good performance for most important traits, or else they would probably not be used widely. Therefore the theoretically expected increases in performance with crossbreeding are relatively small for any one trait. However, heterotic effects are often additive and multiplicative. For example, if the average straightbred pig at 150 days weighs 100 kg and the average crossbred weighs 105 kg, heterosis per pig is 5 kg, or 5 percent. If the average number of pigs raised per litter is 10 for straightbreds and 10.5 for crossbreds, heterosis for litter size would be 5 percent. Expressed as weight produced *per litter* or *per sow farrowing,* straightbreds would average 1,000 kg and crossbreds 1,105 kg. Heterosis is then slightly more than 10 percent or approximately the sum of that for the two components of *litter weight at 150 days.*

Crossbreeding Systems

There are a virtually limitless number of crossbreeding systems which could be used for commercial production. Each of these depends upon the strong foundation breeds included in the crosses. Those most often used include: (1) *Two-breed crosses*—Only the first-cross of F_1 offspring are produced with all sold for slaughter or other commercial use. This system is most useful for situations in which females of a specific breed are well-adapted to a given environment, but whose offspring need inheritance from another source to be suitable for feed-lot or slaughter purposes. It takes advantage of individual heterosis for vigor, survival, growth, efficiency, or other traits. (2) *Backcross*—First crosses are made, and males are sold for slaughter. Crossbred females are mated to males of one of the parental breeds, and all offspring sold for slaughter. This system takes advantage of maternal heterosis and of a part of individual heterosis. It is especially useful in situations where adaptation to a specific environment is required from a particular maternal breed, with crossbred females also being satisfactory in adaptation, but where more than half the additive effects from the other parental breed are desired for feed-lot or carcass traits. (3) *Three-breed crosses*—Males of breed A are mated to females of breed B with the resulting male offspring being sold for slaughter. The A × B females are bred to males of breed C, and all offspring sold for slaughter. The system takes advantage of both individual and maternal heterosis. It also has the advantage that breeds can be used to complement one another. For example, in meat animals, breeds A or B or both can be selected from among those with good maternal abilities while breed C (often called the "terminal" sire breed) can be selected for desired growth, efficiency, and carcass traits. (4) *Sequence breeding*—Systems in which males of two or more breeds are used in *sequence* on crossbred female populations.

A systematic program with a sequence of two breeds is known as *crisscrossing*. Males of breed A are mated to females of breed B to produce the first crossbreds. Females from this first cross are then mated to males of breed B to begin the system, as illustrated in Figure 10-2. In crisscrossing systems a backcross is necessary in the second generation. Nevertheless, in mating crossbred females to males of one of the parental breeds (backcrossing), the first-generation crossbred females can express heterosis for maternal performance. As noted from Figure 10-2 crossbreds in a crisscrossing system move toward having about $\frac{2}{3}$ of their inheritance from the breed of their immediate sire and $\frac{1}{3}$ from the other breed used in the system.

A system of crossbreeding which systematically uses three or more breeds is known as *rotational crossing*. In a system using three breeds, males of breed A are mated to females of breed B. The two-breed-cross daughters are mated to males of breed C. Rotational crossing with three breeds is illustrated in Figure 10-3. As the rotation continues, the offspring will trend toward having 57 percent of their inheritance from the breed of their immediate sire, 29 percent from the breed of their second sire (maternal grandsire), and 14 percent of their genes from the third breed. When n breeds are included in a rotational cross, the expected percentage of inheritance from the immediate breed of sire is given by the expression $50 \times 2^n/2^n - 1$.

The offspring from a two-breed crisscrossing system will be expected to express about $\frac{2}{3}$ of the heterosis exhibited by the initial crossbred offspring of the two breeds. This follows from the fact that matings will be between purebred males and crossbred females with $\frac{2}{3}$ of their inheritance from the other breed. For the three-breed rotational crossing, heterosis would be expected to be about $\frac{4}{7}$ of its maximum. Again this follows since the males of one breed

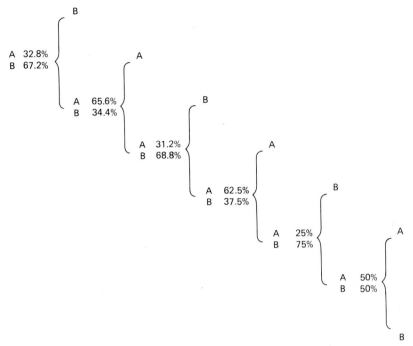

FIGURE 10-2
Illustration of pedigrees of progeny of a two-breed rational crossing system (crisscross-ing) showing the percentage of inheritance expected for progeny of successive genera-tions.

are mated to females with ⁵⁄₇ of their inheritance from breeds other than the breed of their mate.

Much experimental work has made it possible to predict the general useful-ness of many breeds in specific crosses. Carmon[3] and coworkers attempted to predict the performance of progeny of crisscrossing and rotational crossing from the performance of the single crosses and the parental breeds or strains. For predicting the performance of progeny R_2 from crisscrossing they used the expression

$$R_2 = C_2 - \frac{C_2 - P_2}{3}$$

where C_2 = Performance of the single cross
P_2 = Average performance of the two parental breeds

[3] Carmon, J. L., et al. 1956. *J. Anim. Sci.* 15:930–936.

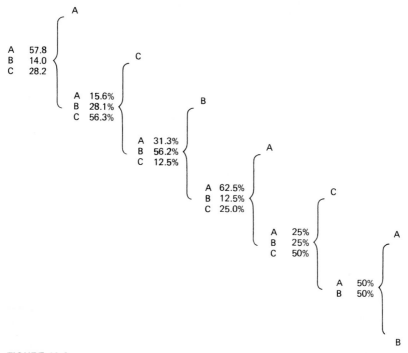

FIGURE 10-3
Illustration of pedigrees of progeny of a three-breed rotational crossing system, showing the percentage of inheritance expected for progeny of successive generations.

Note the reduction in performance from the single-cross performance is ⅓ of the expected heterosis.

In the case of a three-way rotational cross, the expression for predicting the performance of the rotational-cross progeny R_3 is

$$R_3 = C_3 - \frac{C_3 - P_3}{7}$$

where C_3 = Average performance of the three possible single crosses among three breeds
     P_3 = Average performance of the three parental breeds

Note that the expected reduction in heterosis is ⅐ of the average heterosis for the three possible single crosses.

Carmon and coworkers tested these crossbreeding schemes with inbred lines of mice. Single crosses were superior to rotational crosses and these were superior to the parental lines. In theory the performance of the rotational cross is expected to approach the average performance of all the possible single

TABLE 10-1
SUMMARY OF PERCENTAGE OF EXPRESSED HETEROSIS EXPECTED
IN PROGENY FROM ALTERNATIVE CROSSBREEDING SYSTEMS*

	Heterosis	
Mating system	Individual	Maternal
2-breed cross	100	0
A × B		
3-breed cross	100	100
[C × (A × B)]		
Rotational crosses†		
2 breeds	33	67
3 breeds	86	86
4 breeds	93	93

*From Dickerson, G. E. (1973), "Proceedings, Animal Breeding and Genetics Symposium," p. 61, American Society of Animal Science.
†At equilibrium in the rotation.

crosses as the number of breeds included in the rotation increases. However, the breeding system becomes complicated as more breeds are included, and in practice certain single crosses are often far superior to the average of all possible single crosses. Thus a system of crossbreeding utilizing a specific three-breed cross would offer maximum heterosis, including that for maternal performance, when it can be used.

A summary of the percentage of heterosis expected in progeny resulting from some of the more commonly used crossbreeding systems is given in Table 10-1. Individual and maternal heterosis are important to consider for beef cattle, sheep, and swine crossbreeding systems as well as the additive genetic merit of the foundation breeds. This is illustrated by the following expression of the expected performance of crossbred progeny:

Expected performance = mean + sire*/2 + dam*/2 + dam's maternal effect
+ individual heterosis + maternal heterosis

Crossbreeding may not be the best way to go for all situations. Decisions as to which crossbreeding system to use with each species and for specific environmental situations, production systems, and market demands all depend upon many factors. Some of the more important include (1) level of heterosis for important traits, (2) need for complementary breed combinations, (3) availability of superior breeding males of different breeds, (4) percentages of total populations which must be straightbreds to supply females of specific types to the system or availability of crossbred females for purchase, (5) size of herd, and (6) complexity and cost of system relative to

*Additive breeding values as deviations from mean.

expected returns. Each producer must weigh the facts and decide whether crossbreeding will have advantages over other breeding systems for the particular situation.

Crossbred animals usually exhibit a mixture of the breed trademarks of the parental breeds. For some people this reduces the satisfaction from raising livestock to the point that they abhor crossbreeding regardless of its economic merits. In today's competitive and specialized agriculture, however, most commercial producers will make decisions on crossbreeding largely on the basis of whether or not they think it will pay.

Results of Crossbreeding

Results of crossbreeding studies are in general accord with theoretical expectations. Crossbreds have usually exceeded the average performance levels of the parental purebreds by margins which are relatively small on a percentage basis for any one trait. Often on a total performance basis they are large enough to be economically important in terms of total production efficiency. Increases have been most important for traits most depressed by inbreeding and those expressed early in the development of the individual.

The increases from crossbreeding often represent large potential increases in net income from a commercial livestock enterprise; i.e., they can be obtained with little or no additional production costs. Thus, the small increases take on added significance when it is realized that an increase of as little as 3 to 5 percent in total production might well represent an increase of 50 or even 100 percent in net income.

It is also necessary to recognize that increased productivity from heterosis is compounded additively, i.e., improvements in heterosis for livability and growth rate translate into the sum of these in total productivity. Heterosis for maternal performance is additive in relation to individual heterosis, and requires that crossbred females be utilized to gain this contribution to total heterosis (note Tables 10-1, 10-2, and 10-3). Gregory and Cundiff[4] presented results from four generations of data and made projections for increase in calf weight weaned for eight generations for several crossbreeding systems as compared to the average of purebred Angus, Herefords, and Shorthorns. For a three-breed rotation the advantage was 20 percent or more for the first and second and subsequent generations. For a two-breed rotation the advantage was about 15 percent.

Advantages from crossing of breeds of farm animals appear to be of sufficient importance that commercial producers should give serious consideration to the development of systematic crossing programs if near-maximum performance is to be attained. This is particularly true of the species with higher reproductive rates.

---

[4] Gregory, K. E., and L. V. Cundiff. 1980. *J. Anim. Sci.* 51:1224–1242.

Beef Cattle   The Angus, Hereford, and Shorthorn breeds are similar in average mature weight and carcass characteristics. The following is a summary of 16 experiments in which cows of at least two of these breeds produced straightbred and reciprocal crossbred calves:

1 There was an average advantage for crossbred matings of 3.3 percent in calves weaned per cow bred (or exposed for breeding).

2 Crossbred calves were 3.1 percent heavier at birth than the average of straightbreds.

3 There was an average advantage in weaning weight of 5.6 percent for crossbred calves.

4 Due to variation in feeding methods and length of tests, summarization is difficult but crossbreds had an average advantage of about 2 to 4 percent in rate of gain after weaning.

5 Crossbreds apparently had a small advantage in efficiency of gain after weaning. The magnitude of this effect is not well-defined due to the fact that cattle in most experiments in which feed efficiency was reported were fed for time-constant periods. Thus, on average, the crossbreds were heavier throughout the feeding period making measurement of true differences in efficiency very difficult.

6 There are only small differences between crossbreds and straightbreds in either grade of carcass quality or yield of retail trimmed cuts, but with a tendency for crossbreds to be slightly fatter at equivalent ages.

Fertility and maternal qualities of straightbred and crossbred cows of the Angus, Hereford, and Shorthorn breeds have been compared in five experiments. Both straightbred and crossbred cows were bred to produce crossbred calves. Approximate average advantages of 5 percent in calf crop raised and 5 percent in weaning weight of calves have been found for the crossbred cows. A summary of the percentage of heterosis among crosses of European beef breeds is given in Table 10-2.

When compared to most European breeds, the Brahman is more tolerant to high environmental temperatures, is more resistant to some external parasites, and may have a somewhat superior ability to subsist on low-quality forages. As a pure breed it does not grow rapidly nor is it high in reproductive rate. Its carcasses are lower in quality than those of British breeds, and the tenderness of lean is not as highly rated. However, Brahman-British crossbreds show a remarkable level of heterosis for reproductive and growth traits.

Much of the research work with such crosses has been done in the southern part of the United States. It is uncertain how similar crosses would perform elsewhere. Results on Brahman-British crosses from several state and federal stations cooperating in the Southern Regional Beef Cattle Breeding Project show heterosis of approximately 10 percent for preweaning growth rate for both Brahman × British and British × Brahman crosses. There is an additional 4 to 5 percent increase in growth of calves from crossbred cows bred to either type of bull. Carcass data from similar crosses show heterosis for growth rate

TABLE 10-2
SUMMARY OF HETEROSIS AMONG CROSSES OF BRITISH BEEF
BREEDS (ANGUS, HEREFORD, AND SHORTHORN)*

Trait	Percent heterosis†
Individual heterosis	
Calving rate	Slight
Calf survival	2–4
Weaning rate	3–5
Weaning weight	5–7
Rate of gain after weaning	3–6
Feed efficiency after weaning	Small
Carcass traits	Some for fatness
Yearling weight, females	5–7
Mature weight, females	2–4
Age at puberty	2–4
Cow, length of life	Fairly large
Maternal heterosis (crossbred cows with crossbred calves)	
Calving rate	3–5
Calf survival	2–4
Weaning rate	7–9
Weaning weight	8–12
Weight of calf weaned/cow exposed	16–20

*From Long, C. R. 1980. *J. Anim. Sci.* 51:1195–1223.
†Percent advantage over average of parental breeds.

as expressed by carcass weight per day of age and some heterosis for dressing percent. For other carcass traits the crosses tend to be intermediate between the parental types, but somewhat closer to British types.

At least in subtropic situations, Brahman-British crossbred cows have important advantages over straightbreds of either parental type in both percentage of calf crop and weaning weight of calves. Taken together, these resulted in heterosis in calf weight weaned per cow mated for the crossbred cows over the average of the two straightbreds of 47 and 39 percent, respectively, in one study. Thus, crosses of this kind are among the most heterotic known in the animal-breeding world. Objective data are limited on the longevity of Brahman crossbred cows, but observation in many herds indicates that they have real advantages in length of life. In a Texas study Brahman-Hereford crossbred cows were an average of 3 years older when their last calf was weaned, and they had reared three more calves when they left the herd than Herefords. The cross of ¼ Brahman and ¾ Hereford cows exceeded Herefords by 2.3 years and 2.3 calves raised. A summary of the percent of heterosis among crosses of British and Brahman breeds is given in Table 10-3.

Dairy Cattle Several crossbreeding experiments have been completed in the United States. Heterosis for yield has varied considerably in the several

TABLE 10-3

SUMMARY OF HETEROSIS AMONG CROSSES OF BRITISH-BRAHMAN
BEEF BREEDS*

Trait	Percent heterosis†
Individual heterosis	
Calf crop, weaned	1–3
Weaning weight	8–10
Rate of gain after weaning	8–12
Feed efficiency after weaning	Little information
Mature weight, females	10–15
Cow, length of life	Fairly large
Maternal heterosis (crossbred cows with crossbred calves)	
Calf crop, weaned	8–12
Weaning weight	8–12
Weight of calf weaned/cow exposed	8–24

*From Long, C. R. 1980. *J. Anim. Sci.* 51:1195–1223.
†Percent advantage over average of parental breeds.

studies. The U.S. Department of Agriculture study in Beltsville, Maryland, reported no heterosis for Ayrshire-Brown Swiss crosses and 8 to 10 percent heterosis for Ayrshire-Holstein and Brown Swiss-Holstein crosses. In a study in Georgia, 7 to 11 percent heterosis for first lactation milk yields were reported for Holstein-Jersey and Brown Swiss-Jersey F_1's. The Guernsey-Holstein F_1's in Illinois exceeded their parental average by 11.6 percent. In Holstein–Red Dane crosses in Indiana, heterosis was 9.6 percent, and the South Carolina study reported 3.8 to 4.5 percent heterosis for yield traits of Brown Swiss–Holstein crosses. The above results are based primarily on first lactation results. A lower level of heterosis is expressed for later lactations.

Almost without exception, crossbred animals have exhibited higher reproductive fitness than the parent breeds. In the Illinois study, crossbreds completed 14 percent more total lactations than purebreds. Crossbreds also had higher livability, as losses from birth to calving were 17.4 percent for purebreds as compared to 8.2 percent for crossbreds.

Some degree of heterosis has been evident for most traits studied, but superiority over the parental mean may not be a satisfactory practical criterion. Crossbreeding does become more important as increments of heterosis are expressed by several traits, i.e., reproduction, viability, and production. Heterosis of net merit may result even though the cross does not exceed the better parent in any one trait. A summary of the percent heterosis among crosses of dairy breeds is given in Table 10-4.

The average merit of the breeds available for crossing is also a major practical concern. This is undoubtedly one reason why crossbreeding of dairy cattle in the United States is not used more extensively. Milk volume is of major economic significance, and advanced programs of progeny testing in the Hol-

TABLE 10-4
SUMMARY OF HETEROSIS AMONG CROSSES OF DAIRY BREEDS*

Trait	Percent heterosis†
Milk yield, first lactation	4–10
Milk yield, later lactation	Less than in first
Calf livability	5–8
Cow, length of productive life	10–20
Cow, reproductive efficiency	Variable estimates

*From Turton, J. D. 1981. *Anim. Breed. Abs.* 49:293–300.
†Percent advantage over average of parental breeds.

steins have provided many bulls with superior transmitting ability for milk yield. Many times a bull of inferior merit would have to be used if a systematic crossbreeding program were adhered to rigidly.

Much interest and considerable effort is now being given to the possibility of crossing among strains of Holstein-Friesians that have been developed in different countries, notably the United States, the Netherlands, Great Britain, and Germany. Major increases in milk yield have been obtained when sires from the United States have been bred to females from the other countries. Most of the Friesians from Europe used in this country have been used in beef crossing, hence there is little information on the reverse dairy crosses.

Swine   Crossbreeding for commercial production has been practiced more extensively with hogs than with any other class of farm animal. It is estimated that 80 to 90 percent, or even more, of the market hogs in the principal cornbelt states are crossbred. Much experimental crossbreeding work has been done with hogs, and most of it shows economically important advantages for crossbreds. Most workers in the field have recommended the practice.

Although crossing two breeds results in increased production, systems which involve the use of single- and multiple-cross sows are much more practical and widely used. The practice of using crossbred sows was frowned upon in the early days of the purebred era, but a great deal of experimental work has shown conclusively that crossbred sows express heterosis at an economically important degree in their ability to produce large, heavy litters. Using them as mothers permits the producer to take advantage of heterosis in this important character as well as in rate of gain, survival, and efficiency of gain. Furthermore, no one has demonstrated an undesirable decrease in uniformity of type in progeny of crossbred sows.

To take advantage of the heterosis both in the growing pig and in the sow without having to start over periodically with purebred females, workers at the Minnesota Agricultural Experiment Station suggested systems known as *crisscrossing* and *triple,* or *rotational, crossbreeding*. In these systems, heterosis in both the *offspring* themselves *and the dam* is utilized, thus leading to the expectation that total production will be as great as or greater

TABLE 10-5
INDIVIDUAL AND MATERNAL HETEROSIS FOR SEVERAL TRAITS IN SWINE*

Trait	Individual heterosis		Maternal heterosis	
	Actual†	Percent	Actual‡	Percent
Conception rate, %	3.0	3.8		
Litter size, birth	.10	1.0	.40	4.7
Litter size, 21 days	.56	8.0	.64	8.7
Litter size, weaning	.70	10.0	.58	7.7
Pig birth weight, kg	.04	3.1	.02	1.5
Pig 21-day weight, kg	.16	3.1	.20	3.7
Pig weaning weight, kg	.64	4.8	1.20	8.2
Daily gain after weaning, kg	.06	9.4	0	0
Feed efficiency after weaning§	.02	2.3	0	0
Age at 100 kg, days	−12.0	6.5	−2.20	1.2
Carcass length, cm	.04	0	.16	.2
Carcass back fat, cm	.08	2.5	.13	4.4
Carcass *longissimus* area, cm²	.52	1.8	.12	.4

*From Johnson, R. K. 1980. North Central Regional Publication 262.
†Actual difference between two-breed crosses and purebreds and percent of increase of crossbreds over purebreds.
‡Actual difference between three-breed crosses with crossbred females and two-breed crosses with purebred females and percent of increase of three-breed crosses over two-breed crosses.
§Feed efficiency expressed as gain divided by feed consumed.

than in first crosses. Experimental results from these systems have been good. A number of studies have been conducted with results quite similar to those of earlier research.

A summary of twenty different studies of seven different experiment stations participating in the North Central regional swine project, plus results from one Canadian study, give a comprehensive picture of amounts of heterosis to be expected from swine crossbreeding as well as much information on expected performance of several of the more popular breeds in crossbreeding systems.[5] Heterosis estimates are given in Table 10-5. The author summarized the results as follows:

> Heterosis effects are quite large for reproductive and growth traits but relatively unimportant for carcass traits. When absolute estimates of heterosis for conception rate, litter size at weaning, and pig weaning weight are combined, two-breed crosses weaned 15.4 per cent heavier litters at weaning per female in the breeding herd when purebreds and three-breed crosses were 21 per cent above the average of the two-breed crosses. In addition, two-breed and three-breed crosses are expected to be 6.5 to 7.5 per cent younger at 100 kg and to gain 2.3 per cent more efficiently than the average of the purebreds. Carcasses merit for crossbreds, however, is expected to be equal to the average of the purebreds that made up the cross.

[5] Johnson, R. K. 1980. North Central Regional Publication 262.

Duroc sired pigs excelled in growth rate, whereas Chester White sired pigs were well below average. Pigs from Duroc and Spot dams were superior for growth and pigs from Chester White dams were considerably below average. Hampshire and Poland sired barrows were superior for carcass backfat and *longissimus* muscle area. Significant maternal effects existed for efficiency of growth and carcass merit. Crosses involving Yorkshire as the dam breed were more efficient and less fat than reciprocal crosses where Yorkshire was the sire breed.

Much crossbreeding research has been done in Europe with results quite consistent with those in the United States.[6] As well as reviewing experimental results, Jonsson gives considerable information on specific crossbreeding programs used in several European countries.

Sheep Crossbreeding has long had an important place in sheep production, especially that of fat lambs. The systems used in Great Britain are perhaps the most widely known. In England and Scotland considerable stress has been placed upon the use of breeds adapted to particular areas. Similar systems exist in Australia and New Zealand.

A system of this nature long served as the basis for much of the sheep industry of the Pacific Northwest. Lincoln × Rambouillet crosses were made, and the ewe lambs were raised economically to yearling age on rather sparse ranges, usually in Montana or Idaho. These ewes were sold as yearlings to sheepbreeders in better range areas, principally Oregon and Washington, where they were bred to Hampshire or Suffolk rams and all the lambs were marketed. Such breeders bought all their replacement ewes. Their own operations were comparatively expensive, making it cheaper to buy replacements than to produce them.

A number of experiments have been conducted with sheep in which straightbreds of two or more breeds and reciprocal crossbreds were produced. Levels of heterosis observed have been so variable that averaging does not seem to be an appropriate method of summarization. Several studies have shown appreciable levels of heterosis. In one such study[7] conducted by the U.S. Department of Agriculture at Beltsville, Maryland, five straightbred types and all possible crossbred combinations were produced. Lambs were weaned at 2-week intervals as they reached 80 to 90 days of age. Of the 20 crossbred combinations, 15 showed positive heterosis for fertility, 14 for lambs born per ewe lambing, 17 for lambs weaned as a percent of ewes bred (a figure dependent upon both fertility and lamb survival rate), 15 for individual-lamb weight, and 18 for weight of lamb weaned per ewe bred. Clean-fleece weights of 404 yearling ewes representing 7 of the 10 crossbred combinations possible when reciprocals are combined were compared to those of straightbreds. Six of the seven combinations showed positive heterosis.

---

[6] See review by Jonsson (1975) in reference list at end of chapter.
[7] Sidwell, G. M., and L. R. Miller. 1971. *J. Anim. Sci.* 32: 1084–1102.

Crossbreeding for Formation of New Breeds

Early in this century it seems to have been generally accepted by many that no more breeds would be formed. Nonetheless, numerous new breeds have developed since that time—all on crossbred foundations. Sometimes this has been thought to be a new approach, since the extent to which crossbreeding entered into the formation of older breeds often is not fully realized. More than 50 breeds of livestock considered to be established breeds prior to 1920 are described by Briggs.[8] At least 34 of these are said to have been based on crosses, "mingling of blood," or "infusions of blood," of two or more older breeds, strains, or types. The formation of new breeds from crossbred foundations is thus not new, but traditional.

Attempts to form new breeds may be justified if no existing breed or systematic cross satisfactorily meets the needs of a given geographic area or economic situation. Even though selection within an existing breed might eventually lead to the development of a new strain with the desired characteristics, the process often can be expedited by using a crossbred foundation, provided there are existing breeds whose crosses approximate the desired type. Other matters to be considered are the probable productivity of a new breed in relation to the productivity of crosses between the existing breeds and the difficulty of maintaining systematic long-time crossbreeding. Most breeders lack the time, capital, and capability for the experimentation required to develop new breeds. Major obstacles to the development of new breeds are:

1 The time and expense involved.

2 The large number of animals that should be raised to give a broad base of genetic variability for selection.

3 The difficulty that might be encountered in selling the new breed to producers, even if it had real merit.

In spite of the successes some breeders have had in establishing new breeds, many failures have been recorded, and doubtless many more have been forgotten.

Work of the past half century in the formation of new breeds under controlled experiment station conditions has done much to refute ideas formerly prevalent: (1) that the use of crossbred animals for breeding purposes is followed by wide segregation in economically important traits and a general decline in vigor below that of the parental breeds and (2) that it would be impossible to attain the uniformity of an established breed among the progeny of a crossbred foundation within the span of a human lifetime.

Most new breeds founded in the twentieth century have been based upon a cross of only two parental breeds. Parental stocks in a few cases have included three or more breeds. The characteristics of the available parental breeds are a

---

[8] Briggs, H. M. 1969. "Modern Breeds of Livestock," 3d ed., The Macmillan Company, New York.

major determining factor regarding the number of breeds to use as a foundation.

Loss of heterozygosity when crossbred populations are intermated (*inter se* matings) equals $1/n$, where n is the number of breeds or lines if all contribute equally to the population. To the extent that heterosis is directly related to heterozygosity, a breed derived from a two-breed cross would be expected to retain a maximum of ½ of the heterosis of the foundation cross. A new breed based upon a four-breed cross would be expected to retain a maximum of ¾ of the original heterosis—a figure which compares favorably with heterosis retained in systematic rotation or crisscrossing schemes. The loss of heterosis could be greater in both cases if epistatic effects were important. The key consideration, however, is that multibreed foundations would permit retention of more heterosis than a two-breed foundation.

An important factor in the formation of a new breed to be used commercially as a straightbred is the avoidance of inbreeding. To accomplish this, rather large populations of both foundation crossbreds and of the early generations of *inter se* mating are essential. No hard and fast rules on population sizes can be given, but suggestions of 300 or more females and 10 or more males per generation seem desirable. Further, the base should be kept broad by maintaining lines of descent to several males in each generation, even though in some cases maximum selection differentials for desired traits might be attained by retaining the sons of fewer males.

OUTCROSSING

Outcrossing is the mating of unrelated animals within the same pure breed. It is the most common breeding system in use by American seed-stock breeders. For practical purposes matings are considered as outcrosses if mates have no ancestors in common in the first three to four generations of their pedigrees. The common practice of continually selecting the best-available, unrelated sires for use in a herd is an outcrossing system. It has been responsible for a high percentage of the changes in livestock since breeds evolved.

FIGURE 10-4
Sixty-day-old mice illustrating differences in size produced under selection and outbreeding. *Left,* a representative of Goodales Large White. *Center,* an albino bred without selection for either large or small size. *Right,* a representative of MacAuthur's small line. *(Courtesy of Dr. W. H. Kyle, formerly of the Animal Husbandry Research Division, U.S. Department of Agriculture.)*

The pool of genes in a breed has, in general, been large enough to permit outbreeding systems to bring about marked changes in type and other characteristics. Indeed, selection to change the frequency of the various genes and inbreeding to increase homozygosity are the only two breeding tools available to the breeders of most of our modern breeds, since many breed associations prohibit the introduction of outside genes into a breed.

Genetic Aspects of Outcrossing

For traits that are influenced largely by genes with additive effects and with a high heritability, a system of selection and outcrossing would be recommended for most seed-stock herds. High heritability indicates a high correlation between genotype and phenotype. Individual or phenotypic selection is therefore reasonably accurate in locating those animals with larger-than-average numbers of desired genes. The outcross mating of selected animals results in relatively few undesired genes being fixed in homozygous form. A breeding system of this kind brings about immediate improvement and at the same time does not shut the door on future improvement as an intense inbreeding program might do through the fixation of undesired or deleterious genes.

Growth rate in beef cattle is an example of a highly heritable trait for which a breeding system of this kind should be most effective. In fact, the system should be effective for traits with much lower heritabilities, there being no definite lower limit at which we could abruptly say another system was preferable.

For the present, it appears that outcrossing systems should be used in most seed-stock herds. Selection should be based upon selection index principles in which emphasis is on economically important traits with reasonably high or moderate heritabilities. Less emphasis should be placed on traits of very low heritability. For owners of commercial herds that for one reason or another prefer to use only one breed rather than a crossbreeding system, outcrossing rather than inbreeding or linebreeding should always be used. Such herds cannot afford even small reductions in vigor and performance which could result from inbreeding.

Outcrossing is a useful procedure when a drastic change in the type of either seed-stock or commercial herds is desired. The necessity for doing this might arise if market demands change or if previous selection standards in a herd were unrealistic. When sires of the desired type are available in the breed, they can be used to change the type rather quickly. Sires more extreme in the desired direction might be used in the expectation that the "corrective matings" will result in intermediate offspring of approximately the desired type.

GRADING

Grading is the practice of mating purebred sires of a given breed to nondescript, scrub, or native females and their female offspring generation after generation.

First-generation offspring carry 50 percent of the inheritance of the pure breed and, depending upon the quality of the original females, can usually be expected to be a considerable improvement over their dams. The next generation results in offspring with 75 percent of the hereditary material of the pure breed. In subsequent generations the proportion of inheritance remaining from the original scrub females is halved with each cross. After five or six crosses of purebred sires, the grade animals carry 96.9 and 98.3 percent, respectively, of the hereditary material of the pure breed. From a genetic standpoint they are essentially the same as purebreds.

Grading is thus a process by which a few purebred sires can rather quickly transform a nondescript population into a group resembling the pure breed. The quality and productivity of the resulting grades depends almost entirely upon the genetic quality of the purebred sires used—the grading process is not a creative one. Grading can be used to convert a population of one (or several) breed(s) into a new population which genetically will be predominantly the breed of the sire. This use of grading usually occurs when a breed is introduced into an area or a country in small numbers and it is desired that numbers be expanded rapidly. Often in such situations, animals with $\frac{7}{8}$ (three top crosses) to $\frac{31}{32}$ (five top crosses) of the hereditary material of the new breed are considered purebred. This approach was used extensively in the late 1960s and the 1970s to increase populations of several newly imported "exotic" cattle breeds in the United States.

Limitations on Usefulness of Grading

A word of caution should be introduced against the use of purebreds in grading programs in areas where the particular purebreds used are not adapted. Purebred dairy and beef cattle from temperate zones often are unable to perform satisfactorily in tropical or subtropical areas. In most such cases, the first crosses, still having 50 percent native genes plus the genes for high production from the introduced purebreds, will be a distinct improvement over their native dams. The performance of second and later crosses having 75 percent or more of purebred ancestry may deteriorate due to disease and reproductive difficulties. An example of this is shown in Table 10-6. These data, accumulated under variable conditions in several parts of Brazil and India, indicate that animals with ½ to ¾ imported background produced more milk than those with more imported hereditary material, even though the latter animals presumably had more genes for high milk production. Lack of adaptability and unfavorable managemental conditions apparently placed a limit on the production of animals with high proportions of imported ancestry.

For grading to be successful, it is not sufficient to use purebreds indiscriminately. The purebreds used must have evidenced the ability to perform well under the conditions which their offspring will have to meet.

TABLE 10-6
AVERAGE PRODUCTION OF MILK FOR CATTLE IN BRAZIL AND INDIA
WITH VARIOUS PROPORTIONS OF IMPORTED ANCESTRY, MOSTLY
HOLSTEIN*

Fraction improved breed	Brazil, kg	India, kg
0	1,582	1,786
1/8	1,852	2,208
2/8	1,992	2,246
3/8	2,238	2,242
4/8	2,527	2,422
5/8	2,567	2,376
6/8	2,435	2,430
7/8	2,336	2,341
8/8	2,332	—

*From McDowell, R.E. 1985. *J. Diary. Sci.* 68:2418–2435.

Registration of High Grades

Currently, in the United States a few breeds (mostly new or numerically small) have provisions for such registration. From a genetic standpoint, the registration of high grades would have the advantages of

1 Increasing numbers in the breed and thus broadening the genetic base for future selection
2 Increasing the frequency of desirable genes for quantitative traits if the grades registered were higher in merit than the average of the pure breed
3 Permitting the introduction of genes not already present in the breed

Increasing numbers is especially important to breeds which are numerically small. A larger population would permit more intensive selection of sires and wider use of outstanding sires without an increase of inbreeding levels at unduly rapid rates.

The introduction of specific genes into a breed through the registration of high grades can be immediately useful primarily for traits under the control of only one or a few gene pairs. Introduction of a gene for polled into a horned breed would be an example. A danger in registering high-grade animals is the possibility of inadvertently introducing an undesirable recessive gene into a breed not already having it. The possibilities of this occurring could be reduced to a negligible level by appropriate safeguards. From a genetic standpoint, provision for registration of high grades seems to be desirable.

Registration of high grades could be either desirable or undesirable for the business interests of persons already owning and breeding animals of a given breed. In the case of a breed with limited numbers, expansion of these numbers more rapidly than would otherwise be possible could facilitate its becoming of regional or national importance. If the breed were truly valuable, this would likely result in building a demand from commercial producers which would benefit all breeders. For a breed already present in large numbers, creating additional purebreds could oversupply the market and result in lower average prices; it could also be detrimental to the financial interests of existing individuals.

In Europe, livestock breeds have quite often, but not in all cases, maintained provision for registration of animals with four or five top crosses of registered sires. In many European breeds, the registration of any animal, regardless of whether it descends from registered ancestors in all lines of its pedigree or whether it is the result of a grading process, is dependent upon adherence to performance and breed standards as determined by a physical inspection. In some countries, maintenance of herdbooks is a government function, and the inspectors are government officials.

Eligibility for registration in the United States has typically been based on the *closed herdbook* principle. It is required that descent in all lines of the pedigree be from registered animals. This has been true both for European breeds imported to the United States and for breeds developed in this country during the nineteenth century. Except for exclusion from registry of animals with specified defects or deviations from breed color or horn standards as established by each association, any animal with registered parents is eligible for registration regardless of conformation or producing ability. In the United States, establishment of standards is entirely a function of the individual associations. There is no governmental regulation.

INBRED LINE CROSSES FOR COMMERCIAL PRODUCTION

Commercial hybrid corn is produced by crossing highly inbred lines. Where adapted hybrids are available, they have usually considerably outyielded the best varieties produced by older breeding methods. This superiority is usually attributed to heterosis. Much of it is due also to successful choices of inbred lines which bring together favorable gene combinations in the hybrids. These favorable combinations occur uniformly in each hybrid plant owing to the homozygosity of the parent lines. The higher yields of hybrids are due largely to uniform production from nearly all plants rather than to extraordinary yields by some plants.

As we have seen, most breeds of farm animals are slightly inbred, they differ somewhat from one another genetically, and crosses between them usually show a small-to-moderate amount of hybrid vigor. This leads to the supposition that crosses between highly inbred lines might exceed the performance of presently existing types.

Before such an hypothesis could be tested it was necessary to develop inbred lines. This is a much more difficult task with farm animals than with corn or other crops for the following reasons:

1 Self-fertilization is impossible with animals; thus more than twice as many generations are required to develop the same degree of inbreeding in animals as in plants.

2 Higher costs of maintaining and raising each individual (or genetic unit) make the cost of developing inbred lines high.

3 Due to the expense of formation, development, and maintenance, fewer lines can be developed, thus lessening the possible choice among lines for crossing.

In spite of these difficulties, inbred lines of all species of farm animals have been developed, usually to inbreeding levels of not over 30 to 60 percent, although a few lines have been carried to higher levels, particularly in poultry.

1 Productivity is reduced as inbreeding levels increase with some lines being affected more than others. Traits of low heritability are usually affected most drastically. Some lines have been lost due to infertility or high death rates, but it has been possible to maintain many of those started to inbreeding levels of at least 40 to 50 percent.

2 Line crosses have on the average restored the productivity lost with inbreeding, but the average performance of all line crosses is apparently not above that of noninbred stocks.

3 Crosses of specific lines have sometimes performed at levels apparently above those of outbred animals. This suggests the potential usefulness of the technique. However, there is still doubt as to whether results from the selected crosses are better than could have been attained with the same expenditure of time, effort, and money by selection within outbred stocks and the crossing of noninbred strains within breeds or by breed crosses.

SELECTION FOR COMBINING ABILITY IN CROSSES

From the foregoing, the discerning student will have gathered that, at the present state of knowledge, performance of two or more breeds or lines in crosses is somewhat unpredictable. This is true to a degree for all traits but much more so for those of low heritability. Some lines or breeds appear to "combine well," whereas others do not. This can in most cases be determined only by test crosses. Thus, a breeder attempting to produce lines which will combine well with each other presently has few alternatives but to form a large number of lines. The breeder can then test the lines in crosses and find those which give the best combinations. This is expensive, time-consuming, and unreliable.

For highly heritable traits, the straightbred performance of breeds and lines is a fairly good indicator of performance of cross progeny. For the less highly heritable traits, which usually show the most heterosis, no predictors studied to date have been very reliable in estimating cross-progeny performance. As a general rule, lines or breeds most diverse in genetic origin (i.e., the most generations away from common ancestors) express more heterosis in crosses.

Thus far we note that heterosis is useful and that advantage should be taken of it when crosses are found which result in superior performance. By implication it could have been assumed that nothing was known about how to improve cross performance. From a practical standpoint this currently is nearly true with farm animals. However, a brief consideration of research aimed at improving heterosis or cross-progeny performance may be of interest. Obviously, critical knowledge on the subject must await a better understanding of the nature of heterosis.

Breeding Approaches for Combining Ability

On the assumption that overdominance is of importance in heterosis, a breeding system of *recurrent selection* was proposed.[9] In this system, a highly inbred line, presumably homozygous at most loci, is selected as a tester. A large number of individuals are crossed with this line and their progeny evaluated. Those giving the best progeny are subsequently intermated and a large number of their progeny again tested in crosses with the inbred tester. The cycle is repeated over and over.

If heterosis is largely dependent upon overdominance, this procedure should result in the line selected on cross performance becoming homozygous for different alleles than the inbred used as the tester. In other words, where the tester is *aa,* the selected line would be *AA;* where the tester is *BB,* the selected line would be *bb,* etc.

A system termed *reciprocal recurrent selection* also can be followed.[10] In this system, randomly selected representatives of each of two noninbred strains are progeny-tested in crosses with the other. Those individuals of each strain having the best cross progeny are then intermated to propagate their respective strains. Offspring from these within-strain matings are again progeny-tested in crosses with the other and the cycle repeated. Both recurrent selection to a noninbred tester and reciprocal recurrent selection should lead to improved cross performance, whether it is the result of overdominance, epistasis, or only additive effects.

Both these systems involve progeny testing. Due to the increased generation intervals, this would be expected to result in slower progress than other breeding systems for traits in which genetic variation is predominantly additive and which are moderate to high in heritability. They would be expected to be more useful than other breeding systems only if overdominance or other nonadditive types of inter- or intra-allelic gene action are important in heterosis.

Experimental tests of these techniques with laboratory organisms have shown promise for improving performance of crossbred progeny for some highly heterotic traits that are low in heritability.[11] One such trait is egg-laying capacity in *Drosophila.* Several experiments have shown no advantage of either recurrent or recurrent reciprocal selection over usual selection methods

[9] Hull, Fred H. 1945. *J. Am. Soc. Agron.* 37:134–145.
[10] Comstock, R. E., et. al. 1949. *Agron. J.* 41:360–367.
[11] See McNew, R.W., and A.E. Bell. 1976. *J. Hered.* 67:275–283.

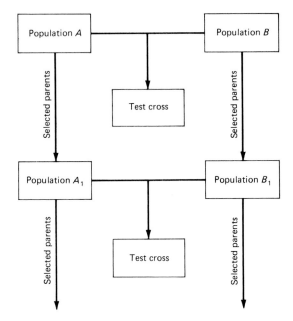

FIGURE 10-5
A simple representation of reciprocal recurrent selection. Males and females of population A are mated to individuals of population B. Parents of the best crossbred progeny are chosen to produce the next generations of populations A and B. Note that parents are selected on the basis of the performance of crossbred progeny, not their individual merit. Actual design of the breeding scheme would be varied according to reproduction rates of species.

for traits of higher heritability even though some of them also exhibit considerable heterosis. Even for the highly heterotic traits with low heritability, some workers have reported that crosses of some highly inbred lines, developed without selection from the same foundations, have equaled or surpassed the productivity of strains developed after many generations of recurrent or reciprocal recurrent selection.

Results of several experiments involving recurrent or reciprocal recurrent selection with swine, sheep, and cattle have been reported.[12] Most of these studies have shown selection based upon cross-progeny performance to be effective to some degree for at least some traits. Not all experiments have included sufficient numbers for critical determinations of comparative effectiveness of selection on cross-progeny performance versus within-breed strain selection based on individual performance. In some cases experimental designs did not permit critical comparisons. It appears, however, that for most traits studied, selection based on individual performance has been as effective or more effective than that based on cross-progeny performance. Litter size in swine and early growth are traits for which cross-progeny selection may be more effective.

[12] Dickerson, G.E., et al. 1974. *J. Anim. Sci.* 39:24–41. Flower, A.E., et al. 1964. *J. Anim. Sci.* 23:189–195. Hupp, H.D., et al. Pp. 23–24, in "Abstracts 69th Annual Meeting," American Society of Animal Science, July 23–27, 1977. Rempel, W.E. 1974. Pp. 849–858, in vol. 1, "Proceedings, First World Congress on Genetics Applied to Livestock."

In some cases[13] sire progenies have ranked in essentially the same orders for economically important traits in progeny tests on females of another strain or breed as compared to progeny tests on females of their own strain or breed. In other cases, changes in rank have been observed. If differences in rank are known to exist among sire progenies depending upon type of females to which bred, potential usefulness of selection on the basis of cross-progeny performance would be suggested if the breed or strain is going to be used commercially in crosses.

However, looking at all the evidence, it must presently be considered as uncertain whether the techniques of recurrent selection have a potential usefulness in farm animals. If they do, it will likely be in breeds or strains in which performance is already high for highly hereditary traits and in which it is desired to improve the potential performance of their cross progeny for the lowly heritable traits related to fertility and livability.

Specialized Sire and Dam Lines

Statistical studies[14] have shown potentially more rapid genetic improvement for breeding systems based on formation and selection of lines or strains for use in specific crosses as compared to selection of each strain for all-around merit. Strains or breeds, for example, with high fertility and other maternal qualities serve as the female parents, and those with high gaining ability, feed efficiency, and desired carcass quality as the male parent. Advantages of specialized sire and dam lines will be greatest in cases of strong negative genetic correlations between traits.

Historically, the system of using strains selected for all-around merit (even in crossing) has been most prevalent in the United States. However, with the increasing availability of diverse germ plasm, more emphasis in commercial production is being focused on crossbreeding systems requiring specialized parental types. To date, emphasis has been based largely upon taking advantage of genetic differences already existing in available breeds.

Increasingly, commercial-meat animal production is based upon crossbreeding. The function of seed-stock breeders is to develop germ plasm for ultimate use in commercial production. They face the question of whether breeds or strains to be used in crossbreeding programs should be intentionally selected for specific qualities for use in two or more breed crossbreeding programs designed to complement each other or whether all should be selected for all-around merit.

The broiler industry is the outstanding example of use of specialized sire and dam lines. Female parent stocks must produce eggs at reasonably high levels in order to keep chick production costs low. Male parent stocks are se-

[13]Koger, M., et al. 1975. *J. Anim. Sci.* 40:230–234.
[14] Smith, C. 1964. *Anim. Prod.* 6:337–344. Moav, R. and W. G. Hill. 1966. *Anim. Prod.* 8:375–390.

lected for their ability to contribute growth and desired carcass characters to the commercial broiler.

The feasibility of specialized sire and dam lines in beef cattle, sheep, and swine must be determined upon the basis of both genetic and economic factors. These basically relate to the costs of maintaining breeding stocks in relation to their potential genetic advantages.

SPECIES HYBRIDIZATION

The widest currently possible kind of outbreeding is the cross of two species. A good example of this is the cross between the jack of the *Equus asinus* species and the mare of the *Equus caballus* species to produce the mule. It seems quite probable that these two species may have descended from common parent stocks far back in the course of evolution. In the dim past, natural selection, working on variation (recombinations, mutations, and chromosomal aberrations), began to set them off into distinct species. They still have many genes in common, but they also have enough different ones to be recognized as distinct species. Up to now they have enough similarity so that the sperm of one will fertilize the ova of the other and produce vigorous offspring. The mule has been valued for centuries for its ability to work under difficult conditions and to withstand abuse.

Genetic differences between the horse and the ass are apparently so great that the male mules are always sterile and the females usually so. The separation of the chromosomes at the time of germ-cell formation is usually abnormal, and gametes are either not formed or are incapable of fertilization. At least four fertile mare mules have been reported, however. In 1920, L. T. Branham of Montalla, Texas, reported that a mare mule in his possession had dropped a living foal to the service of a jack. The contention was supported by affidavits of the owner and some neighbors. The mare was loaned to Texas A&M College and, after failing to conceive when bred to a jack in 1921, dropped a living stallion foal in 1923 as the result of being bred to a Saddle stallion the previous year. Other fertile mare mules have been reported in Indiana, Arizona, and Nebraska.

In all cases, these apparently fertile mare mules have bred as if their fertile ova contained only horse chromosomes. Thus their breeding behavior has been the same as that of mares. They produced foals that resembled mules when bred to jacks and foals that were indistinguishable from horses when bred to stallions.

Chromosomal examination of one allegedly fertile mare mule showed them to be like those of an ass, thus raising the question of whether she was really a mule. A critical review of other reported fertile mare mules has raised questions as to whether any cases have been irrefutably proved.[15]

[15]Eldridge, F., and Suzuki, Y. 1976. *J. Hered.* 67:353–360; and Eldridge, F., and W. F. Blazak, 1976. *J. Hered.* 67:361–367.

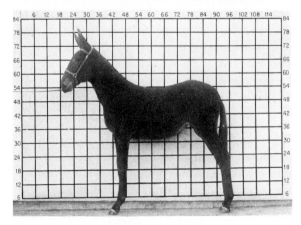

FIGURE 10-6
Top, U. T. Logan Again, a
registered American Jack.
Center, a draft mare. *Below,* a
mule of the type expected from
mating animals such as those
above. *(Courtesy of the late
Professor H. R. Duncan,
Tennessee Agricultural
Experiment Station.)*

TABLE 10-7
SUMMARY OF SOME SPECIES CROSSES IN DOMESTIC ANIMALS

Species crossed	Fertility of hybrid progeny		Remarks
	Males	**Females**	
Ass (*Equus asinus*) ♂ Horse (*Equus caballus*) ♀	Sterile	Sterile	Reciprocal cross known as "hinny." More resemblance to horse
Banteng (*Bos javanicus*) ♂ Cattle (*Bos indicus*) ♀	Sterile	Fertile	*Bos javanicus* X *Bos taurus* offspring also produced
Cattle (*Bos taurus*) ♂ Yak (*Bos grunniens*) ♀	Sterile	Fertile	Often known as "Pien Niu"; excel in hardiness, strength, and milking ability in Himalayas
Cattle (*Bos taurus*) ♂ Gaur (*Bos gaurus*) ♀	??	Fertile	Used as draft and pack animals. Known as "Mathan" or "Gagal"; males occasionally fertile
Horse (*Equus caballus*) ♂ Zebra (*Equus zebra*) ♀	Sterile	Sterile	Docile and resistant to disease and effects of heat
Cattle (*Bos taurus*) ♂ Bison (*Bos bison*) ♀	Sterile	Fertile	Some males of the reciprocal cross are fertile

Several species hybrids which have been produced are summarized in Table 10-7.

The *cattalo* is a term applied to various combinations of domestic cattle and the American buffalo (*Bos taurus* × *Bos bison*). Crosses have been made between domestic cattle and the American bison by many individuals and experiment stations. Male calves from the cross of bison bulls on domestic cows usually die at birth and if they survive are sterile. Females live and are fertile; they produce offspring when bred to domestic bulls. Produce of this and other cross combinations includes some males with a degree of fertility so that it is possible to maintain closed populations descending from the hybrids. The term *cattalo*, as well as several other names, has been used for cattle-bison crossbreds of various percentages of bison and for populations with some fraction of bison inheritance. A considerable amount of exploratory and experimental work has been done, much of it in Canada, on the potential usefulness of these animals for meat production. Initial efforts related principally to cold

FIGURE 10-7
Example of an American Bison–cattle hybrid. This female is ½ American Bison, ¼ Angus, and ¼ Holstein. *(Courtesy of Pierce Rosander, Belle Fourche, South Dakota.)*

climates, but more recently there has been renewed interest for other areas. Most research results have been discouraging, but some commercial use is being made of strains reported to include ⅛ to ⅜ bison ancestry.

The humped cattle of India (*Bos indicus*) are classed as a different species from European-type cattle (*Bos taurus*). Crosses between these two species are commonly used in the southern United States and other subtropical and tropical areas of the world. Their hybrids are fertile in both sexes. There is some doubt that the two groups should be called different species, and their hybrids are usually called crossbreds.

The sheep and goat are related groups, but apparently not so closely related as the above groups. When crosses are made between them, fertilization sometimes takes place and the embryos develop for a time. They die, however, before normal parturition and are resorbed or aborted. When crosses are made between other less closely related species, fertilization apparently does not even occur.

SUMMARY

Outbreeding is a general term applied to any breeding system in which animals mated are less closely related than the average of the population from which they come. Outcrossing combined with selection is a highly useful technique for within-breed improvement for moderately to highly heritable traits. Heterosis is the difference in performance from the average of parental types

...ch is often observed in crosses between breeds, inbred lines, or species. Its genetic and physiological bases are not clearly understood. Among within-species crosses, heterosis is most apparent for the lowly heritable traits related to fertility and viability. Even though it is not well understood, heterosis can be used to advantage in commercial production through crossing breeds and crossing perhaps inbred lines within breeds. Breeding techniques designed to increase heterosis or to improve combining ability of lines or breed combinations may have increasing applicability in domestic animals. Species crosses, of which the mating of the ass and mare to produce the mule is the best known, result in much heterosis when species are related closely enough to be cross-fertile. Fertility is very low or nil among the progeny of most species crosses.

SUGGESTIONS FOR FURTHER READING

Books

Gowen, J. W. (ed.). 1952. "Heterosis." The Iowa State University Press, Ames, Iowa.

Hayes, H. K. 1963. "A Professor's Story of Hybrid Corn." Burgess Publishing Co., Minneapolis, Minnesota.

Lerner, I. M. 1954. "Genetic Homeostasis." John Wiley & Sons, Inc., New York.

Lerner, I. M., and H. P. Donald. 1966. "Modern Developments in Animal Breeding," Academic Press, Inc., London.

Articles

Bowman, J. C. 1959. Selection for Heterosis. *Anim. Breed. Abs.* 27:261–272.

Cockerham, C. C. 1961. Implications of Genetic Variances in a Hybrid Breeding Program. *Crop Sci.* 1:47–52.

Dickerson, G. E. 1977. Crossbreeding Evaluation of Finnsheep and Some U.S. Breeds for Market Lamb Production. North Central Regional Publication 246, Nebr. Agr. Exp. Station.

East, E. M. 1936. Heterosis. *Genetics* 21:375–397.

Johnson, R. K. 1981. Crossbreeding in Swine: Experimental Results. *J. Anim. Sci.* 52:906–923.

Jonsson, P. 1975. Crossbreeding in Pigs in European Countries. *World Rev. Anim. Prod.* 11:16–49.

Long, C. R. 1980. Crossbreeding for Beef Production. Experimental Results. *J. Anim. Sci.* 51:1195–1223.

Sheridan, A. K. 1981. Crossbreeding and Heterosis. *Anim. Breed. Abs.* 49:131–144.

Terrill, C. E. 1974. Review and Application of Research on Crossbreeding of Sheep in North America. "Proceedings, First World Congress on Genetics Applied to Livestock," pp. 765–777.

Winters, L. M., O. M. Kiser, D. S. Jordan, and W. H. Peters. 1935. A Six Year Study of Crossbreeding Swine. Minn. Agric. Exp. Sta. Bull. 320.

CHAPTER 11

PERFORMANCE AND PROGENY TESTING PROGRAMS

The principles of selection, including the use of individual performance, and aids to selection were presented in previous chapters. Many specialized procedures for measuring individual performance and for family, sib, and progeny testing have been developed for individual species to enhance accuracy and economy. The nature of the traits which are economically important, plus the environmental and managemental circumstances in which the class of animals is to perform commercially, are of key importance in performance and progeny testing. Sex-limited traits (milk yield, maternal performance, egg production), traits requiring sacrifice of the animal (carcass evaluation), and traits expressed late in life require special approaches to evaluate breeding values.

The reproductive rate and particularly the number of offspring which a superior male or female may leave influences the investment that may be justified in the performance or progeny testing process. As many as 50,000 offspring[1] can be obtained annually from a progeny-tested dairy bull using artificial insemination; hence, considerable investment to obtain an accurate assessment of the transmitting ability of such an individual can be justified. In contrast, where a male leaves only 20 to 30 offspring per year, a much smaller investment in a performance or progeny testing effort could be justified economically and genetically. Progeny testing of females is much more restricted, even with superovulation and embryo transfer. The number of progeny obtainable is limited, but individual performance data are usually also available as a basis for selection.

---

[1] Over 110,000 inseminations were made during a 5-month spring breeding season to a Friesian bull in New Zealand (*Hoards Dairyman* 121:837, 1976).

The performance of an animal is the resultant of the heredity of the individual and the cumulative impact of environmental circumstances from fertilization until the time of measurement or observation. A major concern in evaluating performance has been to ensure that the phenotypic expressions of the animals or offspring of animals being compared are secured under comparable conditions. When known environmental effects can be identified and their influence assessed, statistical adjustment using conversion or standardization factors may be useful and desirable. Such is the case for age at calving or frequency of milking for lactation milk yields or for age of dam and season of calving for 205-day weights of beef calves.

Performance testing programs should include a sufficient portion of the population to provide a broad base from which breeding stock can be selected. The probability of identifying truly superior individuals increases with the number of animals tested, assuming a reasonably accurate performance assessment can be made. Ideally, every animal in breeding units should be tested to provide a solid basis for both production culling and the choice of breeding stock. The expense and difficulty encountered in obtaining certain data may dictate that information be collected from a reliable appropriate sample of the progeny and/or herd.

Performance and progeny testing programs must balance considerations of the accuracy of the test, the scope of the testing, and the probable increase in generation length. Emphasis on extreme accuracy may utilize so large a portion of the population for each progeny test that the number of animals progeny tested would be severely restricted. The correlation between the breeding value and the measure of progeny test performance may not be increased sufficiently to offset the reduction in selection intensity. More individuals tested with slightly less accuracy may provide a more rapid genetic advance. Time is also a concern. Where performance can be measured on the individual, two cycles of individual selection may be possible during the time required for a progeny test.

Testing programs require investment of funds also. Rationale for evaluating returns to investment in a breeding program has been developed by a number of researchers. Dickerson[2] provided a framework for evaluating economic efficiency of the individual breeding enterprise, and Brascamp[3] has provided a model for evaluating the returns to investment in a progeny test program.

SPECIAL VERSUS FIELD EVALUATIONS

In view of the evident influence of environmental and managemental effects on performance, testing of individuals under standardized conditions at special test stations has been advocated. Danish breeders were among the first to record herd data on performance and use the data for progeny testing. Prog-

[2] Dickerson, G. E. 1970. *J. Anim. Sci.,* 30:849–859.
[3] Brascamp, E. W. 1973. *Z. Tierz. Zuchtungsbiol.* 90:1–15.

eny testing of bulls for milk yield began in 1902 and progeny testing of pigs began in 1907. With the small breeding units and the disruption of World War I, difficulties were experienced in obtaining reliable progeny tests. In the 1930s Danish progeny testing stations were established for swine, and in 1945 dairy bull testing stations began operating. It was hoped that special control of conditions at the testing stations would increase the accuracy of the results by measurably reducing environmental variation.

In the bull testing stations, 18 to 20 daughters of a bull were transferred to the stations in the fall before first calving. Approximately 30 stations were established with from two to nine progeny groups at each location tested simultaneously. Analyses of results showed that environmental differences among years and stations were not eliminated, nor were the environmental differences among progeny groups at the same station. In general, despite rigorous efforts, special test procedures reduced environmental variation much less than was anticipated. Furthermore, an additional question arose as to how accurately the station performance test could predict the on-farm performance of future offspring.

Under some circumstances the test conditions may be rather different from the farm conditions, and the possibility of a genotype-environmental interaction arises. In beef and dairy cattle these have not appeared to be large for the temperate areas. Studies in the United States and Great Britain have shown that sire-by-herd or sire-by-region interactions are not large. Nonetheless, they could be a source of inaccuracy in testing, which could be minimized by testing the progeny in several herds or environments.

In Danish dairy sire tests, the heifers were stabled together in the barn as a progeny group and milked and cared for by the same herder. This introduced environmental similarities among progeny groups which made test results less accurate in predicting future daughters' performance than did progeny information obtained under field conditions.[4] Accuracy of test results in these situations was restricted by the small number of progeny of a sire that could be tested in a station, whereas a much larger sample of progeny could be assessed economically in field tests. Another limitation of central testing stations is that the expense is so high that sufficient males can rarely be progeny tested to optimize genetic improvement for a breed or subpopulation.

An advantage to test stations is that measurements can be made that cannot be obtained in the field. Among the important ones is feed intake to assess feed efficiency. In many situations beef cattle and swine central testing stations provide information that complements what can be obtained from on-farm testing.

Where artificial insemination is not widespread, central testing stations serve to compare and test animals from elite or nucleus breeding herds. In practice the number of animals which can be tested at a station from any one

---

[4] Touchberry, R. W., et al. 1960. *J. Dairy Sci.* 43:529–545.

herd is limited. The largest contribution to breed improvement would be expected from testing males to be used in elite or nucleus breeding herds. Sufficient males cannot be tested to meet the needs of multiplier or commercial herds. With a hierarchical structure, the genetic improvement in the elite breeding stock moves to multiplier herds and from these to commercial producers. Consequently, the central testing of males should be combined with on-farm testing for female selection.

Central test stations can be important educational tools for setting goals or standards of performance, for demonstrating genetic differences, and for providing a record of annual trends in performance. Some limited insight into the genetic differences between herds can be gained, but the small sample of animals tested from each herd restricts the precision of the estimates of these genetic differences.

DAIRY CATTLE

Several circumstances favor the progeny testing of dairy sires. Milk production is a sex-limited trait, and the breeding value of males must be predicted from the performance of close relatives and progeny. The rapid acceptance of artificial insemination and the advancement of techniques for the freezing and storing of bovine semen have greatly extended the use of outstanding progeny-tested sires, once they have been identified.

Planned progeny testing programs, in conjunction with artificial insemination and production testing, offer exceptional opportunities to improve the accuracy of dairy sire evaluation. Use of a sire in many herds under varying managemental conditions provides an "acid test" for a bull's transmitting ability. Some of the advantages of such a program are that

1 A progeny test can be obtained at an early age which can reliably indicate how future sired daughters will perform.

2 The risks of sire proving are spread among many herds, and for the industry as a whole the number of daughters from poorer sires is reduced.

3 Since the daughters of the bull are scattered in several herds, there is less likelihood that herd trends, special environmental circumstances, or genotype-environmental interactions will unduly influence the proof.

Some of the major limitations to such a program are that

1 The cost of holding young bulls awaiting progeny test results may be high.

2 A sound testing program with education and cooperation of the cooperating breeders is essential to its success.

The cost and amount of time involved in any progeny testing program will be large. However, the use of outstanding bulls proved in artificial insemination offers the best opportunity for future improvement, even though no shortcuts are now available to obtain such bulls.

FIGURE 11-1
Interior of a bull barn for progeny-tested sires for a large breeding organization. Facilities are available in several units to handle over 600 bulls in various stages of progeny testing at this location. (*Courtesy of American Breeders Service.*)

Progeny Test Programs

Major efforts in the cooperative progeny testing of dairy sires now involve mating the best cows to the best bulls proven in artificial insemination (AI) to obtain young bulls for progeny testing. Such pedigree-selected young bulls are then sampled at approximately 1 year of age to obtain sufficient daughters for a progeny test.

In the organization of a cooperative progeny testing program, several important questions must be considered to utilize the resources effectively:

1 How many production-tested cows will be available in the breeding unit or population?

2 How many pedigree-selected bulls should be sampled each year?

3 How many tested daughters per bull should be sought for the initial progeny test?

4 What proportion of the females should be bred to pedigree-selected bulls?

The size of the breeding unit is a primary factor in placing limits on the expected rate of genetic improvement. Expected genetic improvement increases, although at a diminishing rate, as the number of animals and herds in the breeding unit increases. When selection is for milk yield only, Skjervold and Langholz[5] have shown that the potential annual genetic improvement in a unit of 20,000 tested females is 50 percent more than for a unit of 2,000 cows.

Determining the number of pedigree selected bulls to sample and the number of tested progeny to be sought for each bull must be done simultaneously. The semen requirements of the AI stud and the number of proven bulls needed to avoid inbreeding must be considered in a closed breeding system. A decision also must be reached as to the proportion of the population which is to be bred to young bulls and the proportion to progeny-tested sires.

This can be seen more clearly from a parallel expression of the equation for genetic change (ΔG) given in Chapter 7:

$$I_{PS} = r_{GO} \, i \, \sigma_g = \frac{h}{2} \sqrt{\frac{n}{1 + (n - 1)t}} \, i \sigma \, g$$

Here I_{PS} is the genetic superiority of the progeny-tested or proven sire derived from the progeny testing program. The value of r_{GO}—the correlation between the breeding value of the individual being progeny-tested G and the offspring average O—increases as the number of progeny in the test increases. However, when the size of the population is fixed, as n increases i decreases. A balance between n and i provides the highest value for I_{PS} for a *specified* size of breeding unit and resources.

In an idealized situation such a balancing of resources can be computed. Skjervold and Langholz[6] have shown that up to 50 percent of the tested females in large populations could be inseminated to young bulls. In large populations this would require the sampling of an extremely large number of pedigree-selected bulls each year to utilize these matings effectively. The cost of maintaining this large number of bulls from sampling until the progeny test is available could be prohibitive.

An assist in practical planning was provided by Robertson[7] who found that sampling at least 4 to 5 times as many young bulls as are needed for future use is a reasonable lower limit. For large units where investment for storing young bulls permits, a higher testing ratio of 1 to 6 or 1 to 10 will more nearly approach optimal improvement. For a more accurate progeny test, most insemination studs should seek at least 50 tested daughters in many herds so that the reliability of the proof would be high enough to use the bull widely on the basis of his initial proof without undue risk.

[5] Skjervold, H., and H. J. Langholz. 1964. *Z. Tierz. Zuchtungsbiol.* 80:25–40.
[6] Skjervold, H., and H. J. Langholz. 1964. *Z. Tierz. Zuchtungsbiol.* 80:25–40.
[7] Robertson, A. 1957. *Biometrics* 13:442–450.

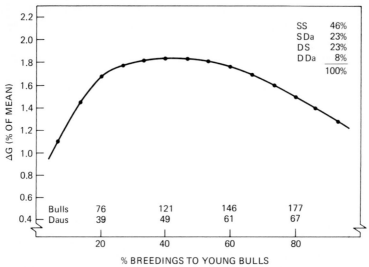

FIGURE 11-2
Relation of expected annual genetic improvement ΔG to percent of breedings
used for progeny testing in a unit of 60,000 tested cows. (*From Skjervold and
Langholz, 1964.*)

Methods of Expressing the Progeny Test

Numerous proposals have been made for expressing the results of the progeny
test. For milk, the lactation yield standardized to a 305-day, twice daily milking (2X), mature equivalent (ME) basis is a *fundamental* unit of measure. If the
number of daughters is large, all daughters are included without selection, and
the feeding and management levels are to be the same for future daughters; the
daughter average can provide a reliable indication of future daughter performance.

Since the daughter average does not account for the merit of the dams to
which a bull was mated, the daughter-dam comparison was preferred by early
breeders. The U.S. Department of Agriculture (USDA) initiated progeny test
summaries using the daughter-dam comparison in 1935.

Unfortunately, the conventional daughter-dam difference could be markedly influenced by changes in herd management. In many cases, the difference
of the daughters compared to their dams directly reflected the yearly
managemental trend in the herd(s). Generally, comments on the daughter-dam
comparisons were prefaced by stating that if the conditions were the same for
the daughters and their dams, the difference reflected the influence of the sire.
But one could not ensure that the conditions were the same for the daughters
and their dams. The dams make their records 2 or 3 years before their daughters. For most herds, only about one-third of the dams make records concurrently with their daughters. Herd opportunity is so subject to change because

of variations in herder, feed supply, and weather and economic conditions that it is practically impossible to provide the same opportunity for daughters and dams.

In recognition of the potential for such environmental influences on the daughter-dam comparison, the USDA-DHIA Herdmate Comparison (HMC) was implemented in 1962. The daughters of a sire were compared to other animals of the same breed which calved in the same herd, season, and year as the daughters being summarized.

Genetic improvement in milk yield was markedly accelerated after the adoption of the HMC through the increased accuracy of the proof and enlarged programs of sire evaluation. However, the transition of the use of frozen semen also made certain of the assumptions underlying the HMC invalid. The most important of these were:

1 Mates, herdmates, and sires of herdmates involved in a bull's summary were not random samples of a breed.
2 No genetic trend in the population.
3 No differential culling among daughters of a bull as compared to their herdmates.

The USDA-DHIA Modified Contemporary Comparison (MCC) method of compilation was implemented in 1974 to replace the HMC and to overcome the invalidity of the above assumptions. Increases in the accuracy of sire evaluation by the MCC necessitated additional complexity in calculations. Information on the background and interpretation of the MCC is provided in the previous edition of this text.[8] The only major change in this procedure since its adoption was the update of the genetic base to that of cows calving in 1982. The new "animal model" method for evaluation was implemented in 1989.

Animal Model   Mixed model linear methods for sire evaluations have been developed by Henderson[9] and they provide a comprehensive approach under a wide variety of situations. This technique combines the desirable properties of the selection index and the capacity of linear model methods with electronic computers to deal with large sets of data with unequal subclass numbers.

The general problem is to obtain the best evaluation of a sire that is regarded as a random individual from a specified population or group. The definition of "best" is the evaluation of breeding value as a linear function of the observations which is unbiased and has the smallest variance for the error of prediction. Procedures derived to meet these criteria have been termed the best linear unbiased prediction (BLUP). While the conceptual framework is more comprehensive, the computational techniques parallel least-squares procedures for fitting constants when unequal subclass numbers are involved.

---

[8] See pages 316–322.
[9] Henderson, C. R. 1974. *J. Dairy Sci.* 51:963–972.

The new USDA genetic evaluation system is based on the use of the performance and pedigree (relationships) of individual animals as the informational and computational unit; hence the animal model. Many earlier procedures have used the performance of progeny groups as the basis of comparison. The MCC is a sire model dealing with sire-progeny groups in a simplified computational approach to gain many of the advantages of BLUP methodology. However, advances in computer technology now have made the animal model feasible. All sources of information, individual performance, parents, and progeny can now be combined in an optimal linear manner.

Analytical and computational procedures begin with the development of a linear mathematical model or equation to describe each lactation record as follows:

$$y_{ijkl} = m_{ij} + a_{kl} + p_{kl} + c_{ik} + e_{ijkl}$$

where m_{ij} = the effect of the ij^{th} management group
$\quad a_{kl}$ = the effect of the genetic merit of the lth daughter of the kth sire
$\quad p_{kl}$ = the effect of the permanent environment on the records
$\quad c_{jk}$ = the effect of the interaction or special influence of her herd i and her sire k
$\quad e_{ijkl}$ = the unexplained random residuals effect on the record

Effects of age, length of lactation, and number of milkings per day are not included, since the lactation records are adjusted for these influences prior to the genetic evaluation.

A management group (m_{ij}) includes records for calvings for a 2-month interval in the same herd, for parity (first or later lactations), and (for Holsteins) for registry status (registered or grade). If fewer than five lactation records are included in a management group, the restrictions are relaxed to include additional records in the following order: inclusion of following or preceding 2-month groups, registered and grades to permit groups to cover 6 months, and first and later parity separation which is maintained by reducing the limiting group size to 3 for a 12-month interval.

The genetic merit of the animal (a_{kl}) includes the merit as evaluated from the animal's performance and the merit of all relatives, progeny, and ancestors. Pedigree information provides relationships to allow records of the individual, progeny, and ancestors to contribute to the genetic merit evaluation. Progeny and ancestor information is used only for sires and other animals without lactation records. Numerators of the relationship coefficients are utilized for all verifiable relationships, as illustrated in Chapter 9, to appropriately weight the information. Parents whose pedigrees are unknown are grouped according to the sex of the parent, the sex of the animal, and the birth year of the animal. For Holsteins, separate groups are defined for animals of U.S. and Canadian origin to account for genetic differences in these base groups.

The inclusion of p_{kl} allows the evaluation to differentiate between the animal's genetic merit and the permanent environmental effects which are independent of the genes of the animal or those genes in common with its relatives.

The value of .25 has been set for the heritability of milk, fat, and protein yields. In appropriately accounting for multiple records for an animal, repeatability of milk, fat, and protein yields is taken to be .55.

Data for the initial animal model evaluations include lactations initiated back to 1960 and available pedigree information. Lactations beyond the fifth are not included, since the records provide little added information regarding the genetic merit and since relatively few contemporaries are available for comparisons. The genetic base for animals born in 1985 is taken as zero. A large number of equations (over 20 million for Holsteins) are required. Solutions must be obtained by iteration rather than by direct matrix procedures. When the solutions for the effects are available, the individual's genetic merit can be computed. These solutions are further compiled to provide the final sire and cow evaluations.

Interpretation of Sire Summaries   The nature of the information provided by the USDA Bull Evaluation and Daughter List is given in Figure 11-3. The predicted transmitting ability (PTA) is an estimate of a bull's transmitting ability for milk, fat, and protein. Transmitting ability is one-half of the animal's average breeding value. Breeding value is measured on the basis of the animal's total average breeding value, whereas transmitting ability is a measure of the average merit of a random one-half of the genotype in a sperm or an egg.

Reliability values (REL) provide an index of the reliability of the PDs for milk, fat, and protein as measures of a sire's transmitting ability. When one chooses between two bulls with the same PDs, less risk is involved when the bull with the higher repeatability value is selected. The standard deviation of the genetic evaluation is approximately $\sigma_g(1 - R^2)^{1/2}$ where σ_g for Holsteins is approximately 500 kg. Nonetheless, it should be kept in mind that the PTA value rather than the reliability determines the long-range genetic improvement. Also because of the sampling nature of inheritance, any one daughter may deviate widely from the performance expected on the basis of her sire's and dam's PTA.

The PTA DOL and CY$ utilizes the milk, fat, and protein PTAs with the economic value of milk in accord with its quality and content. The average price of milk used for the 1987 calculations was $11.20 per cwt for milk with 3.5 percent fat, 3.2 percent protein, and 8.5 percent SNF. Differentials for a change of .1 percent from the above values were 16.8 cents for fat, 11.0 cents for protein, and 7.7 cents for SNF. These are revised annually in accord with market trends.

Although not summarized by the USDA, other traits are evaluated and compiled by the various breed associations and under the auspices of the National Association of Animal Breeders (NAAB). Such evaluations are available on the pedigrees of sires active in artificial insemination studs. The most im-

UNITED STATES DEPARTMENT OF AGRICULTURE
AGRICULTURE RESEARCH SERVICE
ANIMAL IMPROVEMENT PROGRAMS LABORATORY
BULL EVALUATION AND DAUGHTER LIST

JEP CAPT. ASTRO. ROBERT............
HOLSTEIN 1767903[a] BIRTH DATE
NAAB CODE 29H1043[b] 07-14-78

EVALUATION DATE: JULY 1988
BASE: HW90

OWNER
BLUE GRASS ALLIANCE
R 4
CAMPBELLSVILLE, KY 42718
PAGE 1

SIRE NO. H 1458744
DAM NO. H 10986523
MGS NO. H 1427815

PREDICTED TRANSMITTING ABILITY[c]

TRAIT	REL[d]	MILK	%	COMP[e]	DOL[f]	CYS[g]
Fat	89	+1134	-.03	+48	+98	
Protein	57		-.01	+33	+101	+97

PARENT AVERAGE[l]

MILK	COMP.	REL
+1800	+24	43
+69		23

HERD NUMBER 10

TRAIT	HERDS	DAUGHTERS	RECORD PER DAUGHTER	DAUGHTER DIM[j]	FIRST LACTATION CUL[h] %	RIP[i] %	·AGE (MO.)
Fat	89	123	3.2	274	13	11	36
Protein	57	78	2.7	197	13	23	

DAUGHTER AVERAGE

MILK	%	COMP.
18469	3.5	654
17142	3.3	587

DAUGHTER YIELD DEVIATION[k]

NO.	MILK	%	COMP.
22	+225	+.36	+21
17	+212	+.32	+15

CONTRIBUTION TO BULL

WT[n]	MILK (LB)	COMP (LB)
1.0	-4018	-135
.7	-1164	+38

PTA

MILK (LB)	COMP (LB)
+1800	-65
+69	+15

PTA PROTEIN $ THIS RUN HIGHER THAN 57% OF ACTIVE AI BULLS OF BREED IN LAST RUN

Most daughters: 31-12-1234

LACTATIONS INCLUDED FOR BULL

DAUGHTER ID NUMBER	BIRTH MO DAY YR.	HERD CODES FIRST / LAST	LAST LACTATION CALVED MO DAY YR.	AGE	DIM	TRM CD.	CODES	TRAIT	DAUGHTER YIELD NO. REC	MILK (LB)	COMP %	COMP (LB)	YIELD DEVIATION[m] NO.	MILK (LB)	COMP (LB)
H 31ZWY3950	11-03-82	23-17-0127 / 31-23-1944	12-19-85	3-01	257	3	A·B·C / D·E·F	Fat / Protein	2 / 1	14597 / 16982	3.8 / 3.2	465 / 545	18 / 7	-117 / +483	-34 / +59

ELITE COW LISTING

MILK AND FAT

IDENTIFICATION COW	SIRE	BIRTH YR.	MO.	LAST LACTATION HERD CODE	Yr.	Mo.	T DIM	NO. REC	MILK (LB)	FAT %	FAT (LB)	PTA (HW90) REL[d]	MILK	%	FAT (LB)	$$[f]
10873753	1839579	84	12	23-51-2957	89-12	R 126		2	25,294	3.7	992	53	1485	.02	58	177

PROTEIN

NO. REC	PRO %	PRO (LB)	PTA (HW90) REL	%	PRO	PROS	CYS[g]
2	2.9	763	51	-.04	39	168	165

[a] Registration number of bull
[b] AI stud and bull code number
[c] One-half the breeding value (a_{gi}) or the expected difference to be transmitted to the progeny (PTA)
[d] Reliability of the evaluation based on the amount of information included
[e] Components, either fat or protein yields
[f] Dollar value from an index combining values of milk and components
[g] Dollar value of components for cheese
[h] Percent of animals culled before completing first lactations
[i] Percent of first lactation records in progress when evaluation compiled
[j] Average days in milk
[k] The weighted average of daughter's yield deviation adjusted for the genetic merit of their dams
[l] Average PTA of parents
[m] Weighted average yield adjusted for management group, permanent environment, and herd-size effects
[n] Weight given the daughter information in bull evaluation
[o] Predicted transmitting ability (PTA) of daughter

FIGURE 11-3
Format of USDA Bull Evaluation and Daughter List and the Elite Cow Listing with definitions for data. The (HW, 90) indicates the bases for the PTAs with the average of cows born 5 years earlier as zero.

portant additional traits that are evaluated are calving ease and type or conformation. Calving ease provides a measure of the percentage of difficult calvings that are expected when the bull is bred to heifers calving for the first time.

Adoption of the linear type scores for several conformation traits provides for a large number of items for consideration. Unfortunately, few of these have significant relation to performance and utility. Many studs display the several traits in a tabular form, which provides a quick appraisal of strengths and weaknesses in the bull's daughters (Figure 11-4). In recent years much research has been centered on stayability. This is concerned with the percentage of the daughters of a bull that perform and remain in the herd up to, say, 48 months of age. Efforts to refine this functional expression of longevity continue, but it has yet to be incorporated into regular sire summaries.

Cow Evaluations   Although the breeding values for dairy females usually cannot be determined as accurately as those for males whose daughters have records in many herds, cow evaluations are of much value for choosing outstanding dams of sires for progeny testing. Actually a few females with a large number of embryo-transfer progeny could eventually have evaluations with accuracies approaching those of males. This could be especially true with the merchandizing of embryos, so that the cow's daughters have performance records in several herds.

FIGURE 11-4
Example of how the profile for traits other than production for daughters of a sire are displayed by some artificial insemination organizations.

Trait		TRAIT EVALUATION 109 Daughters		TA
Basic Form	Angular		Strong	0.9
Chest Width	Weak		Powerful	0.9
Dairyness	Coarse		Sharp	2.2
Stature	Short		Tall	0.1
Body Depth	Shallow		Deep	1.2
Legs-Side	Straight		Set	0.8
Legs-Rear	Toe Out		No Toe Out	0.4
Foot Angle	Low		Steep	2.0
Rump-Side	Hi Pins		Lo Pins	0.8
Rump Width	Narrow		Wide	1.1
R.U. Height	Low		High	0.2
R.U. Width	Narrow		Wide	0.3
Fore Udder	Loose		Tight	0.3
Udder Depth	Deep		Shallow	0.4
Ligament	Weak		Strong	0.6
Teat Place.	Wide		Close	0.6
Temperament	Problem		Excellent	0.6
Milk. Speed	Slow		Fast	1.3

3   2   1   .25 0 .25   1   2   3

Indexes were previously compiled for the USDA-DHIA Cow Index List utilizing information on the production of the cow and her sire (the cow's paternal sisters). Information on the cow's dam and other maternal relatives was not used, since it was not readily available from the data file. With the introduction of the "animal model" and the use of the relationship matrix, including all known pedigree information, predicted transmitting ability (PTA) values for cows are computed the same way as for bulls. A listing of Elite Cows will include information on the cow's lactation records and her PTAs for milk, fat, and protein with the appropriate reliability and dollar values, as shown in Figure 11-3.

BEEF CATTLE

Research has been under way since the 1930s to examine objective measures for evaluating beef cattle. Performance testing has undergone a comprehensive evaluation by researchers and producers. Most breed associations and many states now sponsor such programs. An all-breed nationwide beef cattle organization was established in 1955 to provide performance and progeny performance and progeny evaluation programs to seed-stock herds.

Nationally the organizations involved in performance testing joined to form the Beef Improvement Federation (BIF) in 1968 to extend and further improve these programs. A major emphasis has been the establishment of uniform procedures for measuring and recording data to permit more widespread use of the results in animal evaluation and selection. The coverage here follows recommendations provided by the Beef Improvement Federation.[10]

On-farm or ranch and central testing are involved and will be presented separately. In either situation effective programs require that animals being compared be given equal opportunities to perform under uniform feeding and management conditions. These conditions should represent nutritional and managemental regimens which are practical and comparable to those under which the progeny are expected to perform. Preweaning and postweaning growth and development are generally handled as two separate phases in line with the general divisions of the commercial beef enterprise.

Farm and Ranch Programs

The Preweaning Phase Weaning weights of the nursing calf are obtained to evaluate differences in maternal performance and growth potential of calves. Calf weights must be adjusted to a standardized age of 205 days and to a mature dam equivalent for comparison. Weights for calves should be recorded as close to 205 days as is practical and within a recommended range of 160 to 250 days.

[10] Beef Improvement Federation. 1986. "BIF Guidelines for Uniform Beef Improvement Programs," 5th ed.

The 205-day weight should be computed on the basis of average daily gain from birth to weaning by the following formula:

$$\text{Computed 205-day weight} = \frac{(\text{actual weight} - \text{birth weight})}{\text{age in days}}$$
$$\times\ 205 + \text{birth weight}$$

If the actual birth weights are not available, the standard birth weight for the breed or breed combination may be substituted. This provides a computed 205-day weight unadjusted for age of dam and sex of calf.

In adjusting the 205-day weight to a mature dam equivalent, the following additive adjustments are made to the computed 205-day weight (Table 11-1). These factors for age of dam adjustment are not precise for all breeds or breeding groups. Higher milk-producing cows such as dairy crosses, Brahmans, and many F_1 crosses produce more milk in relation to their calves' ability to consume. Calves suckled by these higher milking cows actually require less adjustment than is indicated to bring them to the basis of a mature dam.

The performance of an animal relative to the average of all animals in the group often is expressed in terms of "weight ratios" and "gain ratios." For 205-day weight it is

$$\frac{\text{Individual computed 205-day weight} \times 100}{\text{Average 205-day weight of group}}$$

These ratios can provide a useful device for visualizing the relative ranking of individuals in a group. Where sex differences are important, these ratios should be computed separately for each sex. For the ratios to be most meaningful in comparisons, the animals in the group should be contemporaries. While of value, the potential for managemental and genetic differences between herds limit the usefulness of the ratios for between-herd comparisons.

TABLE 11-1

ADDITIVE CORRECTION FACTORS FOR ADJUSTING 205-DAY CALF WEIGHTS TO A MATURE DAM BASIS*

Age of dam class	Additive factors†	
	in lb	in kg
Up to 27 months	52	23.6
27 to 39 months	35	15.9
39 to 47 months	23	10.4
47 to 59 months	9	4.1
59 to 144 months	0	0
Over 144 months	12	5.4

*For both males and females.
†From Beef Improvement Federation (1986), p. 12-5.

Following the procedure suggested in Chapter 7 (see "Repeated Observations") the production of individual cows may be summarized to give the most probable producing ability (MPPA) by the expression

$$\text{MPPA} = \overline{H} + \frac{nr}{1 + (n-1)r} (\overline{C} - \overline{H})$$

where $\overline{H}$ = 100, the herd average
n = the number of calves included in the average
r = .4, the repeatability of weaning weight
$\overline{C}$ = average weaning weight ratio for all calves the cow has produced

Cow	Number of calves	Average weaning weight ratio	MPPA
1	2	90	93.3
2	1	88	95.2
3	4	95	96.4
4	3	110	106.7
5	1	118	107.2
6	4	115	110.9

Note how the rankings on average weaning weight ratios shift when the MPPAs are computed taking account of the number of records.

The Postweaning Phase Yearling weights at 365 days or long yearling weights at 452 or 550 days are important because of their high genetic association with efficiency of gain and yield of retail cuts. Yearling weights should be reported separately for each sex. The postweaning gain test should begin the date that weaning weights are obtained. Age of dam effects on 365-day weights are of the same order of magnitude as age of dam at weaning. Thus the adjusted 365-day weight is computed by the following formula:

$$\text{Adjusted 365-day weight} = \frac{\text{actual final weight} - \text{actual weaning weight}}{\text{number of days between weights}}$$
$$\times 160 + \text{computed 205-day weight adjusted for age of dam}$$

The interval between weaning weight and final weight should be at least 160 days, and the final weight should not be taken prior to 330 days for an individual animal. Adjusted 365-day weights are most useful for herds that develop bulls on a comparatively high level of concentrate feeding. Where bulls are developed less rapidly, adjusted 452-day or 550-day weight should be used in accord with the BIF guidelines.

Weight ratios for adjusted 365-day weight, adjusted 452-day weight or 550-day weight should be computed separately for each sex management group. These weight ratios can be biased downward if lighter calves are culled at weaning. The impact of culling lighter calves at weaning can be adjusted by computing the weight ratio as

$$\frac{W + P}{\overline{Wu} + \overline{Ps}} \times 100$$

where W = Adjusted 205-day weight of individual

P = The postweaning gain of the individual computed from post-weaning average daily gain multiplied by 160, 247, or 345, depending on whether 365-day, 452-day, or 550-day adjusted weight is desired

$\overline{Wu}$ = Average 205-day adjusted weight of all calves weaned contemporaneously with the animal being considered

$\overline{Ps}$ = Average 160-, 247-, or 345-day postweaning gain of all calves tested in a contemporary sex management group

If no calves are culled at weaning, the value for Wu will include the weaning weights for all animals weighed at the older ages of 365, 452, or 550 days.

Central Testing Stations

Feeding and management conditions at central stations can be made much more uniform than for most herd or ranch situations. Central testing stations are used most to (1) compare individual performance of potential seed-stock sires with animals from other herds; (2) compare bulls being tested prior to sale to commercial producers; (3) finish steers or heifers to be slaughtered for progeny test information, and (4) acquaint breeders with records and standards of performance.

When bulls are tested to provide performance information to prospective buyers, the number of bulls assembled per sire or per herd is not a major concern. Nevertheless, a large number of bulls should be tested at a location to provide a sound base for comparison in order to permit culling and still provide adequate animals from which buyers can choose. Such tests are extremely valuable to small breeders who lack facilities and a broad enough genetic base of animals in their herds to make meaningful comparisons.

Specific procedures may vary slightly from one location to another. Even if all test stations rigorously followed precisely a standard set of recommendations, accurate comparisons could not be made between animals at different locations. Many important differences such as nature and quality of the ration, temperature and climatic conditions, potential digestive disturbance and disease, as well as pretest environmental conditions, vary from location to location. The central station test can only offer reliable data for comparison within a specific location and year.

Central testing stations for bulls have been operated for many years. From this experience and the evaluation of test data the following policies and procedures are recommended:

1 Herds from which bulls are consigned should be on herd,testing programs for preweaning performance. The following information should be submitted upon entry to the test station (sire, dam, birth date, actual weaning weight and date, adjusted 205-day weight, within-herd weaning weight ratio, and the number of calves comprising the average ratio.

2 Calves should be at least 180 days and not more than 270 days of age when delivered to test station.

3 There should be an adjustment or pretest period of 21 days or more immediately prior to the test.

4 The feeding test should be 140 days or more in length. Initial and final test weights may be either full or shrunk. If full weights are used, the values should be an average of two weights taken at the same time on successive days. A single initial and final weighing is adequate after a 24-hour shrink.

5 The rations may vary according to locally available feed and test objectives. They should be fed ad lib. Rations of between 60 and 70 percent total digestible nutrients should permit the expression of genetic differences in growth.

6 All bulls sold in a test sale should be examined by a competent veterinarian for reproductive and structural soundness.

Stations vary in the form for reporting test station results. Figure 11-5 shows the basic information which should be reported. The 140-day average daily gain and gain rates are the most important data, since they measure the actual and comparative growth during the period the bulls are together under test conditions. Selection on the basis of 140-day gain should improve weaning weight to the extent that some of the genes which influence feed-lot growth also influence preweaning growth. The gain ratio is computed by dividing the individual animal's gain by the test group average and multiplying by 100. Again this ratio makes comparisons between animals easier in that the average animal has a value of 100, and differences between animals can be easily assessed in percentage units.

Weaning weights and within-herd weaning weight ratios provide good comparisons among bulls which come from the same herd. They are much less useful for comparing bulls from different herds. Although the weaning weight is the resultant of the dam's milk production, the growth potential of the young bull, and other environmental effects, it is the best readily available measure of the dam's milk production. Consequently, it is desirable to have a weaning weight which is above the average of the herd in which the calf was raised (weaning weight ratio, above 100). Actual weaning weight and date weaned are reported to provide information on gain during the period between weaning and initiation of the test. Loss of weight or below average gain during this period may result in higher-than-expected gain from compensatory growth during the test period.

The 365-day adjusted weight combines the adjusted weaning weight and the postweaning gain. The 365-day weight ratio provides a good measure for compar-

Measures recommended for all test stations

Item			Weaning					Gain test						Yearling			
ID No.	Herd tattoo	Birth date	Act. birth wt.	Act. W.W.	Act. 205-day date	Adj. herd day	W.W. ratio w/in herd and No.	Age of dam	(Dte) Init. test wt.	(Dte) Final test wt.	Age in days	ADG	Test gain ratio	Adj. 365-day wt.	365-day wt. ratio	Fat thick	Scrotal circumference
(1)	(2)	(3)	(4)	(5)	(6)	(7)	(8)	(9)	(10)	(11)	(12)	(13)	(14)	(15)	(16)	(17)	(18)

Owner, address, breed, and sire. (insert between sire groups, or in column at the left.)

Each test group (breed and age group) should be listed together on the report and averaged. (Age range in each group should not exceed 90 days and breed should be averaged separately within age group.)

Sire group averages are shown for 3 or more progrmny of same sire.

If sire groups include calves from different age groups, data may be listed together by sires, but with only the average of ratios shown.

Column description

(1) Eartag test number
(2) Permanent identification, such as the ear tattoo
(3) Month/day/year of birth. Example: 2/15/86 for Feb 15, 1986. If all in the same year, may omit year.
(4) Birth weight of the bull
(5) Actual weaning weight
(6) Actual weaning date
(7) Weaning weight adjusted to 205 days and for age of dam according to BIF. If creep-fed, add C after weight.
(8) Adjusted 205-day weight divided by average of all bull calves in same weaning season group and same management coded mutiplied by 100. Minimum entrance requirement is optional with test management. The number of calves making the average is listed in parentheses. Example: 105 (17). This column should be left blank or "ET" printed for embryo transplant calves.
(9) Age of dam at time of calving
(10) and (11) Average of at least 2 full weights taken on different days. May be more than 1 day apart if desired.
(12) Age at end of test
(13) Final weight minus initial weight divided by days on test. Minimum length, 140 days, no maximum.
(14) Average daily gain divided by test group average gain multiplied by 100. (Breed within age group average.)

(15) $$\frac{\text{Final test weight} - \text{Actual birth weight}}{\text{Age in days}} \times 365 + \text{Actual birth weight}$$
$$+ \text{Additive age of dam adjustment for weaning weight}$$

(16) Adjusted 365-day weight divided by test group average of adjusted 365-day weights multiplied by 100. (Breed within age group average.)
(17) Fat thickness may be measured by sonoscope or probe and should be expressed as the absolute value.
(18) Scrotal circumference at time the bull completed test

FIGURE 11-5
Format and recommended measurements for all central bull testing station reports with descriptions for measurements.

ing yearling growth of calves from the same herd. It has a heritability of approximately 50 percent. Nevertheless, care must be exercised in comparing test station results for bulls from different herds, because the weaning growth was not made under comparable conditions. When gain ratios and 365-day weight ratios are nearly the same, the 365-day adjusted weight is a reliable comparative measure of ability to gain to 1 year of age. Weight per day of age also could provide an alternative measure of growth under these circumstances.

Efficiency of feed utilization is expressed as units of feed per unit of gain. Weight of the animal during the test period influences maintenance requirements, and measures of feed conversion should be adjusted for weight.[11] Feed intake is difficult to measure, and most central test stations do not obtain individual feed conversion data. Individual feeding would be required, but group feeding more nearly duplicates feed-lot circumstances and can provide a measure of a sire's transmitting ability when sire progeny are fed as groups in separate pens. Equipment now available to automatically obtain feed intake of individual animals may ease this task in the future. Fortunately, growth rate and gain per unit of feed consumed are highly correlated genetically. Studies indicate that selection for gain alone should result in about 80 percent as much genetic improvement in feed efficiency as would direct selection for reduced feed per unit of gain.

Forage Bull Tests Recently, tests to evaluate bulls when fed primarily forage have developed in the southeastern United States. This approach attempts to measure performance on forage feeding programs similar to those of the commercial operator, rather than on the heavier grain feeding regimen. Bulls are on test for a longer time due to the low-to-moderate levels of energy intake provided by the high forage ration. Due to this additional time on test, the bulls are approaching 2 years of age. They are larger and better conditioned to handle heavier breeding loads and the rigors of commercial operations than yearling bulls taken directly off heavier grain feeding tests.

Ration quality tends to be much more variable on the forage tests, and a longer period on test is recommended. Bulls should be on test at least 168 days. Other recommended differences in the management of the test are provided by the BIF.[12]

National Sire Evaluation

The most important contribution of a National Sire Evaluation Program is in making valid comparisons among sires. The increasing availability of beef sires through artificial insemination has provided a sounder basis for such comparisons. Objective performance tests have been developed with animals being compared to contemporaries to provide accurate within-herd evaluations. Un-

---

[11] Beef Improvement Federation (1986).
[12] Beef Improvement Federation (1986).

fortunately, beef sires are not used as widely in artificial insemination as dairy sires. As a consequence, there are fewer herds where the same sires are used to provide a genetic tie between herds. Nevertheless, several types of sire evaluation programs are being conducted. These range from using field records from seed-stock herds to designated beef progeny tests supervised by a breed organization which designates specific reference sires which are already progeny-tested.

Performance records obtained from regular herd performance programs can be used to estimate expected genetic differences among sires. However, unless a group of sires is used extensively enough in artificial insemination to have many sires used over groups of herds, the accuracy of the comparisons will be low. Newly introduced breeds have encouraged herd performance testing with artificial insemination, and this has given a broad base of information for estimating the comparative genetic merit of sires.

Progeny Testing   All sire evaluation programs involve selection first on the basis of pedigree and own performance for desired traits followed by selection based on progeny performance. Top yearling bulls with outstanding performance and with superior close relatives are candidates for progeny testing. Such testing requires close adherence to the basic principles of any progeny test. All bulls being compared must be mated to a comparable set of cows taking into account factors such as age, breed or cross, and managemental grouping. Within each cow grouping, bulls should be mated randomly. The progeny of all the bulls must be provided an equal opportunity to perform. All test herds must include designated reference sires to provide a standard of reference for within-herd comparisons and also to make broader comparisons of sires across herds.

Progeny tests need to be planned thoroughly and well in advance. A sufficient number of cows should be available for test matings. Unfortunately this is a limiting factor in most single herd progeny testing programs. Management factors which may affect the conception rate also must be optimized to ensure that a maximum number of calves result from the progeny test matings made by artificial insemination.

Reference Sires   Organizations conducting the sire evaluations are responsible for designating the reference sires and for the handling and distribution of the frozen semen for progeny testing. A reference sire for programs using field records should have a large number of progeny (100 to 500) evaluated in a large number of herd groups (10 to 50) in comparison with several (5 to 10) other reference sires. Designated reference sires should be chosen from the top sires tested previously, so that a sire may be used as a reference sire for at least 2 years. Half of the reference sires would be replaced each year, allowing comparisons among the reference sires from year to year.

A compromise must be reached between the number of bulls to test and the number of progeny to secure for each bull. The Beef Improvement Federation has provided guidelines for progeny numbers for reference and test sires. At least 20 progeny should be obtained for each sire being progeny-tested. Reference sires whose performance has been evaluated by progeny testing, should have at least 20 progeny when only one test bull with 20 progeny is being compared. An increase of 5 progeny from reference sires should be provided for each additional test bull up to seven, to a total of 40 progeny from reference sires. Forty reference sire progeny are deemed sufficient even when the number of bulls being tested exceeds seven.

Cooperative multiherd progeny tests are encouraged to provide more accurate progeny tests. These average out environmental influences which may be characteristic of a specific herd, and provide many more sire ties between herds for sire comparisons. Multiherd progeny test programs provide an opportunity to measure genetic change in the herds by comparing the performance of new sires with a base set of reference sires. As artificial insemination becomes more widely accepted and sires are used in many herds, additional alternatives in progeny testing and in measuring genetic trends will become available.

Sire Summaries Organizations responsible for sire evaluation programs should publish a summary including all sires tested regardless of their merit. The summary should include:

1 *Identification:* Complete sire information, including parentage, should be provided.
2 *Own performance:* Record of performance of the sire in his herd of origin as compared to herdmates, including weight and gain ratios should be given.
3 *Progeny performance:* The specific traits which should be evaluated in the progeny test depend upon the objective of the testing, and these are the prerogative of the organization conducting the testing program. Traits suggested for consideration include growth, reproduction, and carcass merit of progeny.

BIF recommends several measures of growth of progeny related to development during related commercial periods, such as weaning weight, feed-lot gain, and yearling weights. Measures of reproduction that may be influenced by the sire and which are important to the beef enterprise include gestation length, birth weights, and calving ease. Ease of delivery for individual unborn calves cannot be predicted accurately. Birth weights are a major factor influencing difficulty of calving among various breeds of sire. A subjective rating of ease of calving and the percent of easy calvings, reflecting the percentage of births without assistance, are desirable and often provided.

Data on carcass merit are difficult to obtain, but they add information on sibs of potential sons of tested sires for future progeny testing. The Beef Carcass Data Service (BCDS), which is a joint USDA–beef cattle industry effort, will assist breeders and feeders in obtaining carcass data. Details are provided in BIF[13] recommendations.

The future progeny of a bull do not perform exactly the same as past progeny, even when a sire is mated to the same cows. Variation occurs partly because of feeding and management, but random sampling in the genes transmitted and variable environmental effects also influence progeny performance. Evaluating a large number of progeny performing at several management units tends to cancel these random effects.

Expected Progeny Difference (EPD) The expected progeny difference should be presented for traits which can be measured quantitatively. The EPD is closely related to the PTA for dairy sires. It is an estimate of how future progeny of a sire are expected to perform relative to the progeny of reference sires or the average of sires included in the test when mated to comparable cows and when the progeny are treated similarly. The progeny difference for a sire is regressed (*b*) toward zero, depending upon the number and distribution of the progeny involved in the difference and the heritability of the trait. For example:

$$\text{EPD} = b \, (\overline{X}_b - \overline{X}_n)$$

where b = the regression coefficient (weighting factor) dependent on the number of progeny and the heritability of the trait
$\overline{X}_b$ = the average of the bull's progeny
$\overline{X}_n$ = the average of the other progeny in the same herd and/or test

Under these simplified conditions the value of b can be expressed for the individual traits as

$$b = \frac{n}{n + (\sigma_e^2/\sigma_s^2)} = \frac{n}{n + [(4 - h^2)/h^2]}$$

where n = the number of progeny
σ_s^2 = the variance due to sires
σ_e^2 = the residual variance or an intraherd or intratest basis for the specific trait

The ratio of the two variances, σ_e^2/σ_s^2, is a function of the heritability of the trait. Table 11-2 gives numerical values of b for selected combinations of n and h^2.

Future Developments Computation of estimated breeding values for weaning and yearling growth began in 1971 and those for maternal performance us-

---

[13] Beef Improvement Federation (1986).

TABLE 11-2
VALUES OF b FOR SELECTED COMBINATIONS OF n AND HERITABILITY, h^2

h^2	.20	.25	.40	.50
n	b	$n/(n + 19)$	$n/(n + 15)$	$n/(n + 9)$	$n/(n + 7)$
5		.21	.25	.36	.42
10		.34	.40	.53	.59
15		.44	.50	.62	.68
20		.51	.57	.69	.74
30		.61	.67	.77	.81
40		.68	.73	.82	.85
50		.72	.77	.85	.88
100		.84	.87	.92	.93

ing data on the calves of the daughters of sires began in 1974. The mixed-model methodology, which can provide best linear unbiased predictions (BLUP) of breeding values, was developed and introduced by Henderson. It is reported that 15 beef breeds have various national genetic evaluation programs in progress. Rapid advances in computer capabilities have paralleled many applications of this methodology for progeny test evaluation and estimation of breeding values. An introduction to the BLUP procedures is provided in the *Guidelines for Uniform Beef Improvement Programs* (pp. 9–3 and 9–19) cited in the "Suggestions for Further Reading."

Increased use of artificial insemination in the future will provide data that will more nearly utilize the assumption of the advanced mixed models and provide a more accurate estimate of breeding values across herds. The acceptance and use of these new approaches will require an extensive educational effort. University personnel, breed associations and other agencies have made an excellent beginning to accomplish this. Breeders and students can obtain information on latest procedures from the Beef Improvement Federation, breed associations, and state extension services.

SWINE

Formal performance testing with swine was initiated by the United Duroc Registry Association in 1938. Subsequently other breed associations developed programs, and a number of states initiated on-farm testing programs sponsored by their agricultural extension services. The first of these was in 1945 in Wisconsin. Several changes in rules and procedures have evolved, but since 1946 there has been a considerable degree of uniformity among these programs.

The first central test station was established in Ohio in 1954, and 43 stations were operating in 25 states of the United States in 1987. During the 1950s most central stations tested only barrows and gilts for performance and slaughter evaluation. The Iowa station began testing boars in 1956, and gradually other stations have shifted to boar testing. Testing of boars is consistent with recent evaluations which lend little support for the continued use of sibs to predict the breeding value of boars at central testing stations (Appendix, Chapter 7). This is also true for body composition where back-fat probes and ultrasonic measurements have been developed.

The National Swine Improvement Federation (NSIF) was formed in 1975 with membership from organizations operating or interested in testing programs.[14] NSIF was organized to promote more uniform central testing procedures and to revitalize on-the-farm testing programs. The purposes of NSIF are to work for *uniformity* of testing procedures, to aid member organizations in *development* of programs, to encourage *cooperation,* to encourage member organizations to conduct *educational* programs relative to the use and interpretation of performance data, and to develop increased *confidence* in the swine industry regarding the economic potential of performance testing. It works through research-extension-industry committees to accomplish its goals. NSIF has published guidelines for testing programs which are drawn upon extensively in the subsequent discussion of on-the-farm and central testing programs.

On-Farm Testing

Most on-farm testing programs are extension-sponsored and vary somewhat in the services provided. Computerized printouts summarizing records and including adjustments to standard ages or weights are usually provided. Records may be submitted to breed associations for their official programs.

In spite of the diversity of formats, programs are all basically aimed at the same objectives. These include providing reliable records on one or more performance traits such as litter size, weaning weight, rate of gain after weaning, or efficiency of gain to market weight. They may include indications of carcass composition from live-animal estimates of back-fat thickness and loin eye area, or actual carcass measurements and cut weights from slaughter of representative animals. The various programs have had increasing participation and have undoubtedly been an important factor in guiding selection programs in the direction of meat-type swine.

The primary purpose of on-farm performance testing programs is to measure the performance of individuals as the basis for selection and genetic improvement within the herd. These programs are designed to assist breeders in evaluating their herds in a systematic manner. Maximum value is gained when the entire herd is enrolled, and all pigs are evaluated. Pigs should be uniformly managed and given an equal opportunity to perform in order that the records be useful for individual pig, sow-productivity, or boar evaluations.

Testing Procedures   Participants in on-farm testing should meet the following minimum criteria:

[14] National Swine Improvement Federation (1987).

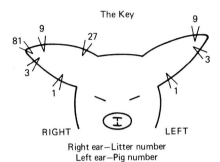

FIGURE 11-6
Ear-notching system to identify litters and pigs in litter. Right ear is for litter identification, and all pigs from the same litter must have same ear notches. "Right" ear is on pig's own right. Left ear is used for individual pig number in the litter. Each pig will have different notches in this ear.

1 *Identification:* A uniform ear-notching system has become standard throughout the swine industry and the NSIF recommends identifying the litter in the right ear and the individual pig in the left ear. The most used system, as illustrated in Figure 11-6, is the 1-3 bottom notch, 9-27 top, and 81 end-of-ear notch. When another notching system is used, a complete description should be stated and attached to the basic document. Supplemental ear tags or tattoos may be useful as optional herd identification.

2 *Birth record:* All pigs should be individually ear-notched; sex, date of birth, and parents should be recorded in an appropriate record book or file.

3 *Sow productivity:* It is recommended that primary selection be based on litter weight at 21 days of age. Basic information should include (*a*) the number of pigs farrowed, including the number alive and dead (but only those fully formed at birth), (*b*) a score for farrowing ease, based on time required for farrowing, (*c*) number of functional teats, (*d*) litter birth weight, and (*e*) number of pigs and litter weight at 21 days.

4 *Postweaning:* Gain tests should begin when the pigs are an average 70 lb, with a minimum range. Final weights should be obtained when the pigs have gained at least 160 lb more than the starting weight and adjusted to 230 lb. Actual daily gain during the test period should be used to adjust to 230 lb.

5 *Days to a constant weight:* The number of days to a desired constant weight provide valuable managemental information. A desired weight of 230 lb is usually specified for market hogs, but other desired weights can be used in the following formula:

$$\text{Adjusted days} = \text{actual age} + (\text{desired weight} - \text{actual weight}) \left(\frac{\text{actual age} - 38}{\text{actual age}} \right)$$

6 *Body composition:* Back-fat thickness should be measured on all pigs between 200 and 260 lb and adjusted to 230 lb. The following formula can be used for adjustment to a 230-lb-weight basis:

Adjusted BF = actual back fat + (desired weight − actual weight)

$$\left(\frac{\text{actual back fat}}{\text{actual wt} - 25}\right)$$

Additional optional records for on-farm testing include determination of feed conversion for groups fed as litters or by sire progeny groups. Loin eye area can be estimated from ultrasonic measurements, and slaughter with complete carcass information is also optional on sibs. Figure 11-7 shows one method of recording results of on-farm testing programs for data analysis and summarization.

N.C. Pork Improvement Program
Swine On-Farm Performance Testing

INDIVIDUAL PERFORMANCE

Owner: _____
Herd Code _____
Litter No. _____
Breed _____
Dam No. _____
Dam Breed _____
Sire No. _____
Sire Breed _____
Date Litter Farrowed _____
No. Farrowed Alive _____
No. After Transfer _____
Sow's Parity _____
21 Day — Litter Weight _____
21 Day — Date Weighed _____
21 Day — No. Weighed _____
Date Weighed & Probed _____

CODES

Pig No.	SEX	Weight	Probes Shld.	Loin

BREED
01 Duroc
02 York
03 Hamp
04 Poland
05 Spot
06 Berk
07 Landrace
08 Chester
09 Crossbred

SEX
0 Gilt
1 Barrow
2 Boar

FIGURE 11-7
Format for collecting data for on-farm test in swine used by the North Carolina On-Farm Testing Program.

Central Station Testing

The primary objective of central stations evaluations is to compare the individual performance of potential seed-stock boars with performance of similar animals from other herds for rate of gain, feed conversion, back fat, and estimates of muscularity. This is of special value for swine because artificial insemination is not used widely. Test stations also provide an educational demonstration to acquaint breeders with performance records and standards of performance. Most arrange for a sale of the tested boars for use in commercial and seed-stock herds.

Testing Procedures   Central station tests seek to provide reliable records for comparison of animals in a specific test and location. Individual pig entry requirements and conditions are:

1 Entry weight from 40 to 65 lb (19 to 28 kg).
2 Entry weight pen day of age from 0.6 to 1.2 lb (275 to 550 g).
3 Minimum of a 7-day adjustment period from entry to station to the beginning of the test.
4 For a pen, the average weight per pig should be 70 lb (32 kg) with a maximum of 80 lb (36 kg). It is recommended that only full-sib boars (littermates) be tested in the same pen. If adequate full sibs are not available, full- and half-sib boars (same sire but different dams) may be included in the same pen. If gilts or barrows are tested, they should be tested in separate pens.

Most central test stations require the following data and information:

1 "On-test" weight, date, age, and weight per day of age for individual pigs
2 Final test weight for pen between 230 and 240 lb (105 to 110 kg)
3 Individual average daily gain
4 Feed conversion on a pen basis
5 Average back-fat thickness adjusted to 230 lb (105 kg)

At the option of the station, loin eye area measurements, other estimates of muscularity and soundness may be reported. However, these are not used in the index.

Environmental differences make it difficult to compare pigs tested at different locations, at different times, and under different forms of management. However, the use of a standard index and index ratios based on contemporary groups provide more valid comparisons. It is recommended that only boars be tested in full-sib groups because of increased accuracy of feed conversion data. If full-sib groups are not used, all boars in a test pen must be from the same sire for the indexes proposed to be applicable. The indexes recommended for boars are given below.[15] For boars fed individually:

[15] National Swine Improvement Federation. 1987. *Guidelines for Uniform Swine Improvement Programs*. Pp. 10–18 to 10–19.

$$\text{Index} = 100 + 70 \, (\text{ADG} - \overline{\text{ADG}}) - 75(\text{F/G} - \overline{\text{F/G}}) - 55(\text{BF} - \overline{\text{BF}})$$

For boars fed in groups with one or more other pigs:

$$\text{Index} = 100 + 80 \, (\text{ADG} - \overline{\text{ADG}}) - 85(\text{F/G} - \overline{\text{F/G}}) - 80(\text{BF} - \overline{\text{BF}})$$

Feed conversion is computed on the basis of an average weight of 230 lb. The actual ratio as well as the boar's ratio compared to the test group should be reported. For boars in on-farm tests, group-fed, and for which no feed conversion data are available:

$$\text{Index} = 100 + 112 \, (\text{DG} - \overline{\text{DG}}) - 120(\text{BF} - \overline{\text{BF}})$$

where the index variables are defined in English units to retain the range of index values now in use:

ADG = Boar's own daily gain on test (lb.)
$\overline{\text{ADG}}$ = Daily gain of all boars on test (lb.)
F/G = Feed conversion (pounds of feed per ꞁ of gain) for individual boar or pen of boars fed
$\overline{\text{F/G}}$ = Average feed conversion of all boars in test
BF = Boar's back-fat thickness (in.)
$\overline{\text{BF}}$ = Average back-fat thickness of all boars on test (in.)

Index values will average 100 for each test or sale group with a standard deviation of approximately 25. This would mean that two-thirds of the boar indexes will be between 75 and 125 index points. About one-sixth will be expected to exceed an index value of 125 points. Individual boar ratios for ADG, F/G, and BF will be of interest, however. They should not be used for independent culling-level selections. Independent culling levels should not be used except in soundness evaluations.

SHEEP

Organized performance testing for sheep has not developed as extensively as with dairy, beef, and swine. Recently, the National Sheep Improvement Program (NSIP) was established to provide systematic procedures for obtaining data to identify high-producing replacement animals, to cull poor-producing animals, and to assess overall management strengths and weaknesses. Procedures incorporate the performance measurements of ancestors, collateral relatives, and progeny to evaluate the genetic merit of animals on a within-flock basis.

Two programs have been developed by the NSIP, one for purebred flocks and one for commercial flocks. In addition, another manual has been developed to provide the background explanations for the analyses which are provided. The owner enrolls in either of the two programs and· records are processed by the NSIP Processing Center at Iowa State University. Flock expected progeny differences (FEPD) are computed for lambs born, pounds of lamb weaned, and weights for up to three different ages (e.g., 60, 90, 120 days) most useful to the owner. If the owner provides wool measurements, these will also be summarized.

It should again be stressed that these evaluations are only on a within-flock basis. They should provide a sound basis for flock genetic improvement for the interested owners.

Owing to the high cost, progeny testing is limited. However, the extensive use of artificial insemination would permit a more widespread use of the best progeny-tested rams and would justify a more substantial investment in progeny testing programs. In Australia and New Zealand, the establishment of nucleus breeding units in large flocks or cooperative breeding schemes utilizing superior females from a number of flocks make use of progeny-tested males in the next layer of flocks for commercial lamb and wool production.

SUMMARY

Programs for genetic improvement should be based on accurate measures of performance which can be obtained on a large proportion of the breeding population. Evaluation of individual performance for progeny testing is discussed in this chapter. Specific methods and programs vary depending upon the nature of the traits and the husbandry methods for the classes of livestock. Much progress has been made in bringing about uniformity of testing procedures and record summarization during the past decade. The initial diversity of programs provided a broad base from which subsequent decisions concerning procedures for measurements, testing conditions, and record summarization could be made. There will continue to be numerous changes in the performance and progeny testing programs. Much effort will be given to providing truly national evaluation programs, so that the broad base of genetic variability can be utilized with increasing accuracy in selecting sires for use in individual breeding units. With widespread use of artificial insemination, progeny testing becomes much more valuable in all classes of livestock, since outstanding sires can be used extensively once they are identified. The best linear unbiased prediction (BLUP) procedures represent the standard for estimating breeding values. Central testing station evaluations will continue to be useful for situations where artificial insemination is not widespread, where breeding units are small, and for the collection of information on feed efficiency.

APPENDIXES
Example of Pedigree Evaluation

A complete pedigree appraisal can be obtained by a multiple regression approach which would weight phenotypic measures of individuals in the pedigree in accord with the heritability of the trait and the genetic relation of the ancestors to the individual. The genetic relationship among as many ancestors as is desired can be taken into account, although generally little is contributed from individuals more remote than $R = .25$. The pedigree index (PI) can be computed

$$PI = b_1x_1 + b_2x_2 \cdots + b_nx_n = \sum_i^n b_ix_i$$

where b_i = Weight given the phenotypic measure for the ith relative or ancestor
x_i = Measure of the phenotypic value for the ith relative or ancestor as deviations from a population mean

The b_i values are partial regression coefficients. They are partial since two or more are involved in multiple regression, in contrast to one coefficient for simple regression. The relationship between the regression coefficient (b) and the correlation coefficient (r) was discussed in Chapter 5. In a similar manner, the partial regression coefficient (b_i) is related to the standard partial regression coefficient (b_i'). Both the correlation coefficient and the standard partial regression have no units per se. In contrast, the actual magnitude of the regression coefficient and the partial regression coefficient are dependent on the ratio of the variability of the value being predicted ($\hat{y}$) or predicted breeding value (PI), and the independent variables (x').

$$b_i = b_i'\sigma y/\sigma x_j$$

The b_i' are unfettered, so to speak, by the variability of the variables and represent the relative predictive values of the several x_j.

Where the same amount of information is available for each individual in the pedigree (e.g., one record), the predictive values of information on each parent and each grandparent for an outbred pedigree would be presented by the values (b_i') in Table 11-3. Note the decreasing relative weights for the grandparents as the heritability of this trait increases.

With the implementation of the "animal model" for estimating predicted transmitting ability in dairy cattle or expected progeny difference for beef cattle, all pedigree relationships are taken into account. The pedigree index (PTA or EPD) of a young animal can be determined by averaging the values for the parents. For example if the bull and cow on which information is presented in Figure 11-3 were to be mated, the expected (PTA) on pedigree index for the offspring for milk would be

$$PI = \frac{-1134 + 1435}{2} = 151 \text{ lb}$$

The calf's PI for protein would be 36 lb and for fat it would be 53 lb. In all three cases the information on the relatives has been weighted in the multiple regression approach.

An approach to pedigree indexing in dairy cattle that can be done with the desk calculator or even by hand is illustrated below.[16] It is assumed that all production records are on a 305-day, 2X, mature equivalent basis. Female lactation records are expressed as deviations from herdmates or contemporaries, and the sire's transmitting ability is expressed as a predicted difference or predicted transmitting ability.

---

[16] Spalding, R. W., et al. 1963. *Cornell Extension Bul.* 1118.

TABLE 11-3
RELATIVE WEIGHTS FOR A SINGLE RECORD OF EACH PARENT AND GRANDPARENT
FOR SEVERAL VALUES OF HERITABILITY

Regression (b_i)	Heritability				
	.10	.20	.40	.67	1.00
Each parent	.15	.20	.28	.35	.50
Each grandparent	.07	.09	.09	.09	0

When the sire of the young bull is extensively progeny-tested there is little practical justification to give consideration to the paternal grandsire or grandam. Female relatives potentially worthy of consideration are the bull's dam, the dam's progeny (the bull's maternal sisters), the dam's paternal half-sisters (progeny of the maternal grandsire), and the maternal grandam. The PI is computed by following the rules enumerated below:

1 The sire's weighted contribution is .50 times the sire's predicted difference or predicted transmitting ability.

2 The maternal grandsire's weighted contribution (dam's paternal sisters) is .17 times the maternal grandsire's predicted differences or predicted transmitting ability.

3 The weighted contribution of each female relative of the young bull is obtained by multiplying the average deviation from herdmates or contemporaries of each female by the relationship coefficient to the young bull (R) and the weights (W) for the number of lactations included in the female's average.

Relative	Relationship coefficient, R
Young bull and dam	.50
Young bull and full sisters	.50
Young bull and grandam	.25
Young bull and maternal sisters	.25

The weighted contribution of each of the relatives is summed to give the young bull's estimated predicted transmitting ability for milk and fat.

Number of lactations	Weighting, W
1	.12
2	.18
3	.20
4	.22
5	.24
6–10	.25

To illustrate the procedure, consider the pedigree in Figure 11-8, and check the rules previously enumerated. Note each female's contribution will be $R \times W \times D$, where R

NAME: ROCKALLI WARS CAESAR-ET   REG. NO.: 2025254

BORN: AUGUST 22, 1986

BREEDER: ROCKALLI FARMS, MT. VERNON, WA

```
                                      GLENDELL ARLINDA CHIEF           1556373
                                      Excellent (93) Gold Medal 7/87
                                      USDA Sire Summary (7/85)
                                      22384 Dtrs. ( 6200 Herds)
SIRE:                    1697572      Pred. Diff.      + 983M -.02% +31F
ARLINDA ROTATE                        Repeatability 99%; Dollar Difference +$104
Excellent (92) Gold Medal 7/87        HFAA Type Summary (7/87)
USDA Sire Summary (7/87)              12637 Class. Dtrs. Avg. 80.2 (Actual)
2944 Dtrs. ( 1433 Herds)             PDT (99% Repty)      +1.63    TPI +557
Pred. Diff.    +1853M +.09% +83F
Repeatability 99%; Dollar Difference +$238
HFAA Type Summary (7/87)             ARLINDA CHIEF ROSE              7370146
876 Class. Dtrs. Avg. 78.5 (Actual)  Good Plus (82)
PDT (98% Repty)    +1.36   TPI +963   2-0  365d  2x  22500M  3.9%   868F
                                      3-2  308   2x  12810   4.0    511
                                      4-1  365   2x  28800   4.1   1176
                                      5-3  312   2x  15670   3.9    615
                                      6-3  365   2x  32080   4.3   1365
                                      7-9  365   2x  27580   4.0   1112
                                      Lifetime:    166,500M  4.0%  6727F


                                      ROCKALLI SON OF BOVA            1665634
                                      Very Good (88) Gold Medal 1/87
                                      USDA Sire Summary (7/87)
                                      3596 Dtrs. ( 1680 Herds)
DAM:                    10982936      Pred. Diff.      +2063M -.21% +34F
ROCKALLI BOVA CINDY *RT               Repeatability 99%; Dollar Difference +$167
Excellent (91) (EX-MS)                HFAA Type Summary (7/87)
Cow Index(1/87): +1946M + 45F  +$179  1824 Class. Dtrs. Avg. 79.6 (Actual)
MC Dev:        +10450M + 237F ( 2 RECS) PDT (99% Repty)    +2.13   TPI +690
2-0  365d  2x  27960M  3.3%  934F
3-2  365   2x  33910   3.5  1175      ROCKALLI APOLLO CINDER        10097435
                                      Very Good (87)
                                      2-1  365d  2x  23200M  3.6%   828F
                                      3-3  365   2x  26400   3.2    839
```

FIGURE 11-8
Pedigree of an outstanding young sire prospect giving complete production information. The pedigree index of this young bull indicates that his daughters should average 3,385 lb of milk above the average animals of the breed, not accounting for the impact of the genetic trend in the population.

and *W* are taken from the above tables and *D* is the average deviation of their records. These computations are summarized below.

Relative	Calculation	Contribution
Sire	1,853 × .50	926 lb
Maternal grandsire	2063 × .17	351
Dam	10,450 × .24 × .50	1,254
		2,531

We note that the average of this young bull's daughters is predicted to be 2,531 lb of milk above the breed average. This does not take account of any genetic improvement in the population during the approximately 5 years the progeny test records are being accumulated. If the genetic change ΔG is 100 lb per year, the pedigree estimate would be discounted by 250 lb ($5 \times \frac{1}{2} \times 100$). The one-half enters the calculation because the cows to which the young bull is mated will transmit only one-half of the accumulated superiority.

Computing limitations and costs have until recently prevented the use of the animal model for BLUP estimation. However, rapid advances in reducing the number of equations that need to be solved have made it useful in many national sire evaluation programs for beef cattle. Currently, BLUP procedures have replaced the national dairy modified contemporary comparison procedure.

Use of Linear Models in Estimating Breeding Values

Mixed linear model methods for sire evaluation were developed by Henderson,[17] and they provide a powerful tool for use under a wide variety of situations. These techniques combine the desirable properties of the selection index and the capacity of linear model methods with electronic computers to deal with large data sets with unequal subclass numbers.

The general problem is to obtain the best evaluation of an animal that is regarded as a random individual from a specified population or group. The definition of "best" is the evaluation of breeding value as a linear function of the observations which is unbiased and has the smallest possible variance for the error of prediction. Procedures derived to meet these criteria have been termed the best linear unbiased prediction (BLUP). While the conceptual framework is more comprehensive, the computational techniques parallel least-squares procedures for fitting constants when unequal subclass numbers are involved.

Actually the modified contemporary comparison for dairy sires was an empirical approach to gain many of the advantages of BLUP principles, even though computationally feasible approaches were not available to cope with the volume of data required for a national dairy sire evaluation when the MCC was implemented. Methodology for beef sire evaluations with performance testing and artificial insemination now draw heavily upon BLUP methods. Certain conditions which make the simple methods of sire evaluation inadequate were identified in the previous discussion in this chapter of the MCC for dairy sires. They include nonrandom distribution of bulls' progeny across herds, genetic trends, and overlapping generations.

Various BLUP models can be developed that are adaptable to a wide range of problems and situations. Appropriate linear models for the individual observations plus matrices of the genetic relationship (numerator of the relationship coefficient) can permit the extraction of information on the comparison of sires across herds or breeding groups when direct progeny comparisons are not available.

[17] Henderson, C. R. 1974. *J. Dairy Sci.* 51:963–972.

Initially, only the records of the progeny of the sires to be compared were included in the data set (sire model), since the volume of data taxed the capacity of earlier computers. The primary objective of the sire model was to obtain estimates of the breeding values of the array of sires. Developments in methodology, paralleling advances in computer technology, have enabled more information to be used and have given increased accuracy. Advancing from the sire model, the sire-maternal grandsire model, now the animal model, with its many variations has become available. The following example will illustrate BLUP as used with the animal model in USDA dairy sire and cow evaluations.

Actual BLUP procedures begin with the development of a linear model or equation to describe each performance record such as was given on page 282:

$$y_{ijkl} = m_{ij} + a_{kl} + p_{kl} + c_{ik} + e_{ijkl} = (Y_{ijkl} - u)$$

The data for this example are protein yields (in kilograms) of cows given in Table 11-4.

From the linear model and the tabular data presented, an equation can be written for each management group (M_{ij})†

$$M_{11} = 5\mu + 5m_{11} + 2c_{11} + 2c_{12} + c_{13} + (a_{11} + \cdots + a_{31}) + (p_{11} + \cdots + p_{33}) = 1700$$

$$M_{22} = 4\mu + 4m_{22} + c_{21} + c_{22} + c_{23} + (a_{13} + \cdots + a_{33}) + (p_{13} + \cdots + p_{33}) = 1120$$

Likewise an equation can be written for each herd-sire effect (C_{ik}).

$$C_{11} = 3\mu + 2m_{11} + m_{12} + 3c_{11} + (a_{11} + 2a_{12}) + (p_{11} + 2p_{12}) = 970$$

$$C_{23} = 2\mu + m_{21} + m_{22} + 2c_{23} + (2a_{33}) + (2p_{33}) = 520$$

† Capital letters for the effects reflect inclusion of the mean, μ.

TABLE 11-4
DATA FOR EXAMPLE TO EVALUATE PROTEIN PRODUCTION

Management group	Herd-sire	Cow ($a_{kl} + p_{kl}$)									
m_{ij}	c_{ij}	11	12	21	22	31	32	13	23	24	33
11	11	300									
11	11		350								
11	12			380							
11	12				270						
11	13					400					
12	11	320									
12	12		360								
12	12			250							
12	13					300					
12	13						280				
21	21							290			
21	22								300		
21	22									320	
21	23										270
22	21							270			
22	22								310		
22	22									290	
22	23										250

Subsequently equations for each animal (A_{kl}) can be written which include a_{kl} and p_{kl}.

$$A_{11} = \mu + m_{11} + c_{11} + a_{11} + p_{11} = 300$$

.   .   .

.   .   .

.   .   .

$$A_{33} = \mu + m_{21} + m_{22} + 2c_{33} + 2p_{33} = 520$$

The mixed-model BLUP equations for these data can be summarized in matrix form and are given below. Values for k_a and k_p modify the Z matrices in the animal equations to estimate $\hat{a}$ and $\hat{p}$.

$$
\begin{bmatrix}
X'X & X'T & X'Z_1 & X'Z_2 \\
T'X & T'T + k_c I & T'Z_1 & T'Z_2 \\
Z_1'X & Z_1'T & Z_1'Z_1 + k_a A^{-1} & Z_1'Z_2 \\
Z_2'X & Z_2'T & Z_2'Z_1 & Z_2'Z_2 + k_p I
\end{bmatrix}
\begin{bmatrix}
\hat{m} \\
\hat{c} \\
\hat{a} \\
\hat{p}
\end{bmatrix}
=
\begin{bmatrix}
X'y \\
T'y \\
Z_1'y \\
Z_2'y
\end{bmatrix}
$$

where X = a matrix relating elements of m to y
T = a matrix relating elements of c to y
$Z_1 = Z_2$ = a matrix relating elements of a and p to y
I = the identity matrix

$k_c = \sigma^2 e/\sigma^2 c = 3.2$, $k_a = \sigma^2 e/\sigma^2 a = 1.8$, $k_p = \sigma^2 e/\sigma^2 p = 2.8$, and A^{-1} is the inverse of the numerator relationship matrix. It can include related individuals, such as the sires or dams, which have no performance records.[18]

The effects of $\hat{m}$, $\hat{c}$, $\hat{a}$, and $\hat{p}$ are vectors identified in the original linear model.

Pedigree information is needed to compute A^{-1}. This is given in Table 11-5. Note that for this example, it is assumed that the dams were unknown. Animals 11, 12, and 13 represent the sires, and their breeding values or PTAs are to be derived from their daughters' performance. For simplicity, it is further assumed that the grandsires and grandams of the three sires are unknown.

The A^{-1} matrix of these relationships for the 10 cows and their sires is given below. An example of the relationship matrix was given in Chapter 9.

c_{11}	c_{12}	c_{21}	c_{22}	c_{31}	c_{32}	c_{13}	c_{23}	c_{24}	c_{33}	s_{11}	s_{12}	s_{13}
1.0	−10.0	−6.0	2.0	0	0	0	0	0	0	−2.0	0	0
−10.0	−6.0	0	0	0	2.0	0	−2.0	0	0	0	−2.0	−2.0
6.0	0	9.0	−3.0	−3	0	0	0	−2.0	−2.0	0	0	0
2.0	0	−3.0	5.0	1.5	−3.0	0	0	0	0	0	0	0
0	0	−3.0	1.5	7.5	−3.0	−3.0	0	0	0	0	0	0
0	−2.0	0	−3.0	−3.0	5.5	1.5	0	0	0	0	0	0
0	0	0	0	−3.0	1.5	7.5	−3.0	−3.0	0	0	0	0
0	−2.0	0	0	0	0	−3.0	5.5	1.5	0	0	0	0
0	0	−2.0	0	0	0	−3.0	1.5	5.5	0	0	0	0
0	0	−2.0	0	0	0	0	0	0	4.0	0	0	0
−2.0	0	0	0	0	0	0	0	0	0	4.0	0	0
0	−2.0	0	0	0	0	0	0	0	0	0	4.0	0
0	−2.0	0	0	0	0	0	0	0	0	0	0	4.0

Regularized procedures to create the complete BLUP matrix given previously and to obtain solutions from the inverse are cumbersome and costly. Schaeffer and Kennedy[19] developed a computational method for solving mixed-model equations by iteration without creating the complete coefficient or incidence matrix. Misztal and Gianola[20] have proposed refinements in the iterative procedures that reduce computational time and expense. This is the essential methodology used in the USDA sire and cow evaluations.

First, the solutions for the managemental groups are obtained; second, from these solutions the other effects are computed. For this example the solutions for the four management groups which include the mean are

$$M_{11} = 338.7, \ M_{12} = 298.4, \ M_{21} = 292.2, \ M_{22} = 279.2$$

[18] Henderson, C. R. 1976. *Biometrics* 32:69.
[19] Schaeffer, L. R., and B. W. Kennedy. 1986. *J. Dairy Sci.* 69:575.
[20] Misztal, I. and D. Gianola. 1986. *J. Dairy Sci.* 70:716.

TABLE 11-5
PEDIGREE RELATIONSHIPS FOR ANIMALS IN EXAMPLE

	Animals	Sire	Dam
	11	11	0
	12	11	0
	21	12	0
	22	12	0
Cows	31	13	0
	32	13	0
	13	11	0
	23	12	0
	24	12	0
	33	13	0
	11	18	19
Sires	12	17	16
	13	14	15
	14	0	0
	15	0	0
Sires' parents	16	0	0
	17	0	0
	18	0	0
	19	0	0

The values as deviations from the mean, using the restriction that the sum of the $m_{ij} = 0$, become

$$\hat{m}_{11} = 36.6, \ \hat{m}_{12} = -3.7, \ \hat{m}_{21} = -9.9, \ m_{22} = -22.9$$

Since $\hat{m}_{11}$ and $\hat{m}_{12}$ represent herd 1, these values indicate that the management was superior to that for herd 2.

The primary concern is to evaluate the a_{ik}. The estimates of these additive breeding values for the cows are

cow 11 = -8.6	cow 32 = -5.1
cow 12 = 6.0	cow 13 = -3.0
cow 21 = 17.2	cow 23 = 4.2
cow 22 = -18.8	cow 24 = 4.2
cow 31 = 9.3	cow 33 = -5.5

The values for the sires are

sire 11 = 1.0, sire 12 = 10.9, sire 13 = 3.8

As related on page 282, unknown parents are identified by groups. Base breeding values are computed for these parents and the parents of the sires. These are appropriately weighted in computing the above breeding-value estimates. Finally it should be remembered that the predicted transmitting abilities (PTA) are reported by the USDA. These transmitting abilities are one-half the additive breeding values computed in the above example.

SUGGESTIONS FOR FURTHER READING

Articles and Bulletins

Beef Improvement Federation, 1986. "BIF Guidelines for Uniform Beef Improvement Programs," 5th ed. North Carolina State University, Raleigh 27695-7621.

Benyshek, L. L., M. H. Johnson, D. E. Little, J. K. Bertrand and Kriese, L. A. 1988. Applications of an Animal Model in the United States Beef Cattle Industry. *J. Dairy Sci.* 71 (Supplement 2): 35–53.

Cassell, B. G. 1988. What Extension Workers Need to Tell Dairy Farmers. *J Dairy Sci.* 71 (Supplement 2): 85–90.

Dickerson, G. E., and L. N. Hazel, 1944. Effectiveness of Selection on Progeny Performance as a Supplement to Earlier Culling of Livestock. *J. Agr. Res.* 69:459–476.

Dickinson, F. N., R. L. Powell, H. D. Norman, L. G. Waite, and B. T. McDaniel. 1976. The USDA-DHIA Modified Contemporary Comparison Sire Summary and Cow Index Procedures. U.S. Department of Agriculture, ARS Production Research report 165.

Henderson, C. R. 1973. Sire Evaluation and Genetic Trends. Pp. 10–41. in "Proceedings, Animal Breeding and Genetics Symposium in Honor of Jay L. Lush." American Society of Animal Science, Champaign, Illinois.

Henderson, C. R. 1974. General Flexibility of Linear Model Techniques for Sire Evaluation. *J. Dairy Sci.* 57:963–972.

Henderson, C. R. 1975. Use of Relationships among Sires to Increase Accuracy of Sire Evaluation. *J. Dairy Sci.* 58:1731–1736.

Lush, J. L. 1936. Genetic Aspects of the Danish System of Progeny Testing Swine. Ia. Agr. Exp. Sta. Res. Bul. 204.

National Swine Improvement Federation. 1987. "Guidelines for Uniform Swine Improvement Programs." U.S. Department of Agriculture Science and Education Administration Extension. Washington, D.C. 20250.

Robertson, A. 1957. Optimum Group Size in Progeny Testing and Family Selection. *Biometrics* 13:442–450.

Robertson, A., and J. M. Rendel. 1950. The Use of Progeny Testing with Artificial Insemination in Dairy Cattle. *J. Genet.* 50:21–31.

Skjervold, H., and H. J. Langholz. 1964. Factors Affecting the Optimum Structure of A.I. Breeding in Dairy Cattle. *Z. Tierz. Zuchtungsbiol.* 80:25–40.

Wiggans, G. R., I. Misztal, L. D. Van Vleck. 1988. Implementation of an Animal Model for Genetic Evaluation of Dairy Cattle in the United States. *J. Dairy Sci.* 71 (Supplement 2): 54–69.

12

DEVELOPING BREEDING GOALS AND PLANS

The material presented in the earlier chapters sought to provide an insight into the fundamental principles available for livestock improvement. These principles must be meshed together and tempered with livestock experience and the economic realities of animal production to develop a productive program. Care, foresight, and persistence are also required to fashion breeding goals which are sound in principle and which will meet the demands of the consuming public. Insensitivity to consumer needs and desires will not be tolerated as substitutes for animal products expand. Concerns about animal products in connection with diet and health have heretofore projected a negative image, but new information now appears to have provided a positive thrust. Animal agriculture must be competitive and capitalize on its inherent strengths. Much new technology will be emerging, which must be judiciously integrated into animal enterprises.

REQUIREMENTS FOR GENETIC IMPROVEMENT

In specific breeds, seed-stock strains or commercial herds, the general requisites for genetic improvement are essentially the same. First, we must assess the genetic merit of our present animals. This requires accurate records of performance for animals of known ancestry. Although tremendous strides have been made in this area in the last decade, much further progress is needed.

In the United States, recognized herd recording schemes for dairy cattle have been in progress since 1906. Approximately 40 percent of the lactating animals are tested for production to assist in dairy herd management. A valuable by-product of this program has been the information for progeny testing.

Beef performance testing began in 1953, and at this writing national sire evaluation programs are in progress for 15 breeds. Although performance testing of swine began in 1936, it has been difficult to develop a truly national program. Sires are used naturally in individual herds with very limited use of artificial insemination. Standardization of swine testing procedures for on-farm and central test stations has been a major advance. A similar on-farm program recently has been initiated for sheep producers.

Second, the influence of animals with the desired genes must be extended and made available to commercial herds. As recording and evaluation procedures have advanced, methods for increasing the number of progeny from superior males and females also have moved forward. Artificial insemination with frozen semen has been most helpful for beef and dairy cattle. Storage of semen in liquid nitrogen has made the best sires available almost anywhere in the world.

Multiple ovulation and egg transfer (MOET) systems can extend the potential impact of superior females. In certain instances MOET programs with nucleus herds should offer the opportunity for advancing the rate of genetic improvement where artificial insemination is not widespread. Even with artificial insemination, it can provide a means for pyramiding genetic improvement in nucleus herds. Surely these techniques, along with cloning, will offer much to extend the influence of superior individuals to enhance future genetic improvement.

BREEDING GOALS

In developing goals, animal breeders must continue to seek out the potential requirements of the consuming public. Insensitivity to consumer needs will be tolerated less and less as competition from substitutes for animal products becomes keener. The expanding United States population, with predictions of over 250 million by 2000, suggests an increase in the demand for food products. If per capita incomes continue to rise, the demand for animal products should continue to be strong. Trends in countries where incomes have risen point out the strong demand for animal protein.

If our future citizens continue to require annually 40 to 45 kg of beef, 25 to 30 kg of pork, 27 to 32 kg of poultry, and the equivalent of 270 kg of milk per capita, increases in production will be necessary. The major portion of the additional meat needs from beef and swine would be expected to come from increased numbers of animals. Efficiencies in production per head would make a measurable but less dramatic contribution to the meat supply. Current trends in dairy cattle production suggest that most of the increased milk needs could be obtained from increasing production per animal. Neither present levels of milk production per cow nor increases in meat output per animal can be maintained without reliance on concentrate feeding. Hence, competition between animals and humans for cereal grains and other concentrated foods may be anticipated in the future.

The surplus of cereal grains we have enjoyed for most recent decades has meant that we have been selecting animals under conditions where the nutritional regimens have included liberal feeding of concentrates. If our future livestock must serve only as harvesters of roughage from nontillable grass lands or as gleaners of by-products which are not satisfactory for human consumption, serious difficulties may result. Would the animals which have been selected under conditions of liberal concentrate feeding respond so poorly under heavy roughage feeding that we would lose much of their productive efficiency? Attention is being given to this point, and in most areas of the world, livestock must continue to perform largely on roughage and by-product rations. In limited experimentation with dairy cattle and beef cattle it would appear that the animals which do best on high-roughage rations are also those which have the best genotype for utilizing rations with a high proportion of concentrates. It is true that the levels of performance vary widely depending upon whether additional concentrated energy is available or whether roughage is the main source of nutrients. However, there is relatively little evidence to suggest that these interactions are of major practical importance.

Large quantities of cereal grains may not always be available for livestock production. Perhaps the change will be reasonably gradual so that available genetic variability for utilization of these foodstuffs may be captured in the selection process during the nutritional transition. Actually the evolution of the bovine has been based primarily on foraging. The advent of high-concentrate feeding is of recent origin, and it could hardly be expected that major genetic changes in the control of the ruminant digestive system would have occurred from such recent indirect selection. Furthermore, ruminants appear to have considerable adaptability to a wide range of different rations without a drastic sacrifice in efficiency of utilization.

In shaping our goals in animal breeding, more emphasis in the future must be given to quality. The livestock industry must produce quality products to stimulate consumption. On the other hand, markets must recompense the producer for the extra quality in order to justify the special effort in breeding and production costs. The contrast of protein versus fat in meat and even in milk is now in open discussion and transition. Increased use of processed and ground meats rather than the complete cuts tends to obscure the marbling and other special features in much of the meat that is marketed. Homogenization of milk destroyed the creamline many years ago, although fat is still in surplus. Both lard and milk fat have been given stiff competition during the past three decades by vegetable oils and fats. There is little evidence that competitive pressures will diminish. Protein from animal sources continues to be in demand when incomes are strong. Substitute products should not have a further depressing effect on animal protein demand at least well into the 1990s. Yet as a new generation of consumers emerges, the stigma against substitute products will dwindle gradually, particularly if their nutritional content is fortified.

Breeding Plans and Programs

Since the 1930s, numerous advances and sophistications in breeding methodology for animal improvement have emerged. Much of the early work of Fisher and Wright was translated by Lush for use with populations of farm animals. Considerable effort has been expended in research areas to determine the genetic variability in numerous economic traits through heritability analyses. Undoubtedly there may be justification for continuing certain aspects of such studies, but they must be more incisive and thorough. Much can be gained from studies with laboratory mammals, and the use of such experiments will continue.

Ideally, an accurate assessment of the nature of genetic variation should provide the basis for devising breeding strategies. Considerable effort has been expended to assess the importance of additive genetic, dominance, and epistatic variance. For most traits in farm animals the importance of variance due to the additive effects of genes can be estimated with reasonable precision. However, knowledge of the magnitude of variance due to dominance and epistasis are woefully imprecise. Thus most breeding plans focus on improving additive breeding merit, utilizing selection with its various aids and modifications.

If simple dominance is the sole reason for departure from additivity in the relation between phenotypes and genotypes, selection can still be a major force in improvement, even though its rate of effectiveness will diminish as the frequency of desired genes exceeds one-half. Significant overdominance would require special cross-population selection to capitalize on these effects. Our knowledge of the physiological nature of gene action at this level is extremely meager. Even more elusive is the determination of epistatic effects. Selection between families or strains would be logical to attain desired gene combinations. The ultimate goal might be perceived as the attempt to produce breeds or strains homozygous for the desired combination of genes. Assuredly, if many loci are involved, it will be almost impossible to embody all the desired genes in one genotype.

Increasing recognition of the value of heterosis or hybrid vigor has resulted in crossbreeding becoming the major commercial breeding system for poultry and swine production. It has also become increasingly important in beef production. Half or more than half of the beef animals marketed are now believed to be crossbreds. Crossbreeding has long been a standard practice in sheep production. Horses needed for special purposes are produced by crossbreeding. The exploitation of hybrid vigor will surely be of concern in the years ahead. It is only in dairy cattle and horses bred for racing that crossbreeding is used to a limited extent. This is basically due to the fact that for these purposes, one breed has been so superior that hybrid vigor is unlikely to be important enough to overcome deficiencies in the additive contribution of the inferior performing breed(s) available for crossing.

Inbreeding does not appear to be destined to make a major future contribution to breeding programs in farm animals. Reproductive fitness rapidly be comes a limiting factor when inbreeding is undertaken in large animal populations. This is particularly true in cattle, where there is limited reserve reproductive capacity. Inbreeding also drastically narrows the base of selection and, hence, the selection differential. Heterosis will be utilized largely through crossbreeding systems, where individual breeds or strains may be selected for criteria which will enable them to meet specific roles in crossbreeding systems. Maternal performance is an important consideration which supports production of special crossbred females as dams for commercial beef and swine production. Utilization of heterosis in the final analysis is dependent upon the development of specialized seed stock—either improved purebreds in the traditional sense or seed stock from new breeding concepts.

The possibility that we may be reaching genetic limits in our farm animal populations during the next 20 years is rather remote. Nevertheless, we should be on the lookout for signs of diminishing rates of genetic improvement. Along with this we will need to find more critical evidence about genetic correlations and especially those negative ones which influence our economic traits. The concept of the selection index has provided a logical framework for planning selection programs, but we still find too limited use of these in practice.

As physiological techniques improve and further understanding of embryo culture is gained, the usefulness of outstanding females will continue to be extended. Ova transfer is now practically operational, but it does not give promise of extending the use of outstanding females to the extent that artificial insemination has done with males. The current use of stored deep-frozen fertilized ova should permit establishing large elite families of breeding females. Perhaps an almost-unlimited number of offspring from a most favorable genotype might be obtained if individual cells could be taken from a culture of the early developing embryo.

Additional information about the basic physiology of the various traits should assist us in genetic improvement. The real question concerns the level at which our inquiry should be directed. It has been proposed that assays of those hormone and enzyme systems which influence the expression of a trait should provide an effective selection aid. Attempts in this area to date have not been very productive.

With a process as complicated as growth or lactation, involving the interaction of an innumerable complex of hormones and enzymes, progress will not be rapid or easy. When the number of enzymes involved approaches the number of loci affecting the trait, we can see the improbability of being able to put the individual genes in their proper perspective. The fact that we presently have difficulty putting the individual traits in their appropriate perspective, even with the mathematical logic of the selection index, points up the magnitude of the challenge.

Attempts to guide genetic change at the chromosome and gene level in the past have generally failed to match expectations. Mutagenic agents have appeared to produce random changes in the genome. Even in plant species where apparently useful artificial mutants have been incorporated into the genome, extensive screening and discarding of untold numbers of undesirable mutants have been necessary.

A deep and abiding conviction exists among researchers that "genetic engineering" may be more directive in the future. The chemical structure of the gene has been determined, and genes have been synthesized. The incorporation of genetic material from two or more species has been accomplished. Viewpoints have been expressed that DNA fractionation, identification, and recombination will obviate the need for the laborious breeding tests now required for gene identification. Experimental verification will be the acid test of such options. Knowledge to control the turning on and off of genetic function still remains elusive. Nevertheless, there are many exciting possibilities ahead of us.

GLOSSARY

acquired character This term applies to the possibility of an environmentally induced change in the body becoming hereditary. It has not been proved to occur.

additive genetic variance Genetic or hereditary variance dependent upon additive gene effects, i.e., a gene has a given plus or minus effect, regardless of which other member of the pair or allelic series may be present.

alleles Members of a pair (or series) of different hereditary factors which may occupy a given locus on a specific chromosome and which segregate in formation of gametes.

analysis of variance A statistical technique for apportioning variance to its sources. In genetics and animal breeding it is used to determine relative influences of heredity and environment on variation in traits.

ancestor Animal of a previous generation that has passed on genes through a line of descent.

androgen A generic term for hormones that stimulate activity of accessory sex organs and sexual characteristics in males. Testosterone is one of these hormones. They are often termed *male sex hormones*.

anestrus A period in which a female does not experience estrus or estrual cycles.

antibodies Specific chemical substances developed by an animal in response to introduction of antigenic substances into the body. Some antibodies occur naturally without introduction of antigens. An example is the naturally occurring antibodies to the human A and B blood groups in individuals not having the blood groups.

antigens Chemical substances, usually complex proteins, which have the capability of stimulating formation of specific antibodies when introduced into animals not possessing them.

artificial insemination Introduction of sperm into the reproductive tract of the female by mechanical means rather than by natural mating.

atavism   Reappearance of an ancestral trait or character after a skip of one or more generations. Also referred to as *reversion*. Usually results from recessive genes being present in the homozygous condition in an individual after having been hidden in ancestors by their dominant alleles.

autosomes   Chromosome pairs which are alike in both sexes.

binomial distribution   A mathematical method for determining probabilities of occurrence of specific combinations of two independent events.

biochemical genetics   *See molecular genetics.*

biochemical polymorphism   A general term usually applied to genetically determined variants of protein, although the term is broad enough to include other substances as well. Polymorphisms of blood and milk have been studied most extensively, but they occur in many other body substances.

biotechnology   The application of biological and engineering technology to animals, plants, and microorganisms. On occasion used in the narrower sense of genetic engineering.

blastocyst   An early embryonic stage beginning about the seventh day after fertilization when cell division has formed a circular hollow ball of cells.

blood groups   Immunologically different blood types due to hereditary differences in antigens or antigenic factors carried on red blood cells.

boar   Male swine (*Sus scrofa*).

breed   Group of animals having a common origin and identifying characters that distinguish them as belonging to a breeding group.

breeding value   Genetic worth of an animal's genotype for a specific trait.

bull   Male cattle (*Bos taurus* or *Bos indicus*). Term is also used for males of several other species.

calf   Young cattle (*Bos taurus* or *Bos indicus*) of either sex to about a year of age. Also used for young of other species.

castration   Removal of the testes of the male.

cattalo   A term applied to animals descending from interspecific crosses of cattle (*Bos taurus* or *Bos indicus*) and American bison (*Bison bison*). They may vary in percentage of inheritance from the bison.

cell   The basic functional unit of all higher living organisms.

chi-square   A mathematical test for goodness of fit of experimental data to expectation. Used in genetics to determine the probability of deviations from expected distributions of phenotypes in offspring being due to chance.

chromosomal aberration   Any deviation from the norm of a species in chromosome number or morphology.

chromosomes   Darkly staining bodies in cell nuclei which carry the hereditary material. They occur in pairs in somatic cells with the number of pairs being characteristic of the species.

cistron   A term used to represent the unit of DNA carrying the information necessary for the formation of one polypeptide chain in protein formation. Functionally, it is equivalent to the gene.

cloning   The process of producing many copies of a single ancestral gene or DNA sequence. In animals and plants it is also the replication of cell lines, tissues, or organisms from somatic or nonreproductive tissue.

codon   A three-letter (three-nucleotide) unit that is part of the genetic code. The codon starts or terminates the series of linked amino acids that constitute a protein molecule and specifies the addition of a particular amino acid at a specific location in the molecule.

coefficient of variation The standard deviation of a trait in all individuals of a population expressed as a percentage of the mean.

collateral relative Individuals not related directly; e.g., aunts, uncles, cousins.

commercial herds Herds maintained for the purpose of producing meat, milk, wool, or other animal products for sale in commercial channels.

conformation Externally visible or measurable variations in shape or body proportions of animals (see also *type*).

correlation Association between characteristics of individuals. The *correlation coefficient* is a statistical measure of degree of association and varies from -1.0 to $+1.0$ (see also *regression*).

covariance Variation that is common between two traits. It may result from joint hereditary or environmental influences.

cow A mature cattle (*Bos taurus* or *Bos indicus*) female. The term is also used for females of a number of other species.

crossbred An animal produced by crossing two or more pure breeds.

crossbreeding Mating systems in which hereditary material from two or more pure breeds is combined.

cytoplasm Nonnuclear portions of the cell.

dam Female parent, the mother of an animal.

deleterious genes Genes which in either the homozygous or heterozygous state have undesirable effects on an individual's viability or usefulness.

deoxyribonucleic acid (DNA) The basic hereditary material of all living matter. It is composed of basic units or nucleotides each of which contains an organic base, a sugar, and a phosphate. It is a chemically complex substance with gigantic molecules in a spiral, double-helix configuration capable of virtually infinite numbers of structural variations.

diploid Cells with two members of each pair of chromosomes. This is termed the $2n$ condition and is characteristic of body cells of all higher animals.

dominance variance That portion of the hereditary or genetic variance over and above that which can be accounted for by additive effects and which is due to dominance. Alternatively, it can be defined as due to *dominance deviations* from a description based upon assumed additive effects.

dominant Genes that have an observable effect when present in only one member of a chromosome pair.

Drosophila melanogaster A small insect (fruit fly) used extensively in genetic research.

ejaculate Semen produced by a male in a single mating act or ejaculation.

embryo Early stage of development of the new individual following fertilization to approximately the time at which organ formation is completed.

embryo transplant Artificial transfer of the embryo from the natural mother to a recipient female by mechanical means.

environment All the external factors within which an animal's genotype acts to determine its phenotypic traits.

environmental variance The variance, in absolute terms, for any character in a population which is due to environmental influences.

enzyme A protein that catalyzes a specific chemical change without being used up in the reaction.

epistasis Genetic effects due to interactions among two or more pairs (or series) of nonallelic genes.

epistatic variance That residual portion of the hereditary variance due to nonallelic gene interactions not accounted for by additive or dominance effects.

estrogen A generic term for substances with biological effects characteristic of estrogenic hormones. Often called *female sex hormones*. They are involved in many reproductive functions including inducing estrus.

estrual cycle A rhythmic or cyclic pattern of reproductive activity in the female. An estrual cycle includes activity from the beginning of one estrus to the beginning of the next.

estrus The period of sexual receptivity of the female. In popular usage it is often called the *heat period*.

ewe A female sheep (*Ovis aries*). Used alone the term indicates a mature female. Also used with other terms to indicate sex of younger animals, e.g., "ewe lamb."

F_1 The hybrid offspring or first filial generation from a given mating.

F_2 Offspring of $F_1 \times F_1$ matings.

F_3 Offspring of $F_2 \times F_2$ matings.

F_n Extension of foregoing.

family Term used to denote relationship. In animal breeding sometimes used to denote a line of descent (similar to family names in people) but more often to represent a group of animals having a genetic relationship.

fertilization Union of the sperm and ovum to produce the fertilized ovum or zygote.

fetus A developing individual between completion of organ formation and birth.

follicle-stimulating hormone (FSH) A hormone secreted by the pituitary gland which stimulates follicle growth in the female and is involved in development of seminiferous tubules and Sertoli cells and with sperm production in the male.

freemartin A sterile, sexually maldeveloped female born as a twin with a male. Found primarily in cattle but may occasionally occur in other species.

full sibs Individuals with the same sire and dam, full brothers or full sisters, or a full brother–sister pair.

gamete Reproductive or germ cell. In animals, the male gamete is the sperm or spermatozoa and the female gamete is the ovum. Gametes carry the reduced or n number of chromosomes.

gene The classical term for the basic unit of heredity. Functionally, it is equivalent to the cistron.

gene frequency The proportion, in a population, of the loci of a given allelic series occupied by a particular gene.

generation interval Average age of parents when their offspring are born.

genetic correlation Association among traits of individuals due to additively genetic influences.

genetic drift Changes in gene frequency in a population due to chance variations in proportions of gametes which are formed carrying specific genes or which succeed in accomplishing fertilization.

genetic engineering Alteration of the genetic makeup of an organism by direct human intervention.

genetic variance See *hereditary variance*.

genetics The science concerned with determining the mode of inheritance or the transmission of biological properties from generation to generation in plants, animals, and lower organisms.

genotype   The complete genetic makeup of an individual.

grade   An animal carrying a preponderance of the hereditary material of a single pure breed but not eligible for registration in the breed's herdbook.

grading   A system of breeding purebred sires of a given breed to nondescript, scrub, or native females and to their female offspring generation after generation.

half sib   One of a pair of animals having one common parent, half brother, or half sister.

haploid   Cells with one member of each chromosome pair. This is termed the *n* condition and often is referred to as the *reduced chromosome number*. The reproductive cells or gametes have a haploid number of chromosomes.

Hardy-Weinberg law   A law which states that in a population mating at random, the proportion of different types of zygotes produced for any allelic pair or series is directly proportional to the square of their respective gametic frequencies.

hereditary   A condition controlled or influenced to some degree by gene action. This is in contrast to characters which are entirely controlled by environmental variables.

hereditary variance   The variance, in absolute terms, for any trait in a population which is due to genetic influences, additive, dominance, and epistatic effects. (See also *heritability*.)

heritability   That fraction of the total variance for any trait in a population which is due to additively genetic effects.

heterosis   Positive or negative differences in performance of progeny from the *average* of the *parental types* (see also *hybrid vigor*).

heterozygote (adj. heterozygous)   An individual in which a given locus in a chromosome pair carries unlike members of a pair or series of alleles.

heterozygote superiority   Gene pairs (or series) in which heterozygous individuals are superior to any homozygote of the pair or series. Technically called *overdominance*.

homozygote (adj. homozygous)   An individual in which both members of a chromosome pair carry the same gene at a specific locus. Homozygotes are therefore genetically pure for a given pair or series of hereditary factors.

hormone   A substance secreted and released into the bloodstream by one ductless gland or organ which acts upon a specific tissue, organ, or another ductless gland.

hybrid   Technically, refers to the offspring of parents which are each genetically pure (homozygous) for one or more pairs of hereditary factors, but with the two parents being homozygous for different members of allelic pairs or series. In practice the term has been extended to include offspring of species crosses, to progeny of crosses of inbred lines, and in some cases to breed crosses.

hybrid vigor   Increased vigor or productivity often observed in hybrid, crossbred, or crossline individuals as compared to that of the *average* of the *parental types* (see also *heterosis*).

inbreeding   A system of mating in which mates are more closely related than average individuals of the population to which they belong.

independent assortment   Refers to behavior at meiosis of genes located on different chromosome pairs.

jack   A male ass (*Equus asinus*).

jenny   A female ass (*Equus asinus*). Also called a *jennet.*

lamb   A young sheep (*Ovis aries*) of either sex to about a year of age.

lethal gene   A gene which results in death of an individual at some stage of life. Lethal genes may be dominant and exert their effect in heterozygotes. Such genes are comparatively rare and difficult to study since they are rapidly eliminated from a population unless their effects occur late in life after affected individuals have produced offspring. Most lethal genes are recessive and exert their effects only when homozygous.

libido   Sex drive necessary for the mating act to occur.

linebreeding   A form of inbreeding in which an effort is made to maintain high relationships in subsequent generations with a favored ancestor.

linkage   Refers to gene pairs (or series), members of which are on the same chromosome, and tend to remain together at meiosis more frequently than would be expected if they segregated independently.

locus   Region on chromosome where a specific gene pair or allelic series is located.

luteinizing hormone (LH)   A hormone secreted by the pituitary gland which controls ovulation in the female and stimulates interstitial cell development in the male.

maternal impression   An old but unproven belief that characteristics of the offspring can be influenced by what a pregnant mother sees, hears, or experiences while pregnant.

mean   Average of all measurements of a given trait in a population.

meiosis   Cell division during germ cell formation in which chromosome number is reduced with each daughter cell receiving only one member of each chromosome pair.

Mendel, Gregor   An Austrian monk who discovered the basic laws of inheritance in experiments with peas. His results, published in 1866, were not recognized as important until 1900.

migration   In a genetic sense, the introduction of genes into a population from a source not hitherto a part of the population.

mitochondria   Bodies in the cell cytoplasm which are rich in fats, proteins, and enzymes and which produce energy for the cell.

mitosis   Cell division in which each chromosome duplicates itself and the daughter cells each have the same number of chromosomes as the parent cell.

mode   Class with the highest frequency when measurements of a given trait in a population are tabulated.

molecular genetics   The science having to do with genetic variation at the molecular level and with the biochemical characteristics of basic hereditary materials. Often called *biochemical genetics*.

monoclonal antibody   An antibody produced by cells developed from fusion of a single antibody producing animal cell with a myeloma cancer cell. Antibodies produced are specific for a given antigen.

monoestrous   Species in which females have only one estrus cycle per year.

mule   The sterile offspring of a cross between the jack (*Equus asinus*) and the mare (*Equus caballus*).

multiple alleles   A series of more than two genes which can occupy a particular locus on a chromosome.

multiple-gene heredity   Hereditary situations in which more than one gene pair (or series) influences a specific character of an animal or plant.

mutation   A sudden, heritable change in genetic material. Chemically, a mutation is due to a change in DNA at a particular point on a chromosome.

muton   The smallest genetic unit capable of change or mutation. A single base in a nucleotide.

nicking   A situation in which offspring are superior to either parent or in which unexpectedly favorable results are obtained from crosses of two breeds or strains. (See also *specific combining ability*.)

normal curve A graphic representation of frequencies of values at varying distances above and below the mean for continuously varying traits measured in large populations.

nucleic acid Molecules composed of purines, pyrimidines, carbohydrates, and phosphoric acid, concentrated in cell nuclei.

nucleotide The basic unit of DNA composed of an organic base, a pentose sugar, and a phosphate.

nucleus The portion of the cell which carries the chromosomes and thus the hereditary material.

oocytes:

    primary oocyte An intermediate diploid cell formed during female gametogenesis. It divides meiotically to form one secondary oocyte and one polar body.

    secondary oocyte An intermediate, haploid cell formed during female gametogenesis. It divides mitotically to form one mature ovum or egg and one polar body.

oogonia A type of cell in the ovary which serves as a progenitor of the female gamete, the ovum.

outbreeding A system of mating in which mates are less related than average individuals of the population being intermated.

outcrossing Mating unrelated animals within the same pure breed. Often, "unrelated" is interpreted to mean no common ancestors in the first four to six generations of their pedigrees.

ovary Principal reproductive organ of the female. Ovaries occur in pairs and have the dual functions of producing both the female germ cells (ova) and the female sex hormones estrogen and progesterone.

overdominance A genetic situation in which individuals heterozygous for a gene pair (or series) are superior in some manner to any homozygote of the pair or series.

oviduct (fallopian tubule) A tubule in the female which carries the ovum from the ovary to the uterus and in which fertilization occurs.

ovulation The release of the ovum from the ovary.

ovum The female reproductive cell of higher animals. It has the reduced or *n* chromosome number.

parity State of female with respect to the number of gestations experienced. *Nulliparity* signifies no offspring born, *primiparity* one gestation, *secundiparity* two gestations, and *multiparity* several gestations.

parturition The birth of young in mammals.

pedigree A record of the animals from which a given individual is descended. The definition is often extended to include animals which are collaterally related to an individual. In animal breeding the term *pedigree information* includes identification of ancestors and collateral relatives and information on their performance or progeny records.

phenotype The external appearance or some other observable or measurable characteristic of an individual.

phenotypic variance Total variance including that due to both environmental and hereditary effects.

pituitary gland A gland located at the base of the skull. It is often called the master gland of the body since it produces hormones which control many functions. In relation to reproduction it secretes three hormones which stimulate and control the ovaries and testes.

pleiotrophy   Genetic situations in which one gene affects more than one qualitative or quantitative trait of an individual.

polar body   A nonfunctional cell formed during female gametogenesis in animals.

polyestrous   Species in which females have more than one estrual cycle per year.

polyploid   A general term applied to cells with three or more times the haploid number of chromosomes. They seldom occur in mammals or birds and are not viable. In plants, polyploids often exhibit above-normal size and vigor.

population genetics   A field of inquiry in which genetics as related to a group or population is considered in contrast to the genetics of individuals.

prepotency   The ability of an individual to stamp its characteristics on its offspring to the exclusion of the effects of genes from the other parent. It is due to homozygosity for dominant genes. Its importance for quantitative traits is often exaggerated by breeders.

probability   Mathematical expressions relating to frequencies with which specific events may occur or fail to occur.

progeny   Young or offspring of given individuals.

progeny test   Estimate of the genetic value or makeup of an individual through measuring or observing the performance, appearance, or other characteristics of a group of progeny.

progesterone   A hormone produced by the corpus luteum of the ovary which is involved in many reproductive functions including preparation and maintenance of the uterus for pregnancy.

prolactin   A hormone secreted by the pituitary gland which brings about secretion of progesterone by the corpus luteum of the ovary and is involved in mammary gland development and function.

puberty   The life stage or age at which animals are first capable of performing all reproductive functions and producing offspring.

purebred   An animal both of whose parents are duly registered in the herd, flock, or stud book of a given breed.

qualitative inheritance   Heredity relating to the traits for which populations can be divided into discrete classes. The phenotypic expression is controlled by environment and by the action of genes that do not obscure the discreteness of classification.

quantitative inheritance   Heredity relating to traits for which populations exhibit a continuous array of variability and whose phenotypic expression is affected by environment and by the action of several pairs (or series) of genes. Effects of individual genes can seldom be detected.

ram   A male sheep (*Ovis aries*). The term used alone normally indicates a mature male but it is often also used in conjunction with other terms to indicate sex in young sheep, i.e., ram lamb.

random mating   A mating situation in which the probability of any male mating with any female is equal to the frequency of individuals of opposite sex in the population regardless of similarity or dissimilarity of appearance, measurable characteristics, or parentage.

recessive   Genes which have no observable effect unless present in both members of a chromosome pair.

recombinant DNA   Modified DNA produced by enzymatic breaking of the molecular DNA chain, after removing, modifying, or adding genes.

recombination   Occurrence in offspring of genetic combinations not found in parents.

recon   The smallest indivisible unit of DNA capable of recombination. A single base in a nucleotide.

registered   An animal recorded in the herd, flock, or stud book of its breed. Usually a registered animal must be the progeny of registered parents, but this is not necessarily true in breeds admitting high grade or specific kinds of crossbred animals to registry.

regression   Amount of change in one trait associated with a unit change in another trait in a population (see also correlation).

related   A term indicating that two individuals have one or more common ancestors or that one is a descendant of the other. In ordinary usage, animals are usually considered to be related only if they have common ancestry in the first four to six generations of their pedigrees.

relationship   The degree to which individuals are more highly related than the average of individuals for the population to which they belong.

repeatability   The tendency for an individual to repeat its performance, e.g., a dairy cow in successive lactations, a ewe in weaning weights of successive lambs, linear measurements, or gains of any animal in successive periods to be similar, etc. In statistical terms, it is the proportion of total variance in a population which is due to similarity of performance of individuals, when all are measured or evaluated more than once.

restriction enzyme   An endonuclease (enzyme) that restricts (cuts) DNA or RNA sequences at a specific location on the nucleotide chain, leading to two or more shorter molecular chains.

ribonucleic acid (RNA)   Substances chemically similar to DNA but having only one strand. RNAs serve a variety of purposes including carrying the genetic code from the nucleus to the cytoplasm and directing protein synthesis in the cell.

ribosome   One of the RNA-rich granules in cell cytoplasm that are sites of protein synthesis.

scrotum   An outpocketing at the posterior area of the abdomen in which the testes are suspended. It serves a thermoregulatory function to keep the testes at temperatures lower than those of the body.

seed-stock herd   A herd maintained for the primary purpose of producing breeding animals for use in commercial herds.

segregation   Separation of members of a pair of hereditary factors at meiosis in germ cell formation.

selection   Any external influence in a population, either naturally or artificially imposed, which enhances opportunities of individuals of some genotypes to contribute genetic material to subsequent generations and thereby to change gene frequencies.

selection index   A system of weighting values for several traits to arrive at a single score or numerical expression for use in determining which of a given group of animals to select for breeding use and which to cull.

selection pressure   The degree or intensity of selection for or against a trait in the selection process.

semen   The fluid in which sperm cells are ejaculated from the body by the male.

semilethal gene   A gene with detrimental effects on the viability of individuals carrying them but which may not cause death in favorable environments.

seminal plasma   Secretions of several glands along the vasa deferentia and urethra which serve as carriers of the sperm cells in semen of the male.

sex chromosomes   Chromosomes which segregate as if they were members of the same pair but which are morphologically different in the two sexes. They, or factors carried in them, are partially or wholly responsible for sex determination.

sex control   Controlling the sex ratio at birth. There are currently no proven methods for accomplishing this.

sex linkage   Inheritance dependent upon hereditary factors located in the sex chromosomes.

sex ratio   The ratio of males to females at a specific life stage such as at birth.

sire   Male parent, the father of an animal.

somatic cells   The body or nonreproductive cells.

somatostatin   A hormone that inhibits the release of growth hormone.

somatotrophin   Growth hormone.

sow   A mature female swine (*Sus scrofa*).

species   A group of animals or plants possessing in common one or more distinctive characteristics and which are fully fertile when intermated.

specific combining ability   Ability of two breeds, lines, or strains to produce specific effects (favorable or unfavorable) in progeny when crossed.

sperm or spermatozoa   The small, mobile male reproductive cell of higher animals. It has the reduced or *n* chromosome number.

spermatid   A haploid cell type which develops directly into sperm cells without division during the process of male gametogenesis.

spermatocytes:
  primary spermatocyte   An intermediate diploid cell which is formed during the gametogenic process in the male. It divides meiotically to form secondary spermatocytes.
  secondary spermatocyte   A haploid cell formed during gametogenesis in the male. It divides mitotically to form spermatids.

spermatogonia   A type of cell in the testes which serves as a progenitor of the male gamete, the spermatozoa.

stallion   A mature male horse (*Equus caballus*).

standard deviation   The square root of the variance for a trait measured in all individuals of a population (see also variance).

steer   A male bovine (*Bos taurus* or *Bos indicus*) castrated before reaching sexual maturity.

superovulation   Production of an increased number of eggs (ova) in mammals at one time as a result of hormone injection.

telegony   An old but unproven belief that characteristics of offspring may be influenced by the sire of a previous pregnancy in a female.

testes (or testicles)   Paired primary reproductive organs of the male. They produce both sperm cells and male sex hormone.

testosterone   The male sex hormone secreted by the interstitial cells of the testes. It controls sex drive or libido and secondary sexual characters and is involved in sperm production.

transcription   The transfer of the genetic message from the DNA in a gene to the messenger RNA that forms the DNA template. Transcription is the first step in transferring the information in the DNA to a cellular protein.

translation   The formation of a specific protein molecule corresponding to the messenger RNA molecule. The protein is formed from cell constituents with the aid of numerous biomanufacturing elements called *ribosomes*.

transmitting ability The average genetic superiority or inferiority which is transmitted by a parent to its offspring.

triploid Cells with three members of each pair of chromosomes. These do not occur in mammals or birds except as an abnormality. Most triploid individuals are not viable.

type A word used in animal husbandry relative to appearance of animals but having several connotations. It is sometimes used more or less synonomously with the word conformation. It is also used to indicate distinctive kinds of animals, i.e., beef versus dairy, large versus small, fine-wool versus coarse-wool, etc. (See also *conformation.*)

uterus The organ of the female in which embryos develop during the period from conception to birth.

variance Average squared deviations from the mean for a trait measured in all individuals of a population.

variation Differences among individuals in measurable or observable traits. Variation may be continuous (quantitative) or discontinuous (qualitative) in nature.

zygote The cell formed at fertilization by the union of the sperm and ovum. A fertilized egg.

INDEX

INDEX